America's First Rocket Company:

Reaction Motors, Inc.

America's First Rocket Company:
Reaction Motors, Inc.

Frank H. Winter

Ned Allen, Editor-in-Chief
Lockheed Martin Corporation
Bethesda, Maryland

Published by
American Institute of Aeronautics and Astronautics, Inc.
12700 Sunrise Valley Drive, Suite 200, Reston, VA 20191-5807

Cover photos (from left to right, top to bottom):

1) Lovell Lawrence Jr. displaying the 6000-C4- rocket engine, ca. 1945. (Smithsonian photo 00166233.)

2) Group closely watching static firing of the 6000C-4 at RMI at Pompton Plains, New Jersey, ca. 1945. (Smithsonian photo 91-13816.)

3) RMI's small team with their first rocket motors, Pompton Plains, New Jersey, 1943. (Photo from Frederick I. Ordway III collection, U.S. Space and Rocket Center, Huntsville, Alabama.)

4) Rocket-propelled ice boat, Lake Hopatcong, New Jersey, February or March 1947. (Photo from Frederick I. Ordway III collection, U.S. Space and Rocket Center, Huntsville, Alabama.)

5) The first full rocket engine test of 6000C-4, Serial No. 1, Pompton Plains, New Jersey, 30 August 1945, though only chambers 1 and 4 worked on this occasion. (Courtesy Orbital ATK; Smithsonian photo 97-16947.)

American Institute of Aeronautics and Astronautics, Inc., Reston, Virginia

1   2   3   4   5

Library of Congress Cataloging-in-Publication Data

On file

AIAA gratefully acknowledges the American Astronautical Society for permission to republish material from *Pioneering American Rocketry: The Reaction Motors Inc. (RMI) Story, 1941–1972*, AAS History Series, Volume 44.

FOREWORD

Reaction Motors, Inc. (RMI) was one of the most important manufacturers of the early rocket and missile age. As the first successful American liquid-propellant rocket-engine company, with roots in the first U.S. spaceflight society, it built the motors that sent Chuck Yeager's X-1 rocket plane through the sound barrier in 1947 and X-15 test pilots into space in the 1960s. Other Reaction Motors engines powered the Viking sounding rocket beyond the atmosphere and helped the Surveyor landers touch down on the moon. Outside the circle of space historians, however, RMI is now often forgotten, because its last iteration Reaction Motors Division (RMD) of Thiokol Corporation disappeared in the early 1970s. The big California companies, Rocketdyne and Aerojet, had conquered the market for large liquid-propellant engines and, as a merged corporation today, continue to overshadow RMI's reputation due to their prominent role in the missile and space race. Yet Reaction Motors' accomplishments are such that it deserves to be remembered.

No one has worked harder than Frank H. Winter to keep the company's memory alive. Together with our late colleague, Frederick I. Ordway III, he published a series of articles about its engine development that were compiled into an expanded book. As longtime Curator of Rocketry at the National Air and Space Museum, Frank worked to understand the artifacts from the American Rocket Society, RMI and RMD that the Smithsonian had earlier accessioned. He also collected more objects from veterans of the Society and the company. The book you are holding is the ultimate product of his efforts: a well-written, interesting, and factually rich account of the origins and history of RMI. Space enthusiasts, engineers, and historians alike will find this book a valuable and enjoyable account of the pioneering era of rocket-engine development.

Michael J. Neufeld
Senior Curator, Space History Department,
National Air and Space Museum

CONTENTS

INTRODUCTION AND ACKNOWLEDGMENTS

Reaction Motors, Inc. (RMI) was founded in December 1941, just a few days after the bombing of Pearl Harbor, as an extremely modest and most bizarre business by four young and idealistic members of the American Rocket Society (ARS). Their single-minded goal: to develop the liquid-propellant rocket for the war effort.

It was bizarre because at the time it seemed utter foolishness that their only experience in this still relatively brand new field of technology had been largely as weekend hobbyists who built and tested their own homemade rocket motors on the ARS's primitive portable test stand. The homemade stand was characteristically set up in remote fields in and around New York and northern New Jersey to avoid prying fire marshals who might threaten to quickly shut down their experimentation as a fire hazard.

Indeed, in those years rocketry was definitely *not* a profession nor any-thing approaching a respectable and bona fide field of engineering. It was no more than an extremely dangerous and frivolous pastime carried out by starry-eyed dreamers. These dreamers foolishly envisioned fanciful futuristic "rocket ships" capable of flights up to the stratosphere, or even to the moon or the other planets, as garishly depicted in science fiction pulp magazines with their interplanetary tales, like *Science Wonder Stories* or "Buck Rogers" comic strips. In fact, the majority of the members of the American Interplanetary Society, which changed their name in 1934 to the American Rocket Society to attract more engineers to their membership, were science fiction writers or avid fans of the genre.

Rocketry was also bizarre because it was then wholly impractical. Typically, an ARS rocket motor burned for mere seconds; some simply exploded immediately. Cooling was the major problem faced by all the experimenters, and they tried the gamut from water jackets to aluminum blocks around the chamber, although without much success. But from 1938 to 1941, ARS Experimental Committee member James H. Wyld evolved a major breakthrough called the Wyld regeneratively cooled rocket motor in which the fuel circulated within a metal jacket surrounding the motor before it was finally injected into the combustion chamber, thereby cooling the motor throughout its run. The Wyld motor was proven on ARS Test Stand No. 2 throughout these same years; most fortuitously the final runs had been made just a few months before Pearl Harbor, in the summer of 1941.

Upon hearing the news of America's entry into war upon the bombing of Pearl Harbor, Wyld, along with fellow members Lovell Lawrence Jr., John Shesta, and Hugh F. Pierce were confident enough in Wyld's breakthrough to use it as a basis to start their business—and for the benefit of the country.

This, then, is the story of the America's first entry into the rocket industry, RMI—America's first liquid-fuel rocket company—and of the incredible major milestones they accomplished thereafter that really did meet some of their earlier fantastic aspirations for the future of the rocket. For one, less than a decade after their founding, they developed the 6000C-4 rocket engine—nicknamed "Black Betsy"—that powered the Bell, the first aircraft to break the sound barrier (at Mach 1) in 1947. A little more than a decade later, in 1960, their XLR-99 Pioneer rocket engine began a series of flights of the X-15 that flew up to Mach 6.72, literally penetrating space.

In the post World War II years, RMI developed the XLR-10 rocket engine for the Viking sounding rocket, which was in fact the first rocket in the world designed, built, and flown for flight into space. Later, in the mid-1960s, they developed the critically important vernier motors for the Surveyor unmanned spacecraft that became the first U.S. soft-landing craft on the surface of the moon.

This book, which has its own interesting history, also helps to fittingly commemorate the 75th anniversary of the founding of this truly pioneering first American rocket company. The seeds of the book were sewn more than 30 years ago when my late, and highly esteemed friend and colleague, Frederick I. Ordway III, and I undertook research toward a three-part series of history papers on the history of RMI, which we subsequently presented at history symposia of the International Astronautical Federation (IAF) Congresses held between 1982 and 1983. The papers, in turn, were expanded and published as a four-part series of articles in the *Journal of the British Interplanetary Society* (*JBIS*) between 1983 and 1987.

Even earlier, in 1977, the present author conducted a telephone interview in which he very fortuitously persuaded John Shesta, one of RMI's four founders, to present a memoir paper for the forthcoming IAF Congress. Shesta agreed to do this with the understanding that I would edit his paper and monitor its progress toward its presentation. John Shesta's paper was subsequently presented in absentia by another longtime friend and colleague, George S. James, at the 12th History Symposium of the 29th IAF Congress, held at Dubrovnik, Yugoslavia, in 1978.

Among the motivations for Fred Ordway and me to undertake our multi-part study on RMI was that Fred, one of the foremost historians of rocketry and spaceflight, had worked for RMI himself very early in his remarkable career, during 1951 to 1953 as a junior engineer. His main responsibility was to incorporate RMI data into reports for the military services and other customers. Another duty was to serve as a liaison between RMI and the Glenn L. Martin Company, the prime contractor for the Viking rocket.

I was motivated to enter that early joint study on RMI's history with Fred because at that time, as the curator of rocketry of the National Air and Space Museum (NASM), I was responsible for dozens of original RMI artifacts. Moreover, during the 1980s, Shesta's invaluable memoir paper was our only knowledge of RMI's history, and he had left that company by the early 1950s.

In our goal of writing those first articles toward a more complete history of RMI, the Ordway–Winter team collected as much documentation as we could on the topic, and also began interviewing RMI "old-timers." Here, Fred had the advantage of interviewing key contacts dating from his own RMI days—including Shesta, but also early RMI managers Edward H. Seymour, Laurence P. Heath, and Harold W. Ritchey, as well as some of RMI's first engineers, including Charles H. "Chuck" Dimmick, Albert G. Thatcher, and Maurice E. "Bud" Parker. Robert "Bob" M. Lawrence, the brother of Lovell and an early RMI treasurer, was also most helpful.

On my side, I was very fortunately able to locate and conduct telephone interviews with RMI old timer Louis "Lou" F. Arata, one of RMI's first employees; later, Lou's widow donated documentation to NASM that came to further help chronicle RMI's early history. I had already known the late Robertson "Bob" Youngquist, a rocket pioneer in his own right, involved in the early Gorgon and Lark missile-motor developments that had used RMI's motors. His work on these early missile engines also proved extremely helpful in our RMI research.

Our RMI writing project in the 1980s also more fully introduced me to RMI's 6000C-4 rocket engine that had powered the Bell X-1; indeed, I considered it my responsibility to learn more and better document this object because NASM has several of the engines it its collection that were then under my curatorial jurisdiction. For this part of our RMI articles, I took the opportunity to not only attend an RMI old-timers biennial reunion held in northern New Jersey, but to also visit and personally interview two key 6000C-4 pioneers residing in the area—William P. Munger and Henry A. Jatczak. Both Fred and I thus came to be grateful to many RMI veterans, most of whom are now sadly longer with us.

The 6000C-4 story itself became a most engrossing topic by itself and I was later led to produce a two-part spinoff IAF paper on this subject. Parts 1 and 2 of these papers were presented at the 40th IAF Congress held in Málaga, Spain, in 1989, and at the 41st IAF Congress held at Dresden,

Germany, in 1990, respectively. For this additional research, I undertook telephone interviews with key RMI engineer James W. FitzGerald and with Benson Hamlin and Ezra Kotcher, to whom I am forever indebted, who played critical roles in establishing RMI's link to the Army Air Corps for what became the Bell X-1 program.

Many years later after our original joint RMI papers, as well as my additional 6000C-4 papers, it became abundantly clear to both Fred and me that the complete story on RMI—and its successor organization, the Reaction Motors Division (RMD) of the Thiokol Chemical Corporation after they merged in 1958—was undeniably worthy of a book, in other words a compilation of our papers. Yet, in order to arrive to this ambitious goal, it was essential to fill in several gaps. For one, a paper on James H. Wyld himself was necessary, and it was a project I had always wanted to tackle anyway, especially because I well knew that NASM had a very rich collection of original Wyld papers. (Some of these papers were donated by his son, Robert, whom I had the good fortune to meet back in the 1970s.) Still other Wyld papers were at the U.S. Space and Rocket Center in Huntsville, Alabama, and I had already made selected copies of some of this material.

I was especially interested to learn and finally document the details of how Wyld came to evolve the development of his groundbreaking regeneratively cooled rocket motor that had been directly responsible for the founding of RMI. This story had never been told before and would make an ideal "Chapter 1" of our projected book. This therefore led to my paper on Wyld presented at the 59th IAF Congress held in Glasgow, Scotland, in 2008; in Glasgow, prior to delivering my paper, I received a surprise phone call on my cell—a message from Robert Wyld, who gave me some additional information, namely about the Scottish roots of his illustrious father.

Still another required paper for the projected book was on the X-15 engine, also pioneered by both RMI and RMD. This paper was presented on my behalf at the 60th IAF Congress held in Daejeon, Republic of Korea, in 2009, by another colleague and friend, Philippe Cosyn.

For the X-15 engine (XLR-99) story, I am most thankful to several more RMI/RMD veterans, notably Robert W. Seaman Jr., RMD's project engineer on the XLR-99; Harry A. Koch, the program manager for that project; Harold Davies; and several others who had served as engineers on the project. Some of the same individuals provided additional help to me with details and insights into yet other RMI/RMD projects throughout the history of the company, which had lasted until 1972. Among them were Harold S. "Sam" Bell Jr., Harry W. Burdett, and Bob Holder. The late Bob Holder is especially appreciated for arranging for the donation of RMI's early example of their key "spaghetti" motor to NASM, which I later arranged to be placed on exhibit at NASM's Udvar-Hazy Center near Dulles International Airport.

Other RMI/RMD old-timers to be thanked are Hartmann J. Kircher and Delwyn L. Olson.

Several of the above-named individuals also provided extremely useful background details toward another IAF paper I authored, covering the history of the technologically very important spaghetti-type rocket-engine development pioneered at RMI from 1947 by Edward A. Neu Jr. That story is integrated into this book. Those who helped me document Neu's breakthrough were Harry Burdett, Sam Bell, Robert Pearlman, and Bob Holder—all of whom had known and/or worked with Neu. The "spaghetti" paper was presented in 2003 at the 54th IAF Congress in Bremen, Germany.

Myron M. Levoy was very helpful on RMI/RMD's later ventures into nuclear and other exotic propulsion. More recently, Ann Dombras helped me toward the present IAF paper I am preparing on the Viking rocket, based upon the findings by Fred and me in our research; Fred had suggested that we do a joint paper on the Viking, and I will therefore certainly dedicate it to him. Ann's role in that project—when she served as perhaps the earliest known U.S woman rocket engineer—is at least partly covered in the present book.

Ken Montanye, an avid collector of all things RMI, is to be especially thanked for his help in so many ways over a number of years. More on Ken's contributions later. Sincere thanks are also due to Dr. Patrick J. Owens, historian at Picatinny Arsenal, who furnished me with some important finds and insights on the Arsenal's connections with both RMI and RMD.

Among the organizations who must be thanked for that first book were NASM and its staff, especially Dr. Michael J. Neufeld, who has always encouraged and supported my research efforts throughout the bulk of my career at NASM and beyond, and ATK (now Orbital ATK) whose communications manager for Missile Products at the time, Ms. Kristin York, furnished me with extremely scarce early RMI/RMD materials.

For producing the first book and nurturing it throughout, I am forever grateful to Robert "Bob" Jacobs, the publisher of Univelt, Inc. of San Diego, and of course, to my co-author Fred, who guided me all the way, as well as providing me with invaluable new interpretations. Most sadly, however, Fred passed away on 1 July 2014 before the galley stage of the book that finally released late in 2015 as *Pioneering American Rocketry: The Reaction Motors, Inc. (RMI) Story, 1941–1972*. This work, of 462 pages, appears as Vol. 44 of Univelt's AAS (American Astronautical Society) History Series.

When it quickly became evident that RMI would soon approach the 75th anniversary of its founding in December 2016, an abridged and popularized history that would also contain new material seemed warranted. This was to include the important opening chapter summarizing the background of the international space and rocket fad of the 1920s–30s that gave rise to the formation of rocket and space-flight enthusiast groups like the American Interplanetary Society, formed in 1930 and later called the American Rocket Society; it, in turn, led to the formation of RMI.

Most happily, David Arthur, the acquisitions and development editor of the American Institute of Aeronautics and Astronautics (AIAA), was immediately supportive of the proposal to publish this book, which thrust me into a most intense and challenging new project that went far beyond the earlier AAS volume produced by Univelt. The end result is that each chapter of the present work contains both new information and new perspectives, although it is written in a more condensed form and more popular tone.

Hence, not only are some of the original contributors to the original, larger work now to be thanked again as given above, but several other newer interviewees and contributors are as well. That is, the newer book opened up a whole new phase of research.

Among the people who contributed towards this phase are Arthur "Art" Sherman, who responded countless times, often in great detail, to my numerous followup questions on his late-1940s–1960s experiences with the company and Harvey Fox, generous with his recollections of one of RMD's last projects, Project ARE (Advanced Research Engine). Thanks also to RMI/RMD veterans Henry C. Pickering and Gerry Braddick, and to Donald Ruggerie for his help toward my better understanding of RMD's early "JumpBelt."

Other individuals who had not been connected with either RMI or RMD are also to be cited for their gracious assistance. Among them is Susan Marczyk, whose very early recollections of "explosive" noises from RMI during their Pompton Lakes days of the early 1940s were most interesting and useful. Norman Joseph Baum, an early resident of Pompton Plains, the site of RMI's second and much larger headquarters and plant (the old "Dunn barn") fascinated me with his recollections of his firsthand knowledge of the building, including its surprising gangster connection—which assuredly had to be included in the new book.

Edward J. Lenik, the New Jersey archaeologist who had discovered the old RMI "blockhouse" at Franklin Lakes and succeeded in having it placed upon the list of National Register of Historic Places in 1978, and who helped me more recently in gaining more background on how this happened, is also to be greatly thanked. On the same topic of the Franklin Lakes site, Jack Goudsward, a lifelong resident and historian of that area, enthusiastically afforded me important historical material on the old Nelson Airport where the Franklin Lakes site was established. Ronald J. Dupont Jr., another northern New Jersey resident and ardent accomplished local historian who had earlier authored an excellent article on RMI, which I used in my first book on RMI, kindly sent me copies of all the original photos taken by

Ed Lenik, one of which is included in the present book. Shea Oakley, the executive director of the Aviation Hall of Fame and Museum of New Jersey, in Teterboro, must certainly be mentioned here for locating an invaluable light-brown-leather three-ring binder full of photos and accompanying data sheets on many of RMI and RMD's various projects over the years. This work, collected and assembled by an anonymous person with an unknown connection to RMI/RMD, contained key material that I used for the book. Shea also kindly furnished me with other material, in answer to one of my historical questions on the Nelson Airport.

I would be remiss if I did not mention the most wonderful and enlightening experience afforded to me by Ken Montanye of a six-hour tour of historic RMI/RMD sites in the environs of northern New Jersey during 9 September 2016. The next day I met and endlessly chatted with several of the company's veterans at their 20th Biennial RMI Old-Timers Reunion, who added more helpful recollections and answers to my historical questions. Among these individuals were Mario Luperi and Carl Kastner.

Much was also gained from pouring through Ken's considerable and impressive collection of RMI/RMD documents and artifacts. Ken also later put me in touch with Glenn Repp, the only known survivor of the terrible accident during routine testing of RMI's XLR-40 rocket engine in 1957. Subsequently, Glenn very graciously filled me in on details of that sad event, the only known fatality in RMI and RMD's more than 30 years of operations. Great appreciation is also extended to William "Bill"/"Billy" Arnold, with RMI since 1946, and who relayed to me his personal story of how he had continued to work for successor RMD on the NASA lifting body program with its use of an RMI engine as late as 1976, some four year *after* RMD had closed. Technically, Bill Arnold appears to have been RMD's last employee, working on their last project. Ironically, I am also most grateful for my earlier contacts with the late Kurt F. Fischer; reviewing his curriculum vitae and other documents for the second book, I discovered that he had been RMI's very first employee, and I now incorporate this important find in the present book.

I am also extremely thankful to my longtime friend and colleague, the eminent aviation historian Dr. Richard P. Hallion, who took time out from his very busy schedule to answer several key questions I had relating to aeronautics for the book.

Among the organizations that greatly helped me with my deeper research into RMI/RMD history is the NASM archives at the Udvar-Hazy Center, with special thanks to Brian Nicklas. Brian should also be cited for unearthing material on the old Nelson Airport as part of the essential story of RMI's establishment of its first test site at Franklin Lakes, New Jersey. Thanks as well to NASM film archivist Mark Taylor for arranging for the multiple showings of early RMI film footage in NASM's collections and to Paul Silberman for his assistance with additional archival needs.

My work at the Udvar-Hazy Center archives entailed numerous revisits into the Lovell Lawrence Jr. papers, in addition to other collections not previously examined, such as the Scott Crossfield papers of the famed X-15 pilot and engineer. These papers revealed new and important finds on Crossfield's role in assessing RMI's earlier XL-30 engine, which had won the NASA–North American Aviation competition for the power plant for the X-15 engine and led to the development of the XLR-99.

Throughout both book projects, Robert "Rob" Mawhinny of NASM's restoration staff has been of enormous and consistent support to me in presenting important new technical interpretations that have covered everything from RMI rocket motors in early RMI drawings and photographs to our joint and multiple review screenings of rare early RMI footage of their 1940s tests at their Franklin Lakes test site in New Jersey.

Stephanie O. Stewart of NASM's Collections Management staff was consistently helpful in arranging and assisting in my new examinations of NASM's varied Gorgon missiles in its collections. I also thank the staffs of the National Archives at both in Washington, D.C. and at College Park, Maryland, in my latest searches in which I was able to ferret out more hitherto unseen documents that shed further light on RMI. A great deal of assistance was also very patiently and generously provided by Cherie Banker, librarian of the Pompton Lakes Library of Pompton Lakes, New Jersey, and elicited several significant new finds.

I certainly also greatly thank Edward Stewart II, the Director of Exhibits and Curation of the U.S. Space and Rocket Center in Huntsville, Alabama, and Carolyn Lawson of his staff for permitting me to pore through the Frederick I. Ordway III and James H. Wyld collections in the Center's Archives and for providing me with copies of materials from these wonderful collections where so many new and remarkable discoveries were made. I am likewise very appreciative to Joshua D. Dinman, director of marketing communications of Orbital ATK, for granting permission to use several RMI and RMD photos for this book.

Lastly, I am extremely grateful to my family—my wife Fe Dulce R. Winter, and our children Ron and Elaine Winter, for their unstinting support, including my latest treatment on the RMI/RMD story—and especially to Elaine for utilizing her amazing computer skills in this endeavor. I therefore dedicate this book to them as well as to the revered memories of the original four founders of RMI: Lovell Lawrence Jr., John Shesta, James H. Wyld, and Hugh Franklin Pierce.

Chapter 1

SETTING THE STAGE: THE WORLDWIDE ROCKET CRAZE OF THE 1920S–30S

"A good rule for rocket experimenters to follow is this: always assume that it will explode."

The Editor (–G. Edward Pendray), "Letters to the Editor," in *Astronautics* (Journal of the American Rocket Society), No. 38, October 1937, p. 8.

THE FOUNDING AND FOUNDERS OF RMI

A mere two weeks after the bombing of Pearl Harbor that thrust the United States into World War II, the first successful American rocket company was founded. The date was 18 December 1941, and the tiny, newborn firm was boldly named Reaction Motors, Inc., more popularly called "RMI."[1] James H. "Jimmy" Wyld, one of the four founders—and only company personnel at the time—had dreamed up the name, envisioning it to become "a kind of General Motors of the rocket world."[2] But as he later recalled, "The analogy was not very close as we had scarcely two nickels to rub together, and our [initial] plant consisted mostly of half of the upper floor of [fellow-founder] John Shesta's brother-in-law's garage in North Arlington, New Jersey, which was about as large as a rather spacious outhouse."[3]

Technically speaking, there had been earlier American rocket companies, such as the American Carrier Rocket Company created in 1886 by Patrick Cunningham, the Ireland-born developer of the Cunningham lifesaving rocket; the Rocket Bomb and Reactive Shell Company formed in New York state about 1918; and the Rocket Airplane Corporation (late 1935–early 1936), its short existence based upon the ill-fated, small liquid-propellant rocket aircraft *Gloria 1* and *Gloria 2*, designed for delivering rocket-mail over short distances.[4] But the Carrier Company was meant to produce strictly solid-propellant (gunpowder-based) lifesaving rockets, the WWI Rocket Bomb concern focused upon gunpowder rocket-propelled bombs, and the Rocket Airplane Corporation did work with liquid-propellant rocket motors. However, all three companies were short-lived commercial failures. Their lack of longevity supports the claim that RMI was the first successful American rocket company.

At the time of RMI's founding, Wyld was 29 and a brilliant although very shy engineer. He was educated at Princeton University, earning a B.S. degree in engineering with high honors in 1935. The Great Depression made it very difficult for young Jimmy to obtain and hold a job, and he may well have been in one of his unemployment modes when RMI was formed.

John Shesta, the oldest of the RMI founders, had recently turned 40. Shesta was born in 1901 in St. Petersburg, Russia (called Leningrad by the U.S.S.R); his original family name was Shestacovsky. In 1915, he accompanied his father, Ivan, to America. The elder Shestacovsky had come to facilitate an arms purchase for the Imperial Russian government's war effort. But after the Russian Revolution of 1917, Ivan and his Sweden-born wife, as czarist loyalists, decided to stay in the United States, and they later Americanized their name to Shesta. John retained a very slight Russian accent throughout his life. The Shestas settled on a farm in the Catskills in upstate New York, and John became an American citizen in 1925. Like Wyld, he became a professional engineer, earning a B.S. degree from the School of Engineering at Columbia University in 1927 and a civil engineering degree from the same school in 1928.[5]

Lovell Lawrence Jr. was the most business minded of the four founders. Even though at 26 he was the youngest, he was made RMI's first president. Lawrence was born in upstate New York, in the village of Port Henry on Lake Champlain. The son of a mining engineer who worked for the Cheever Iron Ore Company, Lawrence, aka "Bun," attended the Montclair State Teachers College (now Montclair State University) in New Jersey, graduating in 1933 with a B.A. degree. He was also technically gifted, and by the time of RMI's formation he had worked for IBM as assistant to the chief engineer of the Radio Type division that developed and produced radioteleprinting devices—precursors to modern fax machines. Lawrence was particularly adept at understanding and drawing electronic-circuitry schematics that would come into excellent use in the intricate designs of RMI's first rocket test-stand circuitry and ignition systems.

Hugh Franklin Pierce, 36 years old at the formation of RMI, was the final member of the founding four and also an anomaly. He was the least educated, probably receiving no more than a high school diploma. Born in Dayton, Ohio, in 1905, Pierce later joined the Navy and enrolled in, although never completed, a 14-month course in U.S. naval aviation, apparently in the 1920s. His Navy career was evidently short lived, perhaps due to health reasons. Pierce wound up settling in New York City, where he found employment as a ticket taker for the subway system.

Pierce too, was mechanically gifted—and better yet, to the benefit of the newly created but ill-funded firm of RMI, he possessed a lathe and other tools that he kept in the basement of his home in the Bronx. A brief 1934 sketch of rocketeers of the day remarked that Pierce was "formerly a naval mechanic, and is clever with tools."[6] Among all four founders, Pierce already had the most experience in handling rocket motors. He had been one of the earliest

Fig. 1.1 RMI founders, clockwise from top left: Lovell Lawrence Jr., John Shesta, Hugh Franklin Pierce, James H. Wyld. (Courtesy Smithsonian Institution, 80-2409, 78-9396, 77-15169, A-4064.)

members of the American Rocket Society (ARS), originally the American Interplanetary Society (IAS), established in New York City in 1930. Pierce had joined by 1932 and was one of its first experimenters. Indeed, he helped construct what later became known as ARS Rocket No. 1, which was not flown but static tested on 12 November 1932 at a farm in Stockton, New Jersey. From then on, Pierce was an active member of the Society's Experimental Committee.

AMERICAN INTERPLANETARY SOCIETY ROOTS

The American Interplanetary Society was founded by a small group (11 men and one woman) of highly idealistic young people who were clearly space minded. In fact, most of them were science fiction writers. The moving force behind the creation of the IAS and its first president, David Lasser was the managing editor of *Science Wonder Stories*. One of the earliest science fiction magazines, *Science Wonder Stories* invariably featured interplanetary stories—then a relatively new genre of popular literature centering on space flight. These tales were usually set in the far future, and involved such thrilling topics as discoveries of new planets or systems with highly advanced extraterrestrials or alien invasions. Mainly, the AIS had banded together to somehow work toward the actual achievement of space flight. At the time this was a vague, wholly naive, and an almost impossibly optimistic notion—especially during these early days of the Great Depression. Understandably, science fiction served as purely escapist literature, and the mere possibility of space flight gave readers hope toward a very distant, but far brighter, future world.

The AIS presented lectures in an attempt to educate the general public on the promises and possibilities of space flight (as well as potential hazards, such as errant meteorites). The small group also raised the issue of rocket experiments as a concrete way to lay the foundations of the necessary technology for space travel.

It is uncertain who first proposed the experiments, but the Society quickly left behind its almost wholly theoretical science fiction-writers period and entered a new experimental phase. The transformation followed the return of AIS vice president G. Edward Pendray, and his wife, Leatrice, or "Lee," to America from Germany early in the spring of 1931. Both Pendrays had been founders of the AIS, Lee the only woman among them. G. Edward was a reporter for the *New York Herald Tribune,* although he also wrote science fiction under the pen name of Gawain Edwards.

In the Berlin suburb of Reinickendorf, the Pendrays had witnessed a static rocket test or two conducted at the Raketenflugplatz ("rocket-flying place," or test area) of the Verien für Raumschiffahrt, or VfR (the Society for Spaceflight). There, vice president Pendray made copious notes and sketches. This was the first time any member of the AIS had ever seen a liquid-propellant rocket motor in operation. The event, Pendray later recounted, "filled us with excitement, and upon our return we reported fully to the Society, on the evening of 1 May 1931, both the method and promise of the German experiments."[7] This report was published in the mimeographed *Bulletin of the American Interplanetary Society* for June–July of that year.

From then on, the American group shifted their whole focus and sought to emulate the Germans in their rocket work. By 1934, the transition became complete when the AIS changed their name to the American Rocket Society to attract more members with needed engineering background. It was also a conscious effort to distance themselves from their earlier, more fanciful, dreamy-eyed science fiction background, which was somewhat looked down upon by the more scientifically minded enthusiasts who wished to advance the cause of spaceflight. The tactic worked, because Wyld joined by late 1934 or early 1935, followed by Lawrence in 1936. It was the ARS that had brought all of the four future RMI founders together, but Germany, particularly the VfR and the experimenters at the Raketenflugplatz, should perhaps be considered as another "root" of RMI.[8]

How then, did the VfR as well as the American Interplanetary Society spring up? What were *their* roots? In a nutshell, they were both products of the spaceflight and rocket fads of the 1920s and 30s.

EARLIEST CONCEPTS OF SPACE FLIGHT

The dream of spaceflight was nothing new by this time; it stretched back countless centuries, as attested by various ancient and fantastic legends found in many cultures of the world. For example, celestial or astronomical bodies figure very prominently in the creation myths of ancient Babylonia, dating back to about 2000 B.C. From that region comes the story of Etana flying to the planet Venus (represented by the goddess Ishtar) on the back of an eagle. But most of the space voyages in ancient myth center around flights to our nearest astronomical neighbor, the moon. One outstanding example is the well-known Chinese folk legend of the beautiful Lady Chang'e, or Chang-o, who flew to the moon by taking a magic elixir. There she met the rabbit who inhabits the moon. This story is so popular in China that an annual Chang'e mid-autumn festival is celebrated the night of the full moon in the eighth lunar month.[9] In 2007, the People's Republic of China launched its first lunar orbiter, a robotic spacecraft named *Chang'e 1*, and on 14 December 2013, the *Chang'e 3* lunar lander/rover successfully touched down on the moon. China thus became the third nation after the former U.S.S.R and the United States to land a spacecraft on the moon.

The topic of spaceflight inevitably entered literature and the other arts as well, although at first the modes of travel depicted were fanciful and the stories themselves were space fantasies. Perhaps the first such fictional description tion was authored by the ancient Greek satirist Lucian of Samosata (~A.D.125–180), whose *Vera Historia* (*True History*) has voyagers unexpectedly conveyed to the moon via a sailing vessel suddenly swept aloft by "a most violent whirlwind" to land after eight days on "a large tract of land" that was "round, shining and remarkably full of light."[10] Lucian's moon was inhabited by strange lunar beings. Yet wholly contrary to the title of his work, Lucian clearly forewarned readers in his preface: "I lie…and…I mean to speak not a word of truth throughout."[11]

Lucian was so pleased with the spaceflight theme that he wrote a second story, *Icaro-Menippus*, although in this tale the moon trip was planned. Like the legendary ancient Greek Icarus, Menippus chose to fashion a pair of wings for himself—a hybrid arrangement that combined a wing from a vulture with

Fig. 1.2   Modern depiction of Lady Chang'e or Chang-o. (Courtesy NASA.)

another from an eagle. After making a number of practice ascents, Lucian's would-be space voyager set out for his destination from the summit of Mount Olympus and eventually reached the moon.

It is not until as late as the Industrial Revolution in the 19th century that real science fiction appears, in which authors attempt to make their literary modes of spaceflight propulsion more believable by incorporating some species of technology. Even so, this strictly literary technology was characteristically quite far-fetched by today's standards. A recent survey by the author of the variety of forms of space propulsion in the science fiction literature of this period revealed that a sizable number of these stories used balloons to get into space; almost the same number used forms of electromagnetism; and a like number used meteors or comets (the heroes riding on passing meteors or comets). By far, the most popular form of fictional propulsion was antigravity. H.G. Wells's classic novel *The First Men in the Moon* (1901) is a prime example of a work that featured this propulsion mode.[12]

THE ROCKET ENTERS THE PICTURE

The same survey revealed that the rocket was an extreme rarity in very early science fiction from this period. The main reason can be summed up in two words: stagnant technology. Throughout the approximately 1,000 years of rocket history, from its apparent origins in Song Dynasty China (circa A.D. 960–1279) up to, say, 1900, the applications of the rocket were limited to a mere firework; an occasional although not usually effective small-range battlefield weapon; a conveyor of signals; a lifesaving device for throwing ropes to stranded ships offshore; whaling rocket harpoons; and a few other miscellaneous applications. Fundamentally, then, throughout those many centuries, the rocket had changed very little internally. With very few exceptions, the propellant remained gunpowder, a solid propellant of very limited power. The so-called Congreve and Hale era, based upon the improvements of the two British inventors, Sir William Congreve and William Hale, did witness a resurgence of the war rocket, although this lasted only from early Napoleonic times until the late 1890s.[13]

Fig. 1.3 Rockets as fireworks, depicted in a 1616 German print. (Courtesy Frank H. Winter collection.)

One of the exceptions in the early science fiction literature that does feature a rocket as the means of propulsion is the short story "Bagley's Inter-planet [sic.] Skyrocket," by the Michigan-born Howard Dwight Smiley (1879–1940), which appeared in 1908 in the British publication *Blue Book Magazine*.[14] It was perhaps the earliest story that unambiguously described a rocket as the primary means of space propulsion. However, as the title implies, the vehicle was essentially a giant firework-type (gunpowder) rocket, not a liquid-propellant type, which had not yet been developed.

In sum, it would have been ludicrous in the 19th century to even think about the rocket applied to spaceflight, much less seriously consider spaceflight itself on a scientific basis. And throughout the preceding centuries, the cause of rocket motion was almost totally misunderstood. The most prevalent theory of rocket motion was that a rocket needed air to "push against."[15] Hence, the rocket would *never* work in the vacuum of space. This belief persisted in the face of Newton's third law of motion, propounded in his *Principia* of 1687: "For every action there is an equal and opposite reaction."[16] The third law, as is well known, correctly explains rocket motion and why the rocket *can* work in a vacuum. In fairness, the *Principia* was very difficult to understand and with some exceptions, its relevance to rocket motion was simply missed, even by many professional artillerists and savants of science.

THE ROLE OF JULES VERNE

Unquestionably, the most significant science fiction space story to appear during the Industrial Revolution was Jules Verne's classic novel *A Trip to the Moon* (1865), followed by its sequel, *Around the Moon* (1870). Verne's mode of propulsion—a giant cannon shooting a hollowed-out giant cannon shell that conveyed his literary space heroes to the moon—simply would not have worked for a variety of technical reasons. But in its day, and for many years thereafter, Verne's novel was the most believable literary work on space travel. It thus exerted an enormous impact upon the general public as well as upon the scientifically minded.

To his great credit, Verne painstakingly worked out many other details of what spaceflight might be like and was remarkably prescient on some: his launch from Florida, his correct description of zero gravity, and the use of the rocket—even if of the firework type—fired in the vacuum of space for diverting the space capsule in *Around the Moon* to avoid crashing into the moon and heading it back to Earth. Moreover, Verne concluded his story with the safe splashdown of his literary space capsule in the Atlantic Ocean.

Little wonder that both Verne space novels led to a flurry of books and space travel stories by a host of followers. These popular-culture spinoffs included

Fig. 1.4 **Jules Verne's fictional cannon-launched space capsule the moment the space travelers electrically fire firework-type rockets to aim the capsule back to Earth. Illustration by Émile-Antoine Bayard from** *Around the Moon* **(1870). (Courtesy Frank H. Winter collection.)**

the internationally acclaimed Jacques Offenbach opera *Voyage dans la Lune,* which premiered in Paris 1875, with a revival in 1877. (It is still performed occasionally.) In 1902, the famous French cinematographer Georges Méliès produced a now-classic silent film with the same title. Méliès's *Voyage dans la Lune* is widely regarded as not only the earliest example of the science fiction film, but the first space film—and one of the most influential movies in cinema history.

Tsiolkovsky

In the meantime, in the small provincial town of Kaluga in Imperial Russia, about 120 miles southwest of Moscow, there resided from 1892 a partly deaf schoolteacher named Konstantin Eduardovich Tsiolkovsky. Tsiolkovsky devoted most all of his available time working out the problem of spaceflight and by the 1880s had already determined that reaction propulsion, which he later identified as the rocket, was the key answer.

Almost devoutly inspired by Verne, Tsiolkovsky eventually produced his seminal article, as translated from the Russian: "The Exploration of Cosmic Space by Means of Reaction-Propelled Devices."[17] It appeared in the popular scientific journal *Nauchnoe Obozriene* (*Science Review*), published in St. Petersburg in May 1903, just seven months before the Wright brothers made their first successful flights in a controllable, manned aircraft at Kitty Hawk, North Carolina. Tsiolkovsky's article contained all the fundamental mathematics, including the so-called Tsiolkovsky rocket equation, or "ideal rocket equation," required for achieving spaceflight, in addition to a detailed description and depiction of a liquid-propellant rocket using the optimum propellants of liquid hydrogen and liquid oxygen.[18] For a number of reasons this article should have been an incredible major milestone in the history of astronautics.

In retrospect, however, the article's importance seems to have been neglected at the time and for a number of years thereafter. One reason was the language problem; far fewer people outside Russia could read Russian than is the case today. Another was the extremely limited circulation of *Nauchnoe Obozriene*, and there is an indication that the magazine was then simply not available at all in America. It is not known with any certainty what its availability situation was in European countries, either. Russia was also considerably more isolated than at present. Transportation options were a lot more limited, a problem confounded by Russia's vast size. In addition, state-sponsored censorship and greatly restricted interpersonal interactions with citizens of other nations likely played a part in limiting distribution of Tsiolkovsky's article.

Thus, as groundbreaking as it was from a modern perspective, Tsiolkovsky's article was obscure and virtually inaccessible to a wider readership in the West at the time. Nevertheless, Tsiolkovsky continued to considerably expand his spaceflight concepts and writings (he paid for publication of various pamphlets out of his own meager salary as a girls'-school teacher), but further evidence shows that his name as a spaceflight pioneer was not really introduced to the West until as late as the mid-1920s, and his earlier work along these lines had even been neglected in his own country [19]

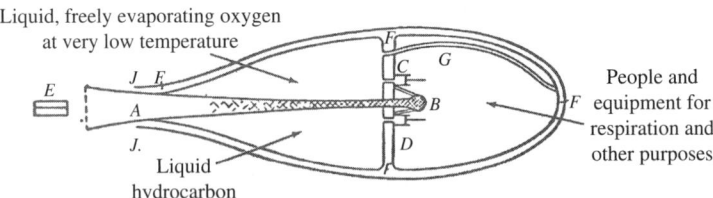

Fig. 1.5   Schematic drawing of Konstantin Tsiolkovsky's spaceship, published in 1903 in *Nauchnoe Obozriene* (*Science Review*). (Courtesy NASA.)

GODDARD

Meanwhile in America, Robert H. Goddard of Worcester, Massachusetts, also caught the space bug. He, too, was enthralled by the Jules Verne space novels as a youngster. But Goddard's thinking about the possibilities of space-flight were far more influenced by the "Mars furor."

The Mars furor originated in 1877, when the Italian astronomer Giovanni Schiaparelli in his initial observations of the planet Mars—in close opposition to Earth that year—observed a dense network of apparent dark markings on the planet's surface that he called *canali* in Italian. Schiaparelli meant "channels," as in possible natural channels of water, but the term was mistranslated into English as "canals," which to many indicated artificial constructions that could only have been made by intelligent beings. This immediately led to waves of sensationalistic speculation in the popular press, as well as in the scientific community, about the possibility of intelligent life on Mars. In turn, it led to the rise of a new subgenre of science fiction literature (and also plays and, later, films) with Martian themes. The furor remained almost unabated until the touchdown and subsequent life-search findings of the Viking orbiter/lander space probes to Mars almost exactly a century later, in 1976. They showed that there was abundant evidence of significant flows of water in the planet's distant astronomical past, first revealed by the Mariner Mars space-craft flybys from 1965, but there were certainly no artificial canals nor any detectable signs of present life, even microbial forms (at least on that part of Mars where the Vikings landed).

While still a teenager, Goddard had been captivated with the newspaper serialization of an American version of the classic H.G. Wells novel *War of the Worlds*, written by Garrett P. Serviss and called "Edison's Conquest of Mars." It appeared in the *New York Evening Journal* and the *Boston Post* in early 1898.[20] Goddard was so enthralled that he experienced a revelation (literally, a daydream) about a trip to Mars. He then vowed to devote the rest of his life to solving the "spaceflight problem."[21]

Goddard later went on to earn a Ph.D. in physics in 1911 and become a professor of physics at Clark University in his hometown of Worcester. All the while, he sought to fulfill his vow and in 1919 produced another seminal work in the history of both rocketry and "astronautics" (as it later came to be called), his 79-page treatise, *A Method of Reaching Extreme Altitudes*. This was published by the Smithsonian Institution, released on 12 January 1920.

Goddard arrived at this point through very many years of brainstorming and theorizing about the most viable means of space propulsion, although he had confined his ever-changing ideas to a number of private notebooks that have never been published in full. He examined all conceivable possibilities for propulsion—from centrifugal force to somehow using radio waves, magnetic reaction, acoustic waves, solar energy, atomic energy ("if it could be harnessed"), to other radiation sources, levitation, ion or electrostatic propulsion, and so on.[22] At the same time, he jotted down other, nonpropulsive spaceflight ideas, like sending cameras to other planets, steering a spacecraft ("the space car," he quaintly termed it) by photoelectric cells, and so forth. But it was not until 1908 that he came to realize that the rocket appeared to be the answer, although this mode needed considerable further development. One of his notebooks shows that on 24 January 1909 he had definitely settled on the rocket. From this time on, young Goddard more fully focused upon the development of the rocket.[23]

After he became Dr. Goddard in 1911, he went much further. He began applying for patents in 1913. His very first, No. 1,102,653, was granted on 7 July 1914 for a "Rocket Apparatus." The patent documentation contained several fundamental advances, such as the use of the de Laval nozzle for achieving far greater efficiency; the staging, or "step" principle; and the use of a much more powerful "double base" (a nitroglycerine–nitrocellulose-based propellant, also termed "smokeless propellant"). Goddard also hinted at, although did not elaborate upon, the use of liquid propellants. The next year, with the help of a $5,000 grant from the Smithsonian, he began initial experiments with solid, smokeless-propellant rockets.

The name of Konstantin Tsiolkovsky was still unknown in the West at the time, and both the Russian and the American worked completely independently of each other but toward the common goal of the realization of spaceflight. But the mid-January 1920 release of Goddard's *A Method of Reaching Extreme Altitudes,* based upon his experiments and including those that proved conclusively that the rocket *can* work in a vacuum, enormously complicated matters for him as well as the history of rocketry and astronautics.

THE "SPACE ROCKET" CONCEPT ENTERS THE PUBLIC CONSCIOUSNESS

Shy by nature and well aware that the notion of spaceflight was then not a respectable subject to be taken up by a physics professor, Goddard had purposely written his treatise for fellow academics and kept it buried within the Smithsonian's *Miscellaneous Collections* of publications. The treatise also advocated the development of the rocket for the exploration of the upper atmosphere, not for spaceflight. And Goddard purposely chose a bland title that did not suggest spaceflight at all. Today, we would say that Goddard's *Method* was a proposal for a "sounding rocket" capable of reaching far higher altitudes than an ordinary sounding balloon with scientific instruments attached.

Yet, buried in his small concluding section titled "Calculation of Minimum Mass Required to Raise One Pound to an 'Infinite' Altitude," he dared to mathematically work out the theoretically maximum possibility of a rocket that *might* be able to reach the moon.[24] Moreover, he suggested his theoretical, multistage, unmanned, solid-propellant rocket could carry a certain amount of flash powder in its nose to explode upon the lunar surface to "signal" its impact upon the moon, an event that could be detected by observers on Earth with telescopes.

Just before the treatise was published, the Smithsonian sent out a press release (unfortunately, the original is missing) to all the major U.S. papers, and evidently to some overseas, in which the theoretical "moon rocket" was stressed. The press release produced astonishing and wholly unexpected results—especially for Goddard.

Overnight, the shy New England professor became front-page news across the country, and also wound up in the papers of countries as far away as South Africa and Australia. His alleged moon rocket caused a sensation, especially as it was inferred (quite incorrectly) by the vast majority of these stories that

Fig. 1.6   Robert H. Goddard next to his rocket prior to its secretive launch on 16 March 1926 in Auburn, MA. (Courtesy NASA.)

Goddard was actually *proposing* a moon rocket. In addition to the stories, which continued for months, there was an outpouring of cartoons, jokes, and criticisms of his plan, as well as a stream of magazine articles (public radio broadcasts were not yet fully established), poems, songs, and films with space-rocket themes. There even emerged a host of volunteers, male and female, who wished to accompany Goddard in his rocket to the moon—or Mars. In essence, Goddard's modest *Method* had unintentionally planted the idea of the space rocket in the public consciousness of millions. It did not matter that the technical details were lost upon, or were unavailable to, the general public; routinely, in fact, many of the stories were very generalized and distorted.[25]

In his own reaction to this unexpected turn of events, Goddard became more reclusive, or rather, intensely more secretive and guarded. Even so, from 1921 he began to experiment with liquid propellants, although none of this was leaked to the press. Nor did word leak of Goddard's monumental and historic launch of the world's first liquid-propellant rocket on 16 March 1926 on his Aunt Effie's farm in Auburn, Massachusetts.

OBERTH

In 1923, there appeared what may be termed the third seminal work in the history of astronautics and rocketry. This was the slim, 92-page book published in Munich, Germany, titled *Die Rakete zu den Planetenräumen* (*The Rocket into Planetary Space*). The author was Hermann Oberth, a 29-year-old physics student who was born to German parents in 1894 in the small town of Nagyszeben (then also called Hermannstadt in Austria-Hungary, although today it is known as Sibiu and is located in Romania).

Oberth also had been significantly influenced by Verne's space novels, encountering them at the age of 11 and then periodically rereading these classics enough to learn them by heart. Like Goddard, Oberth, too, developed his own space ideas independently (of both Tsiolkovsky and Goddard) and allegedly, by 1917, had presented his plan of a liquid-propellant (liquid oxygen/watered alcohol) "missile" to the Prussian Minister of War, but was turned down. (However, the authenticity of this alleged scenario requires further research for verification.)

By 1922, Oberth did propose a doctoral dissertation on the possibilities of the rocket; but it was rejected as "too fantastic."[26] On 3 May of the same year, he wrote in his quaint English the following letter to Goddard:

> Already many years I work at the problem to pass over the atmosphere of our earth by means of a rocket. . .I learned by the newspaper, that I am not alone in my inquiries and that you, dear Sir, have already done much important work at this sphere. In spite of my efforts, I did not succeed in getting your books about this object. Therefore I beg you, dear Sir, to let

me have them. At once after coming out of my work I will be honoured to send it to you, for I think that only by common work of the scholars of all nations can be solved this great problem.[27]

Neither Goddard's reply nor Oberth's copy of it has been found; the latter was likely destroyed in a World War II bombing. But we do know that Goddard duly sent a copy of his *Method* to Oberth that was received by 12 June, and that Oberth promised to add a few words on *Method* to his upcoming book scheduled for publication that autumn.

OBERTH AND GODDARD COMPARED

Goddard must have been totally stunned when he finally went through *Die Rakete*. This little book far surpassed his own *Method* in several highly significant respects. *Method* was a dry, formula-filled scientific treatise on Goddard's solid-propellant experiments, with a just a few passing pages on the possibili-

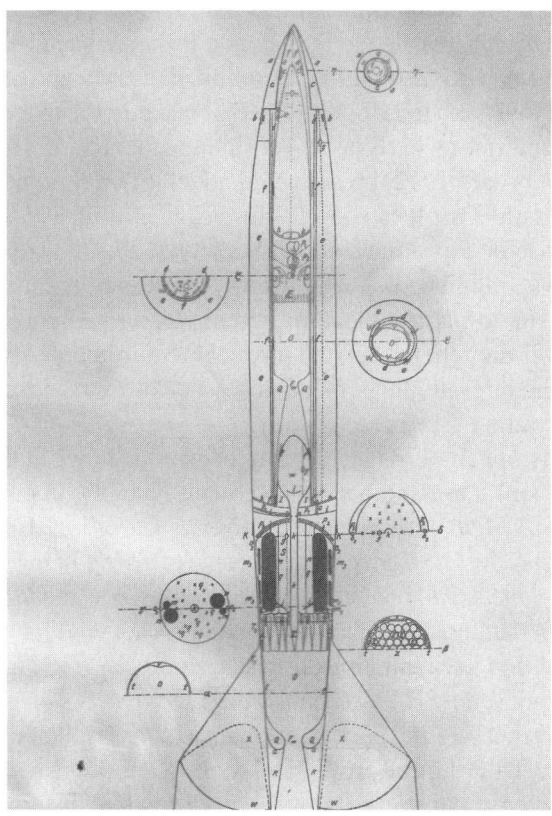

Fig. 1.7   Drawing of Hermann Oberth's Model B liquid-propellant space rocket. (Courtesy Frank H. Winter.)

ties of his theoretical, *unmanned*, solid-propellant, multistage rocket making a one-way trip to the moon. In dramatic contrast, Oberth's work stressed the promise—and technical feasibility—of the development of a liquid-propellant manned space rocket. Beyond this, his book covered a wide spectrum of other aspects of spaceflight, including fine details on liquid-propellant rocket construction (from propellant injectors to cooling); aerodynamics as applied to rockets; guidance and navigation; life-support systems; the possible physical and psychological effects of spaceflight upon space travelers; spaceflight hazards and their possible remedies; reentry and recovery techniques; and telescopic tracking. Oberth additionally outlined how a space rocket could evolve from a basic unmanned prototype rocket through a Model B sounding vehicle up to a modified Model B that could carry men into space. He even included suggestions for diverlike space suits.[28]

ESNAULT-PELTERIE (REP)

It would be remiss to not include another early pioneer who is all too often ignored in favor of the usual triumvirate of astronautics pioneers: Tsiolkovsky, Goddard, and Oberth. This is the Frenchman Robert Esnault-Pelterie (1881–1957), also known as REP. A highly accomplished early aviation pioneer and inventor of the joystick aircraft control, REP became intrigued with the possibility of spaceflight as early as 1908 (independently of Tsiolkovsky and Goddard). In February 1912, he delivered a lecture on this speculation in St. Petersburg, Russia. That was soon followed by a similar lecture before the Société Française de Physique (French Society of Physics) on 15 November 1912, which was published shortly after. "Numerous authors made a man traveling from star to star a subject for fiction," he wrote in his introduction.[29] "No one has ever thought to seek the physical requirements and the orders of magnitude of the relevant phenomena necessary for realization of this idea. . .This is the only aim of the present study."[30]

Unfortunately, much of his lecture had been condensed in the printed version but it was still a remarkable achievement. Notably, in comparison with Goddard's later *A Method of Reaching Extreme Altitudes,* REP had proposed the use of a rocket for the transport and return of living beings from Earth to the moon and planets, and provided—like Tsiolkovsky and Goddard—equations for motion that took into account that the rocket could fly in the vacuum of space, in addition to calculations for escape velocity and flight times to the moon, Mars, and Venus. He also considered factors like thermal problems: particularly, overheating in space from intense solar radiation. REP did not, however, suggest the liquid-propellant rocket, although he did propose a "hydrogen-oxygen mixture" without specifying whether this would be in a solid, liquid, or gaseous form (solid seems implied, as he used kg values for the hydrogen-oxygen).[31] Beyond this, REP proposed the potential energy of radium with theoretically far more power and which he favored over chemical propellants. This made REP perhaps the earliest to publicly suggest a form of atomic propulsion, but at this very early stage no system could be worked out as to how this energy could be tapped. He could go only so far as to merely mention that the power of radium might be considered—nothing more.

There is much evidence that REP's theories reached newspapers by the following year, 1913, and up to the early 1920s, surprisingly even to small-town American newspapers, and undoubtedly must have also appeared in papers large and small in France and other nations. For instance, among these stories was the oft-repeated piece "Airman's Dream" that ran in the *Iron County Register* (Iron County, MO), the *Greenville Journal* (Greenville, OH), and the *Lawrence Democrat* (Lawrenceburg, TN), as well as more obvious treatments like "A Trip to the Moon" in the *Evening Times* (Grand Forks, ND) and the *Aberdeen Herald* (Aberdeen, Washington Territory). Known larger papers that covered REP's spaceflight concepts include the *London Morning Post*, *Evening Star* (Washington, DC), the *New York Herald*, the *Evening Public Ledger* (Philadelphia), and the *New York Tribune*.[32]

For all this—probably because he became more fully involved pursuing his other passionate interest, the advance of aviation—REP made no further headway in developing and publicizing his spaceflight ideas until almost 15 years later, in 1927, when he helped create the REP–Hirsch Prize. The REP–Hirsch Prize was a monetary award and a medal that was awarded annually to the person or institution that had made the greatest contribution to the advancement of astronautics, or spaceflight, the previous year. The award was cofounded by his banker friend, André-Louis Hirsch. REP went on to write important books on the topic, in addition to performing rocketry experimentation between 1932 and 1936, including some on liquid-propellant systems. Without a doubt, REP was France's leading authority on both rocketry and spaceflight throughout all these years.[33]

Then why was REP largely overlooked by both the general public and the scientific community during the pre-World War I period and never afforded anywhere near the acclaim Goddard had certainly received after the release of *Method* in early 1920?

For one, a much shorter paper with an obscure title that was buried in France's *Journal de Physique* could hardly compare with the far more substantial work by Goddard, produced by the world famous and highly respected Smithsonian Institution. But ultimately, it was the uncharacteristically sensationalistic press release circulated far and wide by the Smithsonian that made the difference—this and the fact that Goddard was an esteemed American physics professor and his theories for a flight to the moon *seemed* to be backed up by very solid experimentation that had been supported financially by the Smithsonian. Then, as now, public perception (though very faulty) and the backing of a powerful means of spreading publicity were key

elements in understanding the impact of the first wave of spaceflight publicity and how the then-novel concept of the space rocket entered public
consciousness.

OBERTH AND VALIER TRIGGER THE SPACE FAD

Oberth's thoroughness in describing the liquid-propellant rocket for *manned*
space flight in his *Die Rakete* ultimately made it a far more revolutionary work
than either those of Goddard or REP, and it was consequently very widely
reviewed, as much as or more than Goddard's *Method*. (REP as a contributor
to the international dialogue on spaceflight, meanwhile, remained obscure
until the late 1920s.) And although *Die Rakete* was very technical and full of
mathematics, within a very short time Oberth's book created disciples—
researchers who shared his views, helped disseminate them, and made their
own contributions to what we may call the beginnings of astronautics. One
was the Austrian Max Valier, who was inspired to write a similar work to
explain Oberth's ideas in laymen's language. The result was *Der Vorstoß in
den Weltenraum* (*The Advance into Space*), published in 1924. It was a huge
success, passing through six editions before 1930. The book was accompanied
by Valier's numerous articles and illustrated lectures on the subject of space
travel. In turn, Oberth's *Die Rakete*, widely popularized especially by Valier,
triggered what we know today as the rocket and space fad of the 1920s
and 30s.

Valier, who was not as scientific as he should have been, was nonetheless
a pioneer in his own right. He was famous as one of the most colorful
experimenters of the day, one who built and rode rocket-propelled cars
(using multiple, electrically ignited solid-propellant units), rail cars, and
even ice sleds. The late 1920s was the high point of the fad that also witnessed, especially in Germany, a rash of others who experimented with
rocket cars, rocket-propelled motorbikes, bicycles, gliders, boats, skates—
and the first attempts at "rocket mail." Such stunts were invariably featured
not only in Sunday newspaper supplements but in the newsreels of the day.
Rocket explosions in these so-called experiments were a common
element.

Back in America, Dr. Goddard must have been utterly repelled by all this
activity in which rocketry and spaceflight had suddenly taken on a sensationalistic mantle of showmanship, supported by the press and other public
media of the time. For the same reasons, these nonsense rocket stunts stood
in stark contrast to what the more serious-minded space and/or rocket enthusiasts and leading members of the rocket or spaceflight advocacy groups had
in mind.

Fig. 1.8   Cover of Max Valier's *Der Vorstoß in den Weltenraum* (*The Advance into Space*), 1925 edition. (Courtesy Michael Ciancone.)

Paradoxically, Valier's own rationale had been to bring the *potential* of the rocket to the public. He also conceived of the evolution of rockets up to transcontinental rocket planes that could lead to true spacecraft. Tragically, though, soon after he began to switch from gaseous (carbon-dioxide) to liquid propellants for his rocket cars, Valier was killed in an explosion in 1930—and hailed as rocketdom's first "martyr." He was 35.[34]

Valier was a pioneer in another sense, in helping to create one of the first space-travel advocacy groups, the Verien für Raumschiffahrt (VfR), or Society for Spaceflight, mentioned earlier in this chapter. Founded by about nine men (including Valier) and one woman in the summer of 1927 in the parlor of the *Goldenen Zepter* (Golden Scepter) tavern in Breslau (now Wrocław, Poland), the VfR was preceded by similar groups, but these had been relatively ineffectual and short lived.[35]

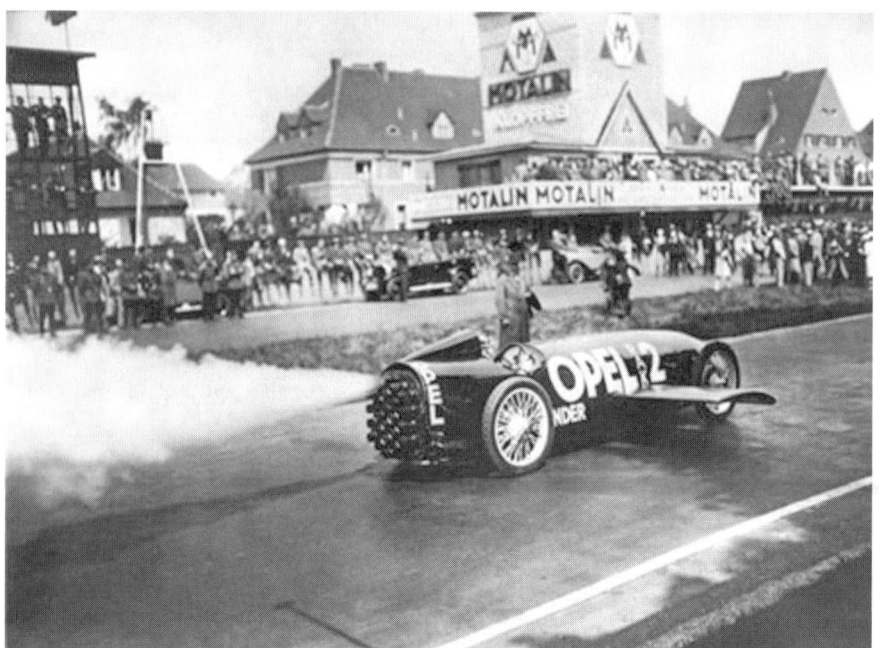

Fig. 1.9   Run of a typical Valier rocket car, the Valier-RAK 2, which reached a speed of 230 km/h (143 mph) on 23 May 1928. (Courtesy Frank H. Winter collection.)

THE SPACE FAD OPENS IN RUSSIA

The first group we know of during the rocket-fad period, however, was the Society for the Study of Interplanetary Travel, formed early in January 1924 as a section of the N.E. Zhukovsky Air Force Academy in Moscow, U.S.S.R.

Indeed, the rocket fad had spread as far as Russia and was therefore truly international in scope. This underscores the tremendous, newfound enthusiasm for spaceflight during that time. Credit must also be given to the Poland-born Yakov Isidorovich Perelman for being the foremost, and perhaps first, popularizer of the works of his countryman Konstantin E. Tsiolkovsky (Poland was under Russian control when Perelman was born in 1882.) But it was not until the mid- to late 1920s, or *after* the emergence of Oberth's seminal first work in 1923, that Tsiolkovsky became more widely known, including overseas.

In addition to Tsiolkovsky, the Russians could also boast of other exceptional early pioneers in the field of rocketry as applied to the potential of spaceflight, notably Friedrikh A. Tsander and Yuri V. Kondratyuk. They were prominent in their country during the 1920s, though at the time they were even less known in the West than Tsiolkovsky.[36]

Among the goals of the Zhukovsky Academy group were to "bring together all persons in the Soviet Union working on the problem," to obtain "full information on the progress made in the West," and to "disseminate and publish correct information about....interplanetary flight."[37] This same group, later renamed the Society for the Study of Interplanetary Communication, or known by its Russian acronym of OIMS, additionally intended to publish a journal named *Raketa* (*Rockets*). But this never came about, and after several very well-publicized public debates (including one about the alleged Goddard moon rocket), OIMS folded by the end of 1924, due in part to a lack of finances—a stark realization of the enormous complexity of the over-ambitious goals they had planned, and the unsettled state of the nation, then just out of the throes of a civil war. A number of other Russian groups followed OIMS, some of which became actively engaged in rocketry experiments during the 1930s.

OTHER SPACE GROUPS AND THE BIRTH OF THE WORD "ASTRONAUTICS"

Small groups also began to appear in Austria from 1926. The American Interplanetary Society emerged in 1930, followed by a number of lesser-known groups in this country, while the British Interplanetary Society (BIS) was formed in 1933 and still exists.[38]

Despite their highly idealistic pronouncements of seeking to "share" their knowledge and ideals internationally toward the cause of spaceflight, for a variety of reasons and with very rare exceptions, these desired interactions by advocacy groups never fully matured. One exception was the the visit by G. Edward Pendray of the American Interplanetary Society to the VfR in which he obtained firsthand details on their rocket hardware and passed them on to the AIS.

Another more subtle yet very important legacy of the space and rocket fad was the introduction of a name for the new science: *astronautics*. The term was invented by the Belgian science fiction writer J.J. Rosny, a pseudonym for Joseph-Henri-Honoré Boex. On 26 December 1927, Rosny, with friends REP, the banker and spaceflight enthusiast André-Louis Hirsch, and others gathered in the home of Hirsch's mother at 47 Avenue d'Iena in Paris to form a committee with the Société Astronomique de France (the French Astronomical Society) to promote spaceflight. At the end of the dinner, they adopted a plan for what became known as the REP–Hirsch Prize, as described previously, to be awarded annually to the person or organization who had made the most notable achievement toward the realization of spaceflight. But a word was needed to describe the subject of the prize. Rosny thought of *astronautics*, an almost literary reference, literally meaning "navigating the stars," although to this day interstellar flight is still very far off.

Nonetheless, the term was appealing and seemed a perfect counterbalance to the discipline of *aeronautics*, meaning "navigation of (or moving through) the air." Astronautics was thus adopted and thereafter, REP helped legitimize the new word in all his future writings on the topic, including the title of one of his landmark books, *L'Astronautique,* in 1930.[39] In 1927, the Hirsch-funded prize itself became known as the REP–Hirsch Astronautical Prize. This was the first known award in the history of spaceflight, though it lasted only from 1928 to 1940 due to the start of World War II.[40] Interestingly, the word *cosmonautics,* preferred by the Russians and other Eastern Europeans, is probably more appropriate, as it means "navigating space" or "flying in space."

GODDARD IN RETROSPECT

As for Goddard, who continued to experiment with liquid-propellant systems almost uninterruptedly until his death in 1945 and was the preeminent authority on rocketry for most of those years, he largely remained very secretive and aloof from ARS and other experimenters. Moreover, a closer examination of his situation reveals that contrary to the widespread popular misconception that the Germans (specifically, the German Army Ordnance) stole knowledge of Goddard's rocket work that led to the development of the infamous large-scale V-2 rocket of World War II, it appears both Goddard and the German Army (Ordnance) conducted their respective rocket programs in secret and entirely independently of each other, often using similar, logical engineering approaches.[41] However, the convergent discoveries and advances by both Goddard and the Germans do not detract from Goddard's greatness and remarkable achievements, which culminated in a total of some 244 patents, many of them taken out posthumously by Goddard's widow. A significant number of his advances also anticipated many features found in later rocket developments.[42]

ENTER THE V-2

To further define the bigger picture, the V-2 (originally designated the A-4), which was the world's first large-scale rocket, had sprung from entirely different roots. In late November 1932, the 20-year-old Wernher von Braun—a young but highly gifted member of the Board of Directors of the VfR—was secretly hired by Army Ordnance to join their nascent rocket program, which had begun in 1929 with solid propellants. What they actually did was to pay him to complete his doctoral thesis on rocket propulsion, allowing him to conduct experiments at the Army Ordnance test range at Kummersdorf, about 15 miles south of Berlin. He eventually was appointed technical director of the program that developed the A-4 (V-2) rocket, which became operational early in September 1944, deployed first against Paris and then London, with other targets in Belgium and the Netherlands.[43]

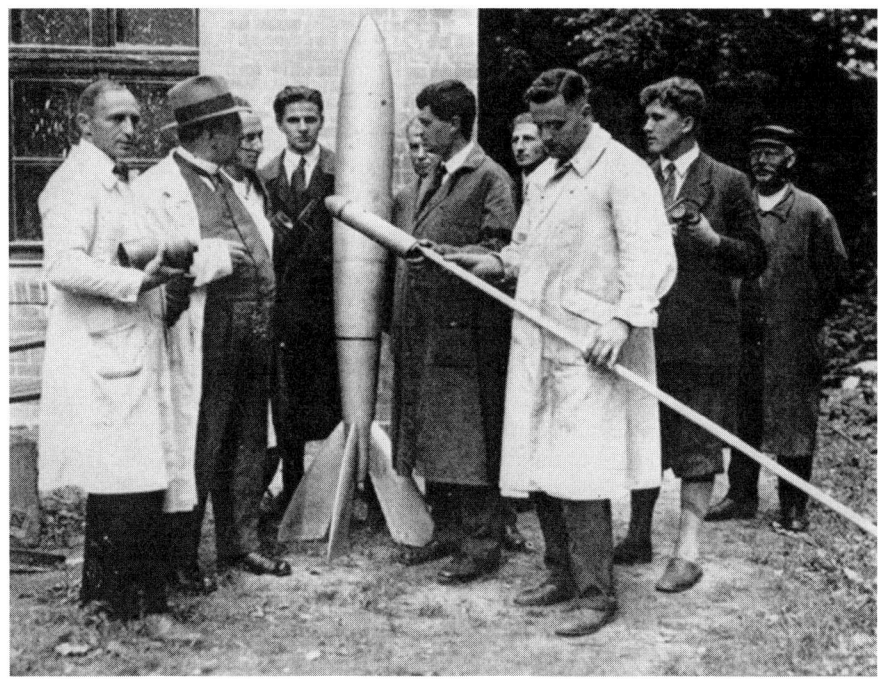

Fig. 1.10   Leading members of the Verien für Raumschiffahrt (Society for Spaceflight), including Hermann Oberth, fifth from left, and a young Wernher von Braun, second from left, in 1930. (Courtesy Frank H. Winter collection.)

The V-2—the world's first large-scale liquid-propellant rocket and direct ancestor of all modern liquid-fuel rockets, including launch vehicles like the giant Saturn V that took men to the moon—thus originated *not* from the work of Goddard, but from roots sown in 1923 by Oberth, which then evolved into the first hardware of the rocket and spaceflight fad that followed. Yet in America, the same fad was surely one of the roots of RMI, which in turn created its own remarkable milestones in the history of rocketry and astronautics.

ENDNOTES

1. It should be noted that RMI actually began their operational history two days earlier, on 16 December 1941, although the company was officially registered on 18 December. (See Frank H. Winter and Frederick I. Ordway III, *Pioneering American Rocketry: The Reaction Motors, Inc. (RMI) Story, 1941–1972,* AAS History Series, Vol. 44, Univelt, Inc., San Diego, 2015, p. 62, Fig. 2.2. The start-of-operations date of 16 December is treated in the latter part of Chapter 2 and at the beginning of Chapter 3 of the current book. In this opening chapter, the generally accepted date of the official foundation of 18 December is used.

2. James H. Wyld, "The Prehistoric Era of the Regenerative Motor," *The RMI Rocket,* Vol. 2, Dec. 1951, p.6.

3. Ibid.
4. Mitchell R. Sharpe, *Development of the Lifesaving Rocket: A Study in 19th Century Technological Fallout,* MSFC Historical Note 4, George C. Marshall Space Flight Center, Huntsville, AL., 1969, p. 49; Winter and Ordway III, *Pioneering American Rocketry*, p. 80.
5. Shesta's early life and later career are nicely summarized in the article by Boonton Herndon, "Rocket Genius with Big Ideas," *Coronet*, Vol. 33, April 1953, pp. 24–28.
6. Ugo Andres (pseudonym for G. Edward Pendray), "Men of Space," *The New Outlook* (New York), Vol. 164, Oct. 1934, p. 28.
7. G. Edward Pendray, "32 Years of ARS History," *Astronautics and Aerospace Engineering*, Vol. 1, Feb. 1963, p. 126.
8. For detailed histories of the AIS and later ARS, as well as the VfR, consult Frank H. Winter, *Prelude to the Space Age: The Rocket Societies: 1924–1940,* Smithsonian Institution Press, Washington, DC, 1983, and Tom D. Crouch, *Rocketeers and Gentlemen Engineers: A History of the American Institute of Aeronautics and Astronautics. . .and What Came Before,* AIAA, Reston, VA, 2006.
9. Variations of the legend of Chang'e appear in the mythologies of Japan, Korea, Indonesia, and other Asian cultures. For coverage of a recently discovered ancient text referring to the legend of Lady Chang'e's flight to the moon and dating to the Zhou Dynasty (1046–256 B.C.), consult Edward L. Shaughnessy, *Unearthing the Changes: Recently Discovered Manuscripts of the Yi Jing (I Ching) and Related Texts,* Columbia University Press, New York, 2014, especially p. 154. For more in general on mythologies concerning the moon, consult David Thomas, ed., *Moon: Man's Greatest Adventure,* Harry N. Abrams, Inc., New York, 1972 and other editions; Hamilton Wright, Helen Wright, and Samuel Rapport, eds., *To the Moon!,* Meredith Press, New York, 1968.
10. Wernher von Braun and Frederick I. Ordway III, *History of Rocketry & Space Travel,* Thomas Y. Crowell, New York, 1969, p. 10.
11. Ibid., p. 9.
12. For this survey of fictional means of achieving space flight, see Frank H. Winter, "The Silent Revolution: How R.H. Goddard Helped Start the Space Age," in Å. Ingemar Skoog, ed., *History of Rocketry and Astronautics: Proceedings of the Thirty-Eighth History Symposium of the International Academy of Astronautics, Vancouver, British Columbia, Canada, 2004,* AAS History Series, Vol. 35, San Diego, Univelt, Inc., 2011, p. 35.
13. For a comprehensive treatment of the Congreve–Hale rocket developments that were international in scope, consult Frank H. Winter, *The First Golden Age of Rocketry: Congreve and Hale Rockets of the Nineteenth Century*, Smithsonian Institution Press, Washington, DC, 1990.
14. For this story see Dwight Smiley, "Bagley's Inter-planet Skyrocket," *Blue Book Magazine*, (London), Vol. 6, March 1908, pp. 1016–1021.
15. For a brief treatment of the history of rocket motion, consult Winter, *First Golden Age*, pp. 225–227.
16. Newton's third law of motion as explaining rocket motion is now very well established. See, for example, von Braun and Ordway III, *History of Rocketry*, pp. 22, 41, 44, who also briefly discuss how it took centuries for this law to become recognized for its application to the rocket for space flight.
17. Ibid., p. 41.
18. Aarushi Gupta, "The Ideal Rocket Equation (Tsiolkovsky Equation)," 4 June 2015, online; David Baker, *The Rocket,* Crown Publishers, Inc., New York, 1978, p. 19. Consult also Arkady A. Kosmodemiansky, "First Works by K.E. Tsiolkovsky and I.V. Meshcherrsky on Rocket Dynamics," in R. Cargill Hall, ed., *History of Rocketry and Astronautics,* AAS History Series, Vol. 7, Part 1, Univelt, Inc., San Diego, 1986, pp. 115–119.

19. Winter, "The Silent Revolution," pp. 6–7, 41 (Backnote 10).

20. Goddard must have read the version by Serviss in the *Boston Post* that was differently titled than the one in the *New York Evening Journal* and was also localized as "Fighters from Mars, or the War of the Worlds in and Near Boston." Consult also Frank H. Winter, "Garrett P. Serviss, the Would-be-astronaut Who Helped Start the Space Age," *Griffith Observer* (Griffith Observatory, Los Angeles), Vol. 74, July 2010, pp. 3–17. Little known is that "Edison's Conquest of Mars" was also serialized in the *Los Angeles Herald* in 1898. This shows the extent of the Mars furor across the country by this time.

21. This well-known, life-changing episode in Goddard's life is covered as the opening chapter of the authorized biography by Milton Lehman, *This High Man: The Life of Robert H. Goddard,* Farrar, Straus, and Company, New York, 1963, pp. 1–26.

22. For a more complete summary of Goddard's earliest investigations of a possible means of achieving space flight prior to his discovery of the rocket, consult Frank H. Winter, *Rockets into Space,* Harvard University Press, Cambridge, MA., 1990, pp. 14–15, 18. The summary from Goddard's "theoretical period" was drawn from an examination of his original notes held at Clark University's Robert H. Goddard Memorial Library in Worcester, MA. For the most complete work on Goddard's life and career, see Esther C. Goddard and G. Edward Pendray, eds., *The Papers of Robert H. Goddard,* 3 Vols., McGraw-Hill, New York, 1970.

23. Goddard's earliest-known notes documenting his realization that the rocket was the solution to space flight are found in Goddard and Pendray, *The Papers*, Vol. 1, pp. 95–99.

24. For a reprint of this section in Goddard's *Method of Reaching Extreme Altitudes,* consult Goddard and Pendray, *The Papers*, Vol. 1, pp. 384–395.

25. For details on the international impact of the release of Goddard, *Method*, consult Winter, "The Silent Revolution," pp. 3–54.

26. Hans Barth, *Hermann Oberth Leben Werk Wirkung* Uni-Verlag, Feucht, Germany, 1985, p. 368.

27. Goddard and Pendray, *The Papers*, Vol. 1, pp. 485–486.

28. For Oberth's contributions overall, consult Hermann Oberth, "My Contributions to Astronautics," in Frederick C. Durant III and George S. James, eds., *First Steps Toward Space*, AAS History Series, Vol. 6, Univelt, Inc., San Diego, 1985, pp. 129–140.

29. Robert Esnault-Pelterie, "Considérations sur les résultats d'un allégement indéfini des moteurs" ("Considerations of the Results of the Unlimited Lightening of Motors"), *Journal de Physique* (Paris), Series 5, Vol. 3, March 1913, p. 218. The full article appears on pp. 218–230, but was fortunately reprinted in Durant and James, *First Steps*, pp. 23–31.

30. Ibid.

31. Durant and James, *First Steps*, p. 26.

32. "Airman's Dream" appeared in the *Iron County Register* (Iron County, MO) of 23 Oct. 1913, the *Greenville Journal* (Greenville, OH) of 23 Oct. 1913, and the *Lawrence Democrat* (Lawrenceburg, TN) for 29 Oct. 1913. "A Trip to the Moon" appeared in the *Evening Times* (Grand Forks, ND for 10 June 1912 and the *Aberdeen Herald* (Aberdeen, Washington Territory) for 3 Feb. 1914. "Motoring to the Moon," originally from *London Morning Post*, appeared in the *Evening Star* (Washington, DC) on 22 March 1913; "Promises Trip to the Moon," (originally from *Westminster Gazette*, presumably *Westminster,* London, England) was printed in the *Evening Star* (Washington, DC) on 25 June 1922; "Journey to the Moon" in the *New York Herald* on 4 June 1922; "Scientists Scoff at Trip to Stars" appeared in the *Evening Public Ledger* (Philadelphia) on 8 June 1922; and "From the Earth to the Moon," in the *New York Tribune* on 16 July 1922. Surprisingly, further research is required to track down publicity on REP's spaceflight ideas in Europe.

33. For a summation of the contributions of REP, see Lise Blosset, "Robert Esnault-Pelterie: Space Pioneer," in Durant and James, *First Steps*, pp. 5–21. For REP's full biography,

consult Félix Torres and Jacques Villain, *Robert Esnault-Pelterie du ciel aux étoiles, le génie solitaire*, Éditions Confluences, Bordeaux, Sept. 2007.

34. For Valier's biography, consult I. Essers, *Max Valier: Ein Pionier der Raumfahrt*, Verlagsantalt Thesia, Bozen, Austria, 1980; and I. Essers, *Max Valier: Ein Vorkampfer der Weltraumfahrt, 1895–1930*, VDI-Verlag GmbH, Dusseldorf, 1968, translated as I. Essers, *Max Valier: A Pioneer of Space Travel*, NASA Technical Translation TT F-664, National Aeronautics and Space Administration, Washington, DC, Nov. 1976.

35. For an overall history of the VfR and many other similar early space-flight-advocate and rocket groups up to 1940, consult Frank H. Winter, *Prelude to the Space Age*.

36. For the works of Tsander Kondratyuk (real name: Aleksandr Ignatyevich Shargei) and the overall early development of the astronautics in the Soviet Union, consult Asif Siddiqi, *The Rockets' Red Glare: Spaceflight and the Soviet Imagination, 1857–1957*, Cambridge University Press, Cambridge, 2010, a work that originated as its author's doctoral dissertation.

37. Winter, *Prelude*, p. 27. Consult this same work for the histories of these other groups up to 1940.

38. For the BIS, see Winter, *Prelude*, pp. 87–97.

39. For the introduction of the term into America, see the editorial "Astronautics," in *The New York Times*, 8 March 1928, p. 24.

40. For a history of the REP–Hirsch Prize, see Frank H. Winter, "The Birth and Rise of 'Astronautics': The REP–Hirsch Astronautical Prize 1928–1940," *Quest,* Vol. 14, No. 1 (2007), pp. 35–40.

41. For an analysis of the alleged German use of Goddard's material toward the development of the V-2, see Frank H. Winter, "Did the Germans Learn from Goddard? An Examination of Whether the Rocketry of R.H. Goddard Influenced German Pre-World War II Missile Development," paper presented at the 65th International Space Congress, Toronto, Canada, 2014, in Marsha Freeman, ed., *History of Rocketry and Astronautics*, AAS History Series, Vol. 46, Univelt Inc., San Diego, 2016, pp. 99–130, and also appearing in *Acta Astronautica*, Vol. 127, Oct.–Nov. 2016, pp. 514–525. The same study and other sources show that German Army Ordnance initially had been influenced by the well-publicized VfR experiment begun in 1931 to begin turning their attentions to the possibilities of the liquid-propellant rocket as a potential weapon. However, they found that there was far too much "showmanship" at the Raketenflugplatz. Thus, Capt. (later Maj. Gen.) Walter R. Dornberger, the future military commander of the A-4 program, and his colleagues decided to undertake their rocket developments entirely in secret and, as much as possible, under "one roof" (that is, inhouse development). Eventually, this effort involved thousands of technicians and scientists, especially after the establishment of the multimillion-Reichsmark rocket research center of Peenemünde, opened in 1936.

42. For a complete listing of Goddard's patents, see Goddard and Pendray, *The Papers*, Vol. 3, pp. 1651–1660, and for a later, more critical biography of Goddard, see David A. Clary, *Rocket Man: Robert H. Goddard and the Birth of the Space Age*, Hyperion, New York, 2003. This work includes an examination of the myths about Goddard, including the one suggesting that his work was used by the Germans to create the V-2.

43. There are a number of biographical works on von Braun. Consult, for example, Michael J. Neufeld, *Von Braun: Dreamer of Space, Engineer of War*, Alfred A. Knopf, New York, 2007.

Chapter 2

JIMMY WYLD'S BREAKTHROUGH AND THE FOUNDING OF RMI

"The greatest obstacle at the present is the lack of a [rocket] motor capable
of withstanding the effects of firing for a sufficiently long time."

– John Shesta, "Report on Rocket Tests,"
Astronautics, No. 31, June 1935.

ENTER JAMES H. WYLD

James Hart "Jimmy" Wyld was an eclectic and highly gifted individual. In
fact, he had been a child prodigy. He practically taught himself to read at the
age of four and later in his youth read the entire 24-volume set of the *Book of
Knowledge* encyclopedia—several times—and was blessed with a very reten-
tive memory.

His interests came to range from aviation (along with an "extensive knowledge of…aerodynamics") to magic, physics, radio, and television (in its early
experimental forms), to photography and astronomy, telescope-making, and
eventually, rocketry.[1] He was also an "amateur optician" and by his own
admission, "a good extemporaneous speaker on technical subjects" and "an
enthusiastic but unskillful accordionist."[2] On top of these talents, he read both
German and French, dabbled in writing prose and poetry, acted in school
plays, and was a devotee of science fiction.

The son of a prominent mechanical engineer, Robert Hasbrouck Wyld,
Jimmy also had an excellent education. He studied under a private tutor from
the ages of 11 to 14, followed by attendance at two prestigious private schools:
the preprep Harvey School in Hawthorne, New York, and the Salisbury School
in Salisbury, Connecticut. Wyld next entered Princeton University, graduating
in 1935 with high honors.

He earned a Sayre Fellowship in electrical engineering and continued
graduate studies at Princeton in 1936. In the meantime, he had been a member
of the American Society of Mechanical Engineers (ASME) since 1933 and
was awarded the Brasher Prize by the Princeton branch of ASME in recognition of two papers read before the branch: "The Modern Telescope Mounting"
(1934) and "The High-Power Rocket" (1935).[3]

Wyld's parents were both prominent high achievers. His father, Robert, was
a graduate of Columbia University, authored a technical article or two (one
titled "Superheated Steam in Industrial Plants" appeared in *Industrial &*

Engineering Chemistry for May 1911), and took out a number of patents, primarily on improved drying apparatuses. Jimmy was thus undoubtedly greatly influenced by his father on the technical side. His mother, Margaret Rebecca Hart, was the daughter of a physician and had obtained a master's degree from the University of Kentucky at a time when few women even entered college.

Wyld's lifelong interest in aviation must have been tremendously stimulated from growing up in an affluent neighborhood in Garden City, Long Island, just a mile from a flying field, most likely Roosevelt Field where Lindbergh took off on the morning of 20 May 1927 for his world-famous solo flight across the Atlantic to Paris. It is not known if Jimmy, then 15, witnessed the takeoff. He later took a course in aerodynamics at Princeton, and in his senior year designed and built a small wind tunnel.

Wyld's other lifelong interest—rocketry—began in 1931 when he read David Lasser's *Conquest of Space*, published that year as the first English-language work on the topic. Thanks to fluency in both German and French, he was also able to read and fully comprehend the works of Oberth, Valier, Esnault-Pelterie, and later pioneers like the Austrian Eugen Sänger. This is well proven in Wyld's own copy of Oberth's *Wege zur Raumschiffahrt* (*Ways to Spaceflight*) of 1929, which he obtained in August 1936, evidently from a book store in Munich while on a post-graduation trip to Europe with college friends. Jimmy's copy, with its well-worn covers, is presently owned by Michael Ciancone, an engineer with NASA's Johnson Space Center, as part of

Fig. 2.1 Cover of David Lasser's *Conquest of Space* (1931), the first English-language book on the topic of spaceflight. (Courtesy Michael Ciancone.)

his impressive collection of similar early classics in astronautical literature. It contains about 200 marginal comments (some of them severe criticisms) on Oberth's text, written in pencil in his very neat and legible handwriting.

JIMMY JOINS THE ARS

Wyld did not become active in liquid-propellant rocketry until 1935, when he joined the American Rocket Society. He related in his diary how this came about.

In the fall of 1934, Wyld read the article "Men of Space" in the magazine *New Outlook* that offered thumbnail sketches of leading rocket experimenters of the day, including some in the ARS. The author of this piece was one "Ugo Andres." Young Wyld then attempted to contact "some of these fellows," but without much success.[4] He then decided to write directly to the ARS and obtained their address from the New York City phone book. His letter, which stated that he was keenly interested in the society's work, "especially the actual experiments," was received and forwarded to G. Edward Pendray. Pendray turned out to be "Ugo Andres."[5]

Jimmy and his new friend William E. "Bill" Rahm Jr. were so eager to get into rocketry that a couple of weeks before Jimmy showed up at the ARS office to sign up for membership, the boys went ahead and built their own "small experimental liquid-fuel rocket."[6] This was Wyld's first exposure to the then-nascent technology. "Rahm's uncle, Professor Lewis F. Rahm of Princeton," he further explained, "gave us some instructions as to nozzle design, and [Bill] Rahm calculated the nozzle dimensions while I worked at the mechanical features. We worked in a basement room in the [University's] new observatory, by permission of [astrophysicist] Prof. [John Q.] Stewart... We eventually evolved a design very much like the German '3rd Mirak' [of the VfR] and this effusion was proudly submitted to Mr. Pendray when I first met him a few weeks later."[7] Interestingly, Professors Rahm and Stewart were among the few academics who supported the new field of experimentation with liquid-propellant rocketry and spaceflight in general, both usually held in low esteem by academics of the time.[8] In another recollection, Wyld wrote that he and Bill Rahm could use the basement room in the observatory but "we were required not to shoot off rockets inside the building!"[9]

The unbridled enthusiasm of Jimmy and his friend was soon dashed, however, when Pendray severely criticized their design. Jimmy omitted the details in his diary except to note that he and Bill "later concocted another design, a 'Repulsor' type similar to the Am. Rocket Society's No. 1, but much smaller (only three feet long)."[10] "As spring wore on," he continued, "Rahm got busy with other matters and I was left to begin the actual construction of the new rocket by myself. I got no further than a bit of fussing around in the [Princeton University] Physics machine shop—I had trouble with the machine work, especially the brazing, and as exams came along I had to give up."[11] Wyld's first venture into liquid-propellant rocketry was therefore very rushed and modest and did not seem very propitious.

These setbacks notwithstanding, a year later in 1936, young Jimmy was appointed the associate editor of the society's journal, then called *Astronautics.* In addition, he was made a member of their Experimental Committee. He then quickly became "deeply engrossed," in his words, "in the Society's early experimental tests…with a crude portable test stand."[12]

EARLIEST ARS EXPERIMENTS

Altogether, the society had built four rockets between 1932 and 1934, although only two flew: ARS No. 2 went up to 250 ft and ARS No. 4 soared to an undetermined altitude and landed about 1,338 ft from its starting point. Pierce and others had started designing No. 5, but this project was abandoned late in 1934 when it was wisely decided to shift to static tests. Very little could be learned from the all too brief and scarce flights, and their Depression-era shoestring budget also made it impractical to continue that course. More could be gained about rocket-motor performance from static tests, leading to better designs and performances.

ARS newcomer John Shesta largely designed and built the stand that afterward became known as ARS Test Stand No. 1. It *had* to be portable because, unlike the society's German counterpart, the VfR (which had folded during the winter of 1933–1934 as the result of an internal political embroilment), the American group did not have the luxury of having a test facility like the Raketenflugplatz; they barely had an office. Rather, the American experimenters were forced to search for this field or that (either in New York state or neighboring New Jersey), then sometimes post lookouts to watch for the approach of any fire marshals and the like so they could conduct their tests in relative peace.

As for the ARS office, Wyld candidly recorded in his diary his surprise to learn that when he visited them to join, the society's secretary was then Dr. Samuel Lichtenstein—who turned out to be a dentist—and "the Society's 'office' consisted of a filing case in his dental studio!"[13] However, ARS meetings were held in more suitable places, like the Academy Room of the American Museum of Natural History, which had afforded them this privilege.

In short, during these difficult times, the ARS was only an amateurish operation, although with a number of highly enthusiastic young and idealistic members eager to learn and advance the so-called state-of-the art of the arcane and relatively brand new field of rocketry. But to these highly dedicated men—people like Pierce, Wyld, and Shesta—the experimentation was an exciting, often dangerous endeavor, that also gave them the opportunity to enjoy a close comradeship.

Shesta's stand was limited to recording thrusts up to only 100 lb and durations of runs using a large clock arrangement that included a prominent second hand for all to see at a safe distance. No temperature measurements were possible. Additional features included a handful of dials that reported pressures of the propellants and pressurizing gas, as the propellants were pressure-fed by nitrogen. To save money and time, Shesta had scrounged tanks, valves, and other parts from ARS Rocket No. 4 to help build the stand.

As for the testing itself, each of the small, hand-held test motors constructed by individual members was simply bolted down within a vertical shaft set within a rectangular steel framework, the nozzle pointing skyward. The shaft was also moveable and thus served as a sort of hydraulic plunger for measuring the thrust, which could be read on a large dial mounted to a nearby detached wooden sawhorse to which were also mounted other dials. (In those years, the term "thrust" was not common; instead, they used the term "reaction.") There were five dials in all, mostly for pressures, and these were linked to the stand proper by feed-off pressure pipes. Of course, one of the dials was a clock, with a very slow second hand, to mark the exact duration of each run. Overall, this was indeed a primitive setup, but it worked. And for each test with the stand, designated ARS observers took notes while crouched nearby, as well as photographs and films, and drew up performance graphs that were later published in the mimeographed journal of the ARS, *Astronautics*.

The baptism of Stand No. 1 took place in Crestwood, New York, (near Pendray's home) on Sunday, 21 April 1935. In it, two different types of nozzles were tried with the same motor in different runs. It was determined by Shesta, who wrote up the test results and finding that were later published in *Astronautics*, that long nozzles were of "no value."[14] "Another point strikingly brought out," he added , "is that the greatest obstacle at the present is the lack of a motor capable of withstanding the effects of firing for a sufficiently long time."[15]

In short, the motor overheated within split seconds, almost as soon as it started. Rapid overheating and inability to maintain longer, steady firings were by far the most severe technical problems that plagued the ARS throughout its entire history of static testing. There were occasional bursts of these small motors, in one instance causing injury to a bystander.

As a result of that serious mishap, the ARS board of directors made sure that for each test thereafter, a special attendance sheet was issued and circulated to all the attendees at the site prior to the testing. At the top of the sheet was a statement, which everyone had to agree to by signing below in the appropriate column. One column was labeled "participants," that is, the experimenters; the other was labeled "spectators."[16] The statement read, in part: "I hereby agree to absolve the American Rocket Society, and its individual experimental personnel, from any claim of any nature whatsoever for injury or damage to persons or property which might occur during the course of experiments conducted this day."[17] Another statement stipulated that each signee also affirmed that they understood "with full knowledge, and having been duly advised, that there are certain hazards involved in being present, as participant or spectator, while experimental work on rocket motors is being conducted."[18]

Naturally, the ARS experimenters had no way of knowing details of any of Goddard's rocket work, or how he solved his own rocket motor cooling problems; nor did either of these parties know how the top-secret German Army Ordnance rocket program handled their own rocket cooling approaches.

ARS Test Stand No. 1 nonetheless served the society very well. The crude rocket motors gradually improved, although overheating remained a perennial issue. The testing apparatus was used until late October 1935, or for only six months, when it had already become obvious that a newer, greater thrust(reaction)-capacity stand was required. This took longer than expected. It was not until three years later, on 22 October 1938, that ARS Test Stand No. 2 became operational for a test in New Rochelle, north of New York City, when Pierce's "tubular" Monel motor was tried.

THE EVOLUTION OF WYLD'S REGENERATIVELY COOLED MOTOR

Meanwhile, a lot had transpired. Jimmy Wyld had evolved his own regeneratively cooled rocket motor that at last offered a promising solution to the vexatious overheating problem.

There are varying accounts as to how Wyld came to choose this approach to cooling. To help commemorate RMI's tenth anniversary in 1951, the company published a historically invaluable special issue of their house organ, *The RMI Rocket,* that featured retrospectives by several RMI pioneers, Wyld included. In his account he offered the following: "I cannot now recall where I picked up the idea of cooling a rocket regeneratively; it was not original with me, of course, and one major source of inspiration was the early work of Eugen Saenger [aka Sänger] at the Vienna Technical College [Wiener Hochschule] in 1933–1934, which I learned about through a fellow member of the [American] Rocket Society, Peter van Dresser."[19]

Another source of inspiration, although Wyld had neglected to cite him, was Oberth. In his classic *Die Rakete* of 1923, according to Irene Sänger-Bredt and Rolf Engel in their landmark history paper, "The Development of Regeneratively Cooled Liquid Rocket Engines in Austria and Germany, 1926–42," Oberth was the first to theoretically propose "a regenerative cooling process."[20] Furthermore, they add that through his *Die Rakete*, Oberth had virtually "influenced the overall development of liquid rockets in Germany

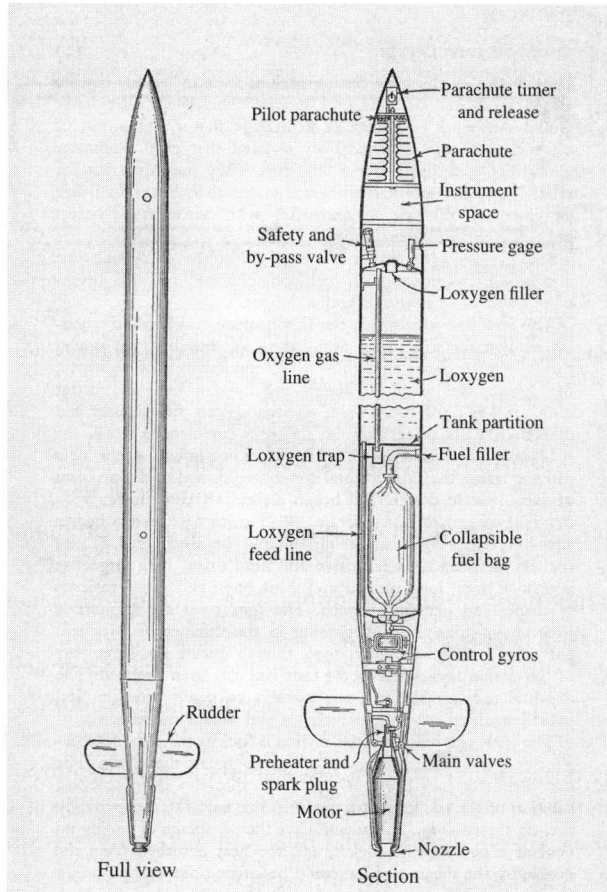

Pilot parachute
Parachute timer and release
Parachute
Instrument space
Safety and by-pass valve
Pressure gage
Loxygen filler
Oxygen gas line
Loxygen
Tank partition
Loxygen trap
Fuel filler
Loxygen feed line
Collapsible fuel bag
Control gyros
Rudder
Preheater and spark plug
Main valves
Motor
Nozzle
Full view
Section

Fig. 2.2   Jimmy Wyld's 1935 design of a stratospheric, or sounding, rocket, (Courtesy Smithsonian Institution, 4561-C.)

and Austria."[21] Wyld, of course, had thoroughly read and consulted this classic book countless times and had made extensive marginal notes in Oberth's expanded work of 1929. What he also should have mentioned is that without a doubt he incorporated Oberth's favorite propellant of alcohol as a rocket fuel which, when mixed with water, helped cool the motor. Jimmy deliberately chose alcohol for his regen motor and therefore, his revolutionary development may claim an additional connection with Oberth.

George P. Sutton's widely recognized standard sourcebook on rocket propulsion technology, *Rocket Propulsion Elements*, has gone through some eight editions since it was first published in 1949. It provides a concise, basic definition of a regeneratively cooled rocket motor: "'Regenerative-cooling,'" Sutton wrote, "denotes the type in which the rocket chamber is cooled by circulating some or all of the liquid propellant (fuel or oxidizer) through a jacket (or coils)

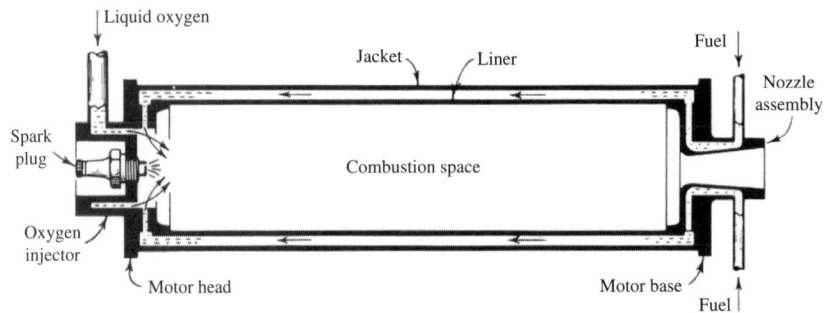

Fig. 2.3 Schematic of Wyld's regeneratively cooled rocket motor. (Courtesy Smithsonian Institution, 4560-E.)

around the combustion device. The heat absorbed by the coolant is therefore not wasted but actually augments the heat content of the propellant prior to combustion."[22] By "augments," Sutton meant that the propellant is preheated, which slightly increases the exhaust velocity of the motor. Regenerative-cooling is therefore an ideal solution to cooling and also increases the efficiency of a rocket motor.

Arriving at this arrangement and making it as simple and as light as possible was vastly more complex than it first appears, and required enormous engineering and mathematical skills that Jimmy Wyld possessed in great abundance. The process involved endless calculations of heat transfers and other thermal analyses, sound knowledge of metallurgy, and on top of these, a mastery of rocket engineering and machining. Testing, retesting, and further retesting were required to verify the design.

"A series of old sketches show that I was evolving the idea all through 1936," Wyld's recollection goes on.[23] "At the time, I lived with two young electrical engineers in a large apartment in Greenwich Village. We...had a small workshop fixed up in an old pantry at the back of the apartment...One of my room mates...had a nine inch South Bend lathe...It was on this machine that my first model regenerative motor, of 100 pounds thrust, was built, mostly in the spring of 1938. The motor was finished by fit and starts, and spent part of the time as an exhibit in the Hayden Planetarium [in New York City]."[24] Curiously, this motor therefore briefly served as a museum piece before it was ever fired.

"At about the same time," Wyld continued, "a new test stand was under construction by John Shesta—a small portable affair, but greatly improved over the original one of 1935."[25] ARS Test Stand No. 2, which still exists and is on exhibit at the National Air and Space Museum (NASM), had mostly been built in Pierce's basement shop in the Bronx. Unlike ARS Test Stand No. 1—modularized in two separate components—No. 2 came as one single arrangement, but was far heavier at 300 lb. A rickshawlike cart with two large wheels and towed by a car was built to haul it to testing fields. Also, the stand's thrust was now doubled to 200 lb. Other marked differences were that each test motor was fired horizontally, not vertically, and all the dials were affixed to the front, or face, of the stand.[26]

Prior to the initial testing of the new stand at New Rochelle in mid-October, Wyld, an avid photographer, took a series of photos of it, probably in Pierce's Bronx basement a few weeks before, when he fit-checked his own motor on it. These rare views reveal not one but two important early American rocketry artifacts in mint condition—the stand, and Jimmy's first regen motor.

Why was the stand hauled way up to New Rochelle for its first tests? Many of the details are obscure, but the answer centered around a little-known smaller sister group of the ARS, the Westchester Rocket Society (WRS) formed in Westchester County, New York, in 1936. One of its founding members, Tucker Gougelman, resided in New Rochelle. New Rochelle offered plenty of room to set up the test stand, and Tucker's father's very spacious home featured a barn or shed where the rickshaw and stand it carried could be safely stored overnight, or longer if necessary. And that is exactly what happened.

Fig. 2.4   Wyld's motor "fit-checked" into the new ARS Test Stand No. 2, a photo taken by Wyld on 13 September 1938, a month before the stand was first used. (Courtesy Smithsonian Institution, 9A05171.)

Initial and Later Tests of Wyld's Motor

The inaugural test of Stand No. 2 was made on 21 October 1938 in an open field, close to a large water tower near New Rochelle, as clearly seen in photos and extant film footage in the NASM's collections. By contrast, Wyld's brand-new regen motor was first tested on the same stand several months later on 10 December, within the stone walls of the foundation of an abandoned unfinished house adjacent to the Gougelman family home. The latter arrangement allowed the use of a long extension cord from Gougelman's house to electrically power part of the stand. (Wyld had originally intended to use a spark plug to ignite his motor, but thought this might be too experimental, and so he opted for a simpler pyrotechnic igniter instead.)

"You can imagine," Wyld wrote in a letter to Peter van Dresser the next day, "how excited I was when we fired up—how excited everyone was, to see how

Fig. 2.5   Wyld, at left with gloves, at the 10 December 1938 test of ARS Test Stand No. 2. (Courtesy Smithsonian Institution, 92-17120.)

this radical and much discussed motor would behave. On opening the valves, a long, crackling, diffuse yellow flame some eight feet long shot out from the nozzle and burnt for three or four seconds, when it suddenly shortened and burned for another two or three seconds with a short blue flame, marked by dark and light Mach waves at regular intervals."[27]

"Amid the greatest excitement," a few moments after the test, he continued, "we examined the motor [and] to our great delight, it had not only not burned out, but was perfectly untouched! Aside from a little sulphur and soot from the fusée [the pyrotechnic igniter] on the outside of the muzzle [the nozzle], it was impossible to tell that the motor had fired."[28] It was therefore decided to give the motor a second trial that same day. That run used the remaining 3.5 lb of alcohol as the fuel and 6.5 lb of the liquid oxygen (LOX) as the oxidizer.

Wyld, who also had a flair for writing long and beautifully descriptive letters, typically on yellow legal pads (this one ran to 14+ pages), dramatically continued his account for van Dresser: "At 3:30, as the grey afternoon was wearing on towards evening, we fired up again, with anxiously beating hearts. And what a run that was! This time…it burned and burned, with a roar fit to knock down the wall."[29]

The official ARS report and its accompanying graph more succinctly summed up the overall results. The maximum thrust was 91 lb, the maximum calculated exhaust velocity was 6,870 ft/s, and the maximum thermal efficiency, about 40%.[30] These might not seem impressive by today's standards, but according to the conclusion of the report, "they represent a great advance on those obtained in former tests, and are among the highest ever recorded… and definitely proves the feasibility of the regenerative method of cooling. The Experimental Committee is now preparing for another series of tests."[31]

The next opportunity to try Wyld's regen motor was long in coming. These were the years of the Great Depression, and despite his many skills and brilliance, Wyld found it very difficult to maintain a job, much less improve his motor. He had always entertained the idea of using this motor in a sounding rocket but this, too, could not be accomplished then, although his plans along these lines were published in *Astronautics*.

It was not until as late as 8 June 1941—a full two and a half years after the triumphs at New Rochelle—that he was able to again prove his vaunted motor. This took place in Midvale, New Jersey, and a new ARS member was present, Lovell Lawrence Jr., who was assigned the duty of operating the ignition switch. Jimmy's motor had undergone minor changes, but was otherwise identical to the original version. This time the propellants were alcohol and liquid air instead of LOX. The report in *Astronautics* for August 1941 started in a dramatic tone: "The air in the valley vibrated, as did the spectators for these tests are exciting to witness. About the middle of the run there occurred a series of chugging sounds and fluctuations were visible in the exhaust flame. This seemed to have little effect on the thrust intensity which

hovered between 80 and 85 lb for the better part of the 26 second run."[32]
Once more, the "chugging" aside, the Wyld regen motor had amply proven
itself.

On 1 August 1941, a little more than two months later, the regen motor was
again secured in Test Stand No. 2 in Midvale, and this time it was subjected to
three additional separate tests. The October issue of *Astronautics* briefly
summed up the results. The first test "saw the motor fire for 21.5 seconds. The
violet flame stretched out a distance of…three feet from the nozzle. The deep
roar was interspersed with sudden detonations spaced out about five seconds
apart [that is, "chugging"]. This phenomenon occurred in all three of the runs,
but did no harm to the motor, though shaking the test stand and the cars [of the
participants]. The second run lasted for 23 seconds, while the final one, with a
leaner [propellant] mixture than previously used, lasted for a surprising time
of 45 seconds" and the thrust "at times reached 135 lbs."[33] Experimental
Committee member Roy Healy proudly concluded in his report that Wyld's
little motor "proved conclusively that a reliable motor for [an] aerological
sounding rocket has at last been designed, built, and tested."[34] Wyld's groundbreaking motor was modest in size. It was a handheld affair, 8 in long and
weighing less than 2 lb.

**Fig. 2.6 Wyld holds his regen motor during a break in ARS testing on 8 June 1941.
(Courtesy Smithsonian Institution, 83-2821.)**

Fig. 2.7 Wyld's motor firing in test of 1 August 1941, also in Midvale. (Courtesy Smithsonian Institution, A4319B.)

THE BEGINNING OF THE WAR AND THE START OF RMI

By this juncture, World War II had already opened in Europe and many Americans were rightly fearful their nation would soon be dragged into it. "The darkening war clouds of 1941," Wyld later recalled, "led Lovell Lawrence to think that our motor work would be of some interest to the U.S. Government, and sometime early in the fall of that year Lawrence, Pierce, Shesta and myself met to discuss plans for a rocket motor company"[35]

"Within a few weeks," Jimmy continued, "Lawrence had succeeded in wangling a visit from a U.S. Navy representative to witness a test run of the regenerative motor [sic.]...in a hidden spot in the woods near Wanaque [in Passaic County, New Jersey, a few miles northwest of Pompton Lakes]. It was a good run, and he departed quite enthusiastic."[36]

We now know that the Navy representative was then-Lt. (later Cmdr.) Charles Fink Fischer. "Fink," as he preferred to be called, had been taught to fly at age 14 in 1927 by a World War I pilot, then joined the Navy at 16. Later, he entered the U.S. Naval Academy and in 1939 enrolled at the University of California at Berkeley on a scholarship toward both a B.A. and M.A. in aeronautical engineering. Fascinated with the potential application of the rocket to boost aircraft, he had met Goddard in the same year on his way to California and became one of his staunchest supporters—and was one of the few whom Goddard trusted. Fink afterwards served as one of the Navy's earlier liaison officers with Goddard when Goddard obtained a wartime Navy contract to help develop JATOs (Jet-Assisted-Take-Off) units to boost heavily loaded seaplanes for the Pacific campaign. Fink was to do the same for RMI.

The U.S. Navy's interest in rocketry began earlier and can be traced to a young naval officer named Robert C. Truax. Truax had been bitten by the rocket bug during the rocket fad when still a high school student in the late 1920s and early 30s. He devoured newspaper accounts about Goddard and the rocket-car stunts in Germany, in addition to "Buck Rogers" comic strips. Later, as a midshipman at the U.S. Naval Academy at Annapolis, Maryland, Truax undertook serious rocket experiments in whatever free time he could muster, with both liquid and gaseous propellants, including a test of one of his motors on the ARS Test Stand No. 2.

By May 1941, based upon Truax's very persuasive recommendations and backed up by his experiences in rocketry, an inhouse JATO project was initiated at the U.S. Naval Engineering Experiment Station (NEES) in Annapolis, seven months before RMI entered the scene. Lt. Truax was thereby designated the officer-in-charge of the Navy's first rocket program, beginning the lifelong professional activities in rocketry that became wrapped up with his naval career. At about the same time, Lt. Fischer, then temporarily in the Naval Reserve, was named to fill Truax's one-man rocket desk job in Washington, D.C., and in effect, became Truax's number-two man. Lawrence's approach to the Navy a few months later proved incredibly timely, and became an even more important endeavor after the bombing of Pearl Harbor early that December. It was then generally believed that JATOs ultimately had the potential to reduce takeoff distances by as much as 60%.

Meanwhile, Wyld had produced a neatly handwritten 17-page report titled, "Wyld 'M-15' Regenerative Rocket Motor — Sketches, Notes, & Discussions," dated 25 September 1941.[37] Whether this report was turned over to the Navy *before* the secret demonstration of Jimmy's motor in the woods of Wanaque in front of Lt. Fischer, or *after*, is not known. As seen from further evidence presented later, Lawrence's letter to Shesta of 4 December 1941 was probably prepared before the demonstration. In fact, Wyld's M-15 report seems to have been a significant early step in the chain of events leading up to the founding of RMI.

The M-15 designation in the title of the report is significant by itself in the development of both Jimmy's motor and the subsequent early history of RMI. Specifically, it is safe to conjecture that "M-15" simply meant "Motor 15" and that Wyld had gone through about that many design iterations before the final model of 1941. Notably, as he related in a talk he gave on his motor to ARS members in their December 1938 meeting, Wyld said he went through "over a dozen preliminary designs" before arriving at his final version.[38]

Another fascinating and recently discovered piece of evidence around developments leading to the formation of RMI is a partial, though key, letter sent by Lawrence to Shesta. It was written on 4 December 1941—just three days before Pearl Harbor.

> I had another very delightful discussion with Lt. Fischer today," the letter begins.[39] "He certainly is being extremely helpful. We have two methods of of procedure in approaching the Navy department. One: to propose to the Navy department that we develop a motor of certain proportions…for a sum of money dealt out to us as we need it. This puts the department in almost complete control of the development…At the end of the development period we would have no control over the production of the motor and the Navy could contract it to any concern willing to manufacture it…Two: that we sell to the Navy department immediately (the motor as it now stands now) for a sum of money set large enough to cover our initial expenditures and then to propose to develop a motor of large proportions for delivery to the Services at the end of development"[40]

Three remarkable things stand out in that passage. First, the opening of the letter strongly suggests that Lawrence had only recently met Lt. Fischer—perhaps in the previous month of November, or October at the earliest. This would certainly fit the trend if Wyld had indeed sent his M-15 report in late September or early October. (Most likely, Lawrence first met Fischer in Washington, as the former was still assigned to the Army's Message Center there during these months.)

Second, but most important, Lawrence conspicuously did *not* mention the formation of a company in this letter, and only alluded to expected prolonged negotiations with the Navy for the rights to produce and sell the Wyld motor to them. It may be that the idea of a company centered around the commercialization of Wyld's motor was entertained earlier—or rather, was considered all along—but at this point the idea was simply put aside, as it is strikingly absent in the letter. Therefore, it seems the sudden and wholly unexpected news of Pearl Harbor really did trigger the true formation of RMI. Last, the letter proves that Lawrence took the lead in promoting Wyld's breakthrough rocket motor. The most business-minded among the four men, Lawrence clearly spearheaded the formation of the company and was to become its first president.

In another little-known and consequential RMI background story, a few months before Pearl Harbor—evidently from August or September—Lawrence had been sent to Washington, D.C., for a six-month tour of duty on behalf of his employer, IBM. (He was assigned to the Signal Corps at the United States Army War Department Message Center in the Munitions Building in Washington to install and guide the operation of IBM teleprinting circuits on the War Department's major communicating networks.). While in Washington, Lawrence took the opportunity to personally visit and speak with George W. Lewis, the director of the National Advisory Committee for

Aeronautics (NACA). It was Lewis, Lawrence later revealed, who solicited interest in the potential of the Wyld motor within the Navy's Bureau of Aeronautics (BuAer) [41]

While Lawrence was on assignment in Washington for IBM, he must have started the whole process in the first place that culminated in the founding of RMI. The second major step was when Lawrence requested Wyld to write his report. As seen, Jimmy completed this task on 25 September. In turn, the submission of the report probably opened up Lawrence's *official* negotiations with the Navy that led to the BuAer ordering Lt. Fischer to New Jersey by October or November to witness the secret test that was conducted at Wanaque with most, if not all, of the future RMI founders also present.

Lawrence's daily work journal for his IBM duty in the Signal Corps War Department Message Center still exists within the Lovell Lawrence Jr. Papers in NASM Archives. They show he was on duty on 7 December 1941 when the disaster of Pearl Harbor struck and the next day he noted that "[The] Honolulu [teleprinting station] was busy on that frequency and we had to sit tight until they were finished."[42] Lovell was therefore not present in New York City with any of his fellow ARS members when the news of Pearl Harbor was broadcast over the radio, nor during the next few days when RMI was finally formed. Shortly after, he may have taken a short leave to travel to both New York and New Jersey to attend meetings that officially established the company and commenced its registration.

RMI SETS UP SHOP AND GLIMPSES OF ITS EARLY OPERATIONS

Matters moved breathtakingly quickly from here on. Within two weeks after Pearl Harbor, on 18 December, RMI registration papers were filed with the state of New Jersey, although operations had actually begun two days earlier; the first page of RMI's handwritten *Journal* starts on 16 December and records the following: "Reaction Motors, Inc., this day begins business at 280 Wanaque Ave., Pompton Lakes, New Jersey."[43] Lovell Lawrence Jr. was named the president; Pierce the vice president; Shesta, the treasurer; and Wyld, who had no inclination for business, was evidently perfectly content to be named the secretary. This was then the sum total of RMI's "staff."

Why Pompton Lakes, New Jersey, and why that address? In a nutshell, there are several interrelated reasons. First, for all intents and purposes, Pompton Lakes in Passaic County was Lawrence's hometown and he was very familiar with it. He had been born in the village of Port Henry, New York, in 1915 and attended school at nearby Mineville—so-named because of the iron-ore mines where his father, Lovell Lawrence Sr., worked as a mining engineer. But in the mid- or late 1920s, when his father obtained a new position as an assistant manager with the Ringwood iron mines in northern New Jersey, the family moved to Pompton Lakes, situated about a dozen miles from the mines.

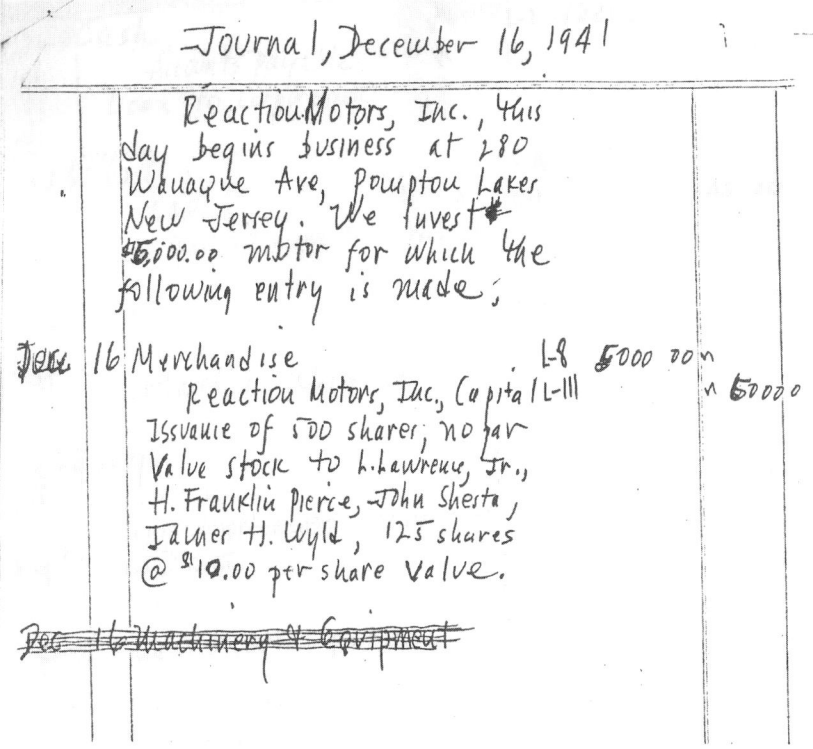

Fig. 2.8 First page of RMI's handwritten *Journal*, showing that the company really started business on 16 December 1941. (Courtesy U.S. Space and Rocket Center, Huntsville, AL.)

Lawrence Jr. thus grew up in Pompton Lakes, attending both junior and senior high school there, and had very fond memories of the place. He knew, too, that the location was relatively remote and ideal for establishing a "secret" rocket company that was also conveniently situated just 20 miles from New York City— and that was also not far from Washington, D.C.—where he could easily maintain close connections with the Navy's BuAer. The Navy had not only helped create the company and assisted them financially, but was RMI's first, and chief, customer. Lawrence must have likewise been well aware of the old and long-abandoned Du Pont explosives powder-mill facility situated in nearby Haskell, a few miles north of Pompton Lakes, that could serve as an ideal site for their initial testing. Historically, this had been known as Du Pont's Haskell plant. More about Haskell later.

Another reason for this part of northern New Jersey as RMI's base was that John Shesta's brother-in-law, Hugo W. Reese, resided in North Arlington, in adjacent Bergen County (as did bachelor Shesta himself), and Reese allowed Shesta and his new business partners to use part of his garage as the company's initial workshop. According to the 1940 U.S. Census, the house and garage of

Hugo W. Reese, which no long exist, stood at 79 Schuyler Avenue in North Arlington. This site was definitely a kind of second headquarters/workshop for RMI, as this address appears on early (1942) RMI stationary, including sample invoice forms.

It did not matter that the total floor space available to them in Reese's 20 × 40-ft garage amounted to only a few square feet in the upper half of the building. For efficiency's sake, this part of the garage was partitioned into two halves: one for administration and drafting, the other for the machine shop and laboratory. Both Pierce and Shesta had also brought in their respective lathes for their RMI work, in addition to an assortment of tools that included a drill press.

And the 280 Wanaque Avenue, Pompton Lakes, address? This was the first headquarters, or official address of RMI.

For some months before the RMI foursome moved there in mid-December 1941, the site had been an empty store on Wanaque Avenue, the main street of Pompton Lakes, and was fully available for rental. The previous tenants had been the family-owned Bauer's Auto Service that apparently had originally moved there in 1935 when they became established.[44] Six years later, according to an item in the Pompton Lakes *Bulletin* for 26 June 1941, Bauer's announced they would move on about 15 July to 736 Hamburg Turnpike in Pompton Lakes. It turned out that the store remained unoccupied for some months thereafter, as evidenced by another item in the *Bulletin* for 2 October, reporting that the King's Daughters of the Pompton Reformed Church were to have a rummage sale there the following day. Lawrence, who probably kept an eye open for such opportunities in his adopted home town, must have picked up on this, and a couple of months later the RMI crew was fully ready to move in.

Another early story about the 280 Wanaque RMI site concerns bicycles. The late Howard Lee Ball Jr., a well-known Pompton Lakes newspaper man and radio personality, often recounted his boyhood recollection of what happened to him at about age 12 when he went to that site in 1941 or 1942 to buy a bicycle. It looked like a bicycle shop, he said—except that "you couldn't buy bicycles there and the guy [inside] said, 'We don't sell bikes.'" "Well," young Howard asked, "what you got them in the window for?" "That's none of your business," came back the response.[45] In other words, as Howard later learned, the bicycles in the front window were a ruse, a "front," because the rocket work they were doing was then secret. A later account by RMI pioneer Robert "Bob" Holder backs up the story: "They put old bikes in front to confuse people about the nature of the company, but the noise of the rocket engines spoiled the ruse."[46] The latter claim may be debatable, however. Another source says that the actual rocket construction work was carried out in the basement of the shop at 280 Wanaque Avenue.

Yet another story about the 280 Wanaque Avenue site maintains that when RMI occupied the premises, some small-scale testing was occasionally done out in the back, on the bank of the Wanaque River. A recent visit to the site by the author showed that the narrow Wanaque River is situated about 300 to 350 ft away, down a slope stretching from the back of 280 Wanaque Avenue. The river is further secluded by trees along its banks, making the area suitable for, say, igniter tests that usually produced relatively low noises (as quick pyro-technic bursts or sizzles). It might have also been felt useful to test near water for safety's sake, in case water would be needed quickly to help douse any fires that might erupt if did not go well.

Some early testing is also said to have been carried out on one of the banks of nearby Lake Inez a little bit to the north and off the Wanaque River. It is impossible to determine where these tests might have taken place, as the lake no longer exists. The small and narrow Lake Inez was created with the build-ing of a dam in 1885; when the Lake Inez Dam was breached by flooding in 1984, the water reverted back to the river. Today, the former Lake Inez is merely a stream of the Wanaque River.

Larger-scale testing of the motors themselves produced far more audible sounds, of course, particularly if a motor burst, and evidently some of their very earliest motor tests were indeed conducted near their shop, which also could have meant at Lake Inez. In the very beginning, the bursts were frequent. Pompton Lakes police chief William F. "Bill" Charles was the only person who knew the real reasons for these explosions—RMI's wartime work on rockets—although he dutifully never revealed these to the local populace. Rather, he vaguely assured RMI's neighbors that the situation was in "good hands," or similar wording, in his efforts to both calm them down and shift attention away from RMI.[47] Susan Marczyk, who was raised in the area, remembered that when she was a little girl the townspeople would blame Du Pont for the periodic explosive or "popping-like" noises, as she puts it.[48] This would have been the second, and then the only *active* Du Pont facility in the area, situated in town near Lake Inez and known as Du Pont's Pompton Lakes Works.[49] This particular Du Pont plant was, in fact, not that far from RMI's shop at 280 Wanaque Avenue.

For certain, the work *was* secret because it was for the war effort. During the time RMI was organizing, BuAer was preparing a contract and the U.S. Navy became RMI's first customer. Under the terms of the contract, RMI was obliged to 1) deliver the existing Wyld motor to the Navy; 2) develop and successfully demonstrate a similar 100-lb-thrust motor, operable with aviation gasoline and LOX; 3) develop a comparable 1,000-lb motor using the same propellants; and 4) demonstrate repeated starts and throttling down to half thrust. All this was to be accomplished within 180 days. That is, RMI was expected to make an exponential leap in the technology from the 100- to 1,000-lb thrust level in the incredibly short time of just six months.

The main aim was develop JATOs for the Navy for potential deployment on heavily loaded seaplanes for the Pacific Theatre of the war involving missions from Pacific islands with short runways. Under these complicated circumstances, the JATOs took on a far greater and strategic significance than is generally believed and had to be accomplished as rapidly as possible. For the same reasons, the Navy soon came to simultaneously support as many as five separate JATO development programs, including one conducted by Goddard. To help his country in this critical time of great need, Goddard put aside his high-altitude research rocket work in lieu of contract projects with both the Navy and Army Air Corps for JATO developments. Sadly, he did not live to return to his research work.

Upon receiving their first payment of $5,000 from the Navy for turning over the original Wyld motor, RMI purchased a quarter-ton Ford truck in early March for a little over $1,100 to haul equipment, and proudly painted the name "Reaction Motors Inc. Pompton Lakes, N.J." on the side of it.[50] The same truck was used to retrieve the ARS Test Stand No. 2 from the ARS in New York City when the society's board of directors voted to loan it to the fledging company for their initial static tests. It is most curious, incidentally, that despite the RMI founders wishing to maintain secrecy over what they were doing, their Ford truck openly advertised their identity. This was perhaps not the smartest thing to do at the time, although such boldness was no doubt due to their sheer youthful exuberance in starting the business. It is even more interesting that this name on the truck continued to show up in early film footage of RMI's testing up to 1943.

Because RMI's top priority was the development of the 1,000-lb motor, a very intensive, increasingly louder testing program had to be carried out—under as much secrecy and as far away from any population area as possible.

Fig. 2.9   RMI's first truck with its boldly painted company name. (Courtesy Smithsonian Institution, 82-5386.)

Yet even the quiet, small-town environment of Pompton Lakes was now found unsuitable. According to Lawrence's brother Robert "Bob", who came aboard the company in early June 1942 to take over the treasurer job, "Because of the noise, a particular area could be counted on for one test only [that day]."[51] It thus appears that as Lake Inez was the closest available area for testing, it was very likely the earliest (if not RMI's first choice) for the main testing of their motors. But the downside of the Lake Inez location was that it was in town and therefore closer to the population center, and the testing noise always made it more risky in attracting unwanted local public attention.

This is why, as Lawrence probably had intended originally, he and his cofounders later agreed to relocate their main testing area a few miles farther north, to the old, deserted Du Pont facility at Haskell, a short distance outside of town. There the noise was not only far less likely to disturb anyone, and the RMI experimenters could also be more concealed.

Fine details of this earliest operational phase of RMI's history are lacking, but it is known that Du Pont's former Haskell plant just outside of town had originally been a smokeless powder mill, started in 1898 and named after its company president, Jonathan Haskell. In 1902, it was sold to Du Pont who also used it as a powder mill until ceasing operations there in 1926. As in their ARS days, all the small RMI team really needed was a wide open and remote space on which to place their portable test stand, as far away from a population center as possible—the Haskell location entirely fitted this bill. The RMI crew may have utilized the remains of a dilapidated Du Pont building or an explosive barrier as a suitable shelter during their static firings.

It will never be known when their testing was shifted to Haskell, nor how extensive it was there, nor other details. Current information suggests that testing must have been carried out at this place as late as September, when RMI at last opened their first permanent test site at Franklin Lakes in neighboring Bergen County, as described in Chapter 3. As mentioned, in addition to the Haskell site there was the intown Du Pont Pompton Lakes Works, located on the aptly named Cannonball Road. Opened in 1902, this plant produced blasting caps and other explosive devices throughout both world wars before it closed in 1994. Whether Lawrence first sought permission from this plant or from Du Pont corporate headquarters in Wilmington, Delaware, to use the abandoned Haskell site for their rocket tests, or whether they secretly used Haskell without telling anyone is another small historical mystery. Most likely it was the latter situation. This would have been in character and the easiest approach, especially because all the founders were very pragmatic and experienced experimenters in rocketry.

FIRST EMPLOYEES AND FIRST BUDGET

The RMI team grew, though not by leaps and bounds. For its first few years RMI remained quite small, then saw a dramatic increase in numbers from 1946. It is difficult to establish the identity of their very first employee, but he seems to have been Kurt F. Fisher, a 24-year-old New Yorker who was another active ARS member and who had witnessed the testings of Jimmy's motor in 1941 at Crestwood. For sure, Fisher was employed by RMI from March to November 1942, when he left to enlist in the Navy. During his nine months of RMI employment, Kurt served as a design engineer, helping design their early motors and involved in early testing.

Interestingly, as part of his wartime Navy duties, Kurt went on to carry out various assignments connected with JATOs. Amazingly, there is photographic evidence in the Reaction Motors, Inc. collection of photos at NASM that Fisher's assignments included a return to RMI on one or more occasions in his Navy uniform, to continue to provide extra manpower to handle the motors. He may have also been involved in early Aerojet work for the Navy. In 1948, Kurt returned to employment with RMI and stayed until 1950; he was known as an aerospace engineer throughout the remainder of his career with other companies connected with rocketry. He went on to make his own technical contributions, including several for the Project Gemini manned space program.

Back in the early1940s when RMI was a very small and truly pioneering operation, *all* early RMI men rolled up their sleeves and participated in the testing that often occupied extremely long, exciting, and challenging days and evenings.

Louis F. "Lou" Arata, another very early RMI employee who arrived in September 1942, later remembered that Shesta "was a Designer, Toolmaker, Machinist, Welder, Pressman, Pipe Fitter and Truck Driver."[52] It is relevant to add that back then all four founders shared the same tiny office with a single telephone, and all engineering work requests were issued verbally. Extreme informality marked the early character of RMI. In the absence of even a solitary clerical secretary, documents from the early RMI years are often handwritten, not typed. The early 1942–1943 RMI *Log Book* with recorded test results was written entirely by hand and in different handwritings—proving that this was a shared task.

In addition to testing, other significant aspects of RMI's first months of operations involved working out budgets and logistics toward the development of their then-single product—a powerful JATO.

In another recently unearthed early RMI document, a letter from Lawrence to Shesta dated 24 January 1942, Lawrence reviews Shesta's proposed budget plan for the remainder of the year, which was also America's first year of the war. This document reveals Lawrence's own attempt to streamline supply ordering. "If we should order fuel or liquid ox.[ygen]," he explained, "we simply make up an invoice with the order number: MF-12[,] M- for material F for fuel and the 12 designates the number of orders we have made up to that time."[53] The designation MT-5 would similarly apply to materials used in testing. "By looking at any invoice," he continued, "we can tell how much was spent for Fuel, Test stand material, [and] general motor material, etc."[54]

In the same letter Lawrence noted: "Saw Fischer yesterday and he is doing his best to get the contract underway."[55] As for their very first proposed budget, for 1942, Lawrence estimated there would be at least 120 "small tests" and 55 large ones, and that by the the year's end, a new test stand would be needed, costing an estimated $1,000.[56] Of course, projections for numbers of tests could never be accurate and were only guesses; the true numbers may have far exceeded these estimates. Motor construction, stock, etc., was to cost about $2,000 in 1942 dollars.

Very shortly thereafter, the Navy contract *did* arrive and was sealed. The stage was therefore now fully set for the greatest pioneering adventure Lawrence, Shesta, Wyld, and Pierce would ever know in their lives.

ENDNOTES

1. James H. Wyld, "Personal Information Regarding James Hart Wyld," in Edward G. Pendray Papers, Princeton University, p. 1; Albert G. Ingalls, ed., *Amateur Telescope Making,* Willmann-Bell, Inc., Richmond, VA, 1996, Vol. 2, pp. 175–176; George F. Bush, "Princeton's Rocketry Pioneer," *Princeton Engineer*, Vol. 29, Dec. 1968, p. 9; Interview with Robert Wyld, son of James H. Wyld, by Frank H. Winter, with Frederick C. Durant III and Tom Crouch in attendance, Washington, DC, 24 Feb. 1976, typed notes in "James H. Wyld" file, NASM.
2. Ingalls, *Amateur Telescope Making*, pp. 175–176.
3. Wyld, "Personal Information," p. 1.
4. James H. Wyld, diary, in James H. Wyld Collection, NASM, n.p.
5. Ibid.
6. Ibid.
7. Ibid.
8. Just a few years earlier, in a much-publicized speech before the Brooklyn Institute of Arts and Sciences on 11 April 1930, Professor Stewart had predicted—although he was way off the mark in several respects—that it would possible one day to fly to the moon, perhaps by the year 2050, by means of a huge metal sphere propelled by ionized hydrogen, the sphere manned by a crew of 60 men and a dozen scientists. Such a wild-eyed pronouncement, especially by an academic of Professor Stewart's standing, only served to further place the notion of spaceflight in the realm of fantasy. (To his credit, Stewart, with Henry Norris Russell, had co-authored *Astronomy: A Revision of Young's Manual of Astronomy.* This became the standard astronomy textbook for about two decades and had been published in several editions of 1926–27, 1938, and 1945. It gave Stewart renown in the field of astronomy. He was very excited about the prospects of spaceflight, but had no real knowledge of rocketry, which was then in its infancy. He was even criticized by Goddard for his bizarre choice of propellant, although Goddard was entirely secretive about his own work along these lines.)

Nonetheless, Stewart's speech and his followup bylined article in the Science section of *The New York Times* for 22 June 1930 became the topic of highly colorful Sunday supplement features across the country that reached even rural papers such as the *Vernon Daily Record* (Vernon, TX). Stewart broadcasted his predication over the radio, in a speech over the Columbia Broadcast System, perhaps one of the first uses of radio to discuss or promote spaceflight.

This kind of fleeting sensationalism typified the general attitude and lack of real knowledge or sophistication about the prospects of rocketry, especially as applied toward spaceflight, during these dizzy years of the rocket and spaceflight fad. Stewart's prediction even reached more respectable publications, such as the *Science News-Letter* and the *Princeton Alumni Journal,* and underscored the poor standing of and attitudes toward rocketry when serious-minded young men like Wyld were embarking upon their own experiments in this new field.

As for Wyld, he is less known—at least publicly—to have been intrigued with the possibilities of spaceflight, than with the sheer technical (mechanical) challenges of working with the rocket as a new power source for transportation. This could be either terrestrial in nature or for space flight, although at this stage in his thinking, the potential applications were stronger for the former. Notably, in his résumélike write-up of 30 August 1937, Wyld added that he had been "Interested in internal combustion engines over many years . . . of many varieties, especially automotive and aviation types." Additionally, he took a "course in engine design at Princeton, and had much experience in engine test work. Also have done much tinkering with automobiles and motorcycles from time to time."

Winter, *Prelude*, p. 74; Esther C. Goddard and G. Edward Pendray, eds., *The Papers of Robert H. Goddard,* 3 Vols., McGraw-Hill, New York, 1970, p. 739; "Scientist Visions Trips to Moon by Year 2050 In Rocket Ships Making 50,000 Miles an Hour," *New York Times*, 12 April 1930, p. 1; John Q. Stewart, "A Flight to the Moon in a Rocket-Driven Ship," *New York Times*, 22 June 1930, Part 9, p. 2; "Shooting to the Moon and Back Again at 2300 Miles a Minute," *Vernon Daily Record* (Vernon, TX), 9 July 1930, p. 10; "All Aboard for the Moon," *Courier-Journal* (Louisville, KY), 8 Feb. 1931, p. 53; "Travel to the Moon by the Year 2050," *Science News-Letter*, Vol. 17, 19 April 1930, p. 243; "Predicts Moon Flight in Hydrogen Rocket," *Popular Science*, Vol. 117, Aug. 1930, p. 31; *The Princeton Alumni Weekly*, Vol. 30, 13 June 1930, p. 926; "Moon Flight Broadcast," *Bulletin of the American Interplanetary Society*, No. 5, Nov.–Dec. 1930, pp. 7-8; Wyld, "Personal Information."

9. James H. Wyld, "The Prehistoric Era of the Regenerative Motor," *The RMI Rocket*, Vol. 2, Dec. 1951, p. 5.
10. Wyld, diary, n.p.
11. Ibid.
12. Wyld, "The Prehistoric," p. 5.
13. Wyld, diary, n.p.
14. John Shesta, "Report on Rocket Tests," *Astronautics*, No. 31, June 1935, pp. 4–5.
15. Ibid.
16. Copy of American Rocket Society typical signed testing statement, issued for ARS test of 22 June 1941, untitled, in "American Rocket Society" file, NASM.
17. Ibid.
18. Ibid.
19. Wyld, "The Prehistoric," p. 5.
20. Irene Sänger-Bredt and Rolf Engel, "The Development of Regeneratively Cooled Liquid Rocket Engines in Austria and Germany, 1926–42," in Frederick C. Durant III and George S. James, eds., *First Steps Toward Space*, AAS History Series, Vol. 6, Univelt, Inc., San Diego, 1985, p. 218.

21. Ibid., p. 220.
22. George P. Sutton, *Rocket Propulsion Elements: An Introduction to the Engineering of Rockets,* 2nd ed., John Wiley & Sons, Inc., New York, 1956, p. 128.
23. Wyld, "The Prehistoric," p. 5.
24. Ibid. The Wyld regen motor was among several "historic rocket motors" on display at the Hayden during April and May 1938; the others were from ARS rockets No. 2 and 4, along with the complete ARS No. 3. This may have the first time anywhere that a functional "modern" (that is, liquid-propellant) rocket and motors were ever exhibited in a museum.
25. Ibid.
26. For a detailed and illustrated history of both ARS Test Stands No. 1 and 2, see Frank H. Winter, "Rocket History through an Artifact: American Rocket Society (ARS) Test Stand No. 2 (1938–1942)," in Kerrie Dougherty, ed., *History of Rocketry and Astronautics: Proceedings of the Forty-Fourth History Symposium of the International Academy of Astronautics, Prague, Czech Republic, 2010*, AAS History Series, Vol. 41, Univelt, Inc., San Diego, 2014, pp. 243–274.
27. Letter from James H. Wyld to Peter van Dresser, 11 Dec. 1938, in "James H. Wyld" biographical file, NASM.
28. Ibid.
29. Ibid.
30. John Shesta, Franklin H. Pierce, and James H. Wyld, "Report on the 1938 Rocket Motor Tests," *Astronautics*, No. 42, Feb. 1939, pp. 6–7.
31. Ibid.
32. J.[ohn] Shesta and R.[oy] Healy, "Report on Motor Tests of June 8, 1941 at Midvale," *Astronautics*, No. 49, Aug. 1941, pp. 2–5. The choice of the location of Midvale, in Passaic County, may well have been made by Lawrence. Located just north of Pompton Lakes where he was raised and lived up to this time, it was remote and very woody with plenty of open space, yet not far from New York City. Midvale was also conveniently close to Lawrence's home.
33. R.[oy] Healy, Wyld Motor Retested," *Astronautics*, No. 50, Oct. 1941, p. 8.
34. Ibid.
35. Wyld, "The Prehistoric," p. 6.
36. Ibid.

In the meantime, even before the 1941 tests, according to an anonymously written article titled "Rockets," appearing in *Fortune* magazine for Nov. 1950, Pierce and Shesta had been approached in 1940 by an unnamed British firm for rocket development but this never came about. Whether this incident planted the seeds of the idea of a rocket "company" among some or all the future founders of RMI therefore cannot be verified. This story might even be apocryphal. "Rockets," *Fortune* magazine, Vol. 42, Nov. 1950, p. 120.37. James H. Wyld, report, handwritten, "Wyld 'M-15' Regenerative Rocket Motor—Sketches, Notes, & Discussions," 25 Sept. 1941, with six internal drawings, each dated 24 Sept. 1941, copy in "James H. Wyld" biographical file, NASM Archives.

Shesta's own recollection is that *after* the demonstration at Wanaque, Lt. Fischer asked that a report on the motor be sent to him. It is likely that *another* report was written by Wyld and sent directly to Fischer. Within the biographical file of Wyld in NASM is a copy of Wyld's undated, 10-page report titled, "Experiments on a Regenerative Liquid Fuel Rocket Motor," believed to have been written in 1941 and donated to NASM in May 1987 by Robert M. Lawrence, one of Lovell's brothers. Significantly, in the back of this report is written, in freehand printing, the words "Navy Department—Shesta & Wyld." This undated report was certainly written at a point when the future RMI founders were negotiating with the Navy. But this report is not as detailed as the one of 25 Sept. 1941 on the development of the M-15 motor, nor does it contain sketches. It is possible this document from Robert Lawrence is a copy of a report sent by Lovell to the Navy, via Lt. Fischer, soon before or just after the demonstration at Wanaque.

38. "The Regenerative Motor Opens New Possibilities," *Astronautics*, No. 42, Feb. 1939, pp. 15–16.

39. Letter from Lovell Lawrence Jr. to John Shesta, 4 Dec. 1941, in "Lovell Lawrence, Jr." file, NASM.

40. Ibid.

41. The latter important anecdote by Lawrence, mentioning his talk with the NACA's George W. Lewis, is found in the retrospective piece "Success Story," in the *RMI News*, Vol. I for June 1949, p. 1. It is likewise mentioned in the Nov.1950 article "Rockets," in *Fortune* magazine, cited in note 36. To date, no written record has been found of Lawrence's meeting with Lewis; it may have been a verbal, face-to-face meeting, with followup phone calls.

42. Lovell Lawrence Jr., "Running Log of Service and Operation [of station] WVQ-WAR on Radioteletype [at Signal Corps, United States Army, War Department Message Center, Washington, D.C.]," 6–11 Dec. 1941, n.p., Lovell Lawrence Jr. Papers, NASM Archives, box 11, folder 14.

43. Reaction Motors, Inc., *Journal*, copy in U.S. Space and Rocket Center, Archives, Huntsville, AL, p. 1.

     The original building at 280 Wanaque Avenue in Pompton Lakes allegedly still stands, although this cannot be confirmed by documentation. Today on this spot is an Edible Arrangements store. The adjacent, adjoining building to its left, with the address of 284 Wanaque Avenue and presently occupied by R & M Hardware, was constructed around 1900. Another building was later added on as part of the same complex and was assigned the address of 280, although the year is unknown. But again, it cannot be determined with certainty whether the present building at 280 Wanaque Avenue is exactly the same building once occupied by RMI from 1941 to 1943.

44. See the Bauer's Auto Service ad in *The Bulletin* (Pompton Lakes), 24 June 1941, p. 4.

     Incidentally, the Bauer's company, now a gas station, still exists at the same Hamburg Turnpike location.

     Wyld's own recollection of the former tenants of 280 Wanaque Avenue was a little faulty. As he reported in the *RMI Rocket* for Dec. 1951, he believed the site had been occupied by "Pat's Tailor Shop." In truth, Pat's Tailor Shop had occupied 280 Wanaque Avenue in 1950, or a year before Jimmy wrote his recollection. (See in the Pompton Lakes Library, an ad for Pat's Tailor Shop in the *Directory of Pompton Lakes*, 1950, p. 180.)

45. Oral history interview with Howard Lee Ball Jr., 29 Feb. 2008, Rutgers Oral History Archives, Rutgers School of Arts and Sciences, available online. See, Ball, Howard Lee, Rutgers Oral History Archives.

46. Holder quote, Sharon Sheridan, "Having fun was part of the [RMI/RMD] job description," *The Star Ledger* (Newark, NJ), 19 June 1997, p. 2.

     Ball intimates that much later, a couple of years after the war, he learned of RMI's real business through his former teacher, Helen Lindsley, who dated Jimmy Wyld in the early 1940s and later became his wife. See for example Howard Lee Ball's columns available online, such as "It happened at Pompton Lakes," *Suburban Trends* (Morris County, NJ), 7 May 2012;"Other Views: 'Weird guys' and the space age," *Suburban Trends*, 22 March 2009; "From Pompton Lakes to outer space," *Suburban Trends*, 14 May 2012.

47. Oral history interview with Howard Lee Ball; Ball, "It happened"; and similar columns by Ball.

48. Telephone interview with Susan Marczyk by Frank H. Winter, 22 Aug. 2016.

49. For an interesting item on the operation of the Pompton Lakes Du Pont plant during the war, see "Du Pont's Adopts 40-Hour Week," *The Bulletin* (Pompton Lakes, NJ), 6 Sept.1945, p. 1.

50. Robert Lawrence, "Excursion," p. 7.

51. [Louis F.] Lou Arata, "How Many Remember?" *RMI Rocket*, Vol. II, Dec. 1951, p. 7.

52. Letter from Lovell Lawrence Jr. to John Shesta, 24 Jan. 1942, in "Lovell Lawrence, Jr." biographical file, NASM.

53. Ibid.

54. Ibid.

55. Ibid.

56. Ibid.

Chapter 3

THE WAR YEARS

"We would test a rocket cylinder in the morning and spend the rest of the afternoon trying to extinguish the fire we started in the surrounding woods. Everyone from the President [Lovell Lawrence] on down joined the bucket brigade."

– Lou Arata, in *The RMI Rocket*, December 1951, p. 7.

RMI STARTS OPERATIONS

As revealed in the rare copies of first issues of RMI's *Journal*, the founders must have been working like beavers since 16 December 1941 when they opened their shop for business at 280 Wanaque Avenue in Pompton Lakes, two days before they officially registered their tiny company. The *Journal* also shows that among the four of them, the founders had been able to scrape together $5,000 as their starting capital, which helped pay the $45 a month for the empty store they rented. They further paid for advance rental on the shop, their incorporation fee, a related state fee, prepaid insurance, stationary and postage, miscellaneous unspecified supplies, and their own payrolls for that month.

During this early time, all the tools and equipment belonging to Pierce and Shesta, including two small lathes, a drill press, and an oxyacetylene welding outfit, were also gathered into Shesta's brother-in-law's garage at 79 Schuyler Avenue in nearby North Arlington. The garage served as RMI's first workshop; the empty store in Pompton Lakes functioned as their main headquarters.

Still another revelation appears upon closer examination of Lawrence's few existing IBM employment records in the Lovell Lawrence Jr. Papers in the Archives of the National Air and Space Museum (NASM). They show that he continued to work for IBM until as late as mid-March 1942. RMI's first four months must therefore have been an especially hectic period, because only three of the founders were regularly around, although in all probability Lawrence helped them during weekends and holidays, and may have also taken regular leave when he could. Lawrence submitted his letter of resignation to the Radiotype Division of IBM on 2 March to "enter a National Defense research position," but the resignation did not come into effect until 15 March.[1]

From these dates it seems that Lawrence must have anticipated the approximate time frame of the signing of RMI's first contract with the Navy. This turned out to be 23 March, which was fairly close to Lovell's estimate. It explains why RMI made the truck purchase that month and began hiring their first worker: Kurt Fisher, another former ARS colleague. Bob Lawrence, Lovell's brother, was likely the second RMI employee, joining in early June. He was followed by Joseph M. "Joe" Porter, who came in late July. Lester "Les" W. Collins arrived in October, and Louis F. "Lou" Arata came aboard in September. Bob did not remain for the duration of the war, however, as he soon signed up for the Army Air Corps. He was appointed a lieutenant in the 317th Fighter Control Squadron that served in China with Col. Clair Lee Chennault's famous "Flying Tigers," although he rejoined RMI after his military duty and remained the company's chief financial officer for years thereafter.

It is uncertain whether the four main tasks for RMI stipulated in the Navy's first contract were already known to the founders prior to the signing. It is likely that they at least received an initial draft of the Navy's expectations, and appear to have operated diligently toward these exceptionally challenging goals. There were, after all, 1001 things to do using Jimmy's handheld small regen motor as a very modest starting point toward the rapid design and

Fig. 3.1 Earliest photo of RMI's employees, summer 1942. Standing, left to right; Pierce, Shesta, and Wyld; crouching: Joe Porter and Kurt Fisher. Lovell Lawrence Jr. not present. (Courtesy Kurt Fisher.)

development of the followup engine requested by the Navy, before leaping ahead to a 1,000-lb-thrust version.

It was not until as late as 16 April that the ARS voted to approve the loan of their Test Stand No. 2 to RMI to conduct their first tests. Crude as it was, this piece of technology served as crucially invaluable test equipment without which the early RMI would not have been able to function or complete their first contract. But ARS Test Stand No. 2 also presents its share of historical mysteries.

In the first place, it must be established that the last usage of the stand by the ARS was their series of tests on 1 August 1941 in Midvale—that was also Wyld's last public demonstration of his revolutionary new motor. It is believed that there were no further official ARS runs on it, and that the series also marked the closing chapter of the rocket motor static-testing phase that the society had begun in 1935. The apparatus was now fully available for any other use by the ARS Experimental Committee. No details are known as to where the bulky stand was stored after the 1941 Midvale tests, but the ARS had a small cramped office in New York City, which would hardly have been suitable for storing this unwieldy object. It was more likely kept in a place like Pierce's basement in the Bronx or in Shesta's brother-in-law's garage at North Arlington, New Jersey.

As related in Chapter 2, it is not known if the stand was taken out again, perhaps in October or November, for at least one secret demonstration of Wyld's motor near Wanaque, New Jersey, for the benefit of Navy Lt. Charles F. Fischer. The stand was turned over to RMI after they were formed in December, but the date of its physical delivery is unknown. It is entirely conceivable it may have been acquired by RMI long before the ARS formalized the loan of this equipment—perhaps as a secret transfer, considering its crucial need for the war effort, coupled with RMI's extremely tight deadline. (It also would not have made any sense for the stand to have been delivered to RMI as late as mid-April.) A secret early transfer of the stand would have been very easy to arrange. If that were the case, RMI's initial tests with the stand may well have started as early as late December 1941, or some time during the following month.

When Lovell Lawrence Jr. finally joined his RMI team full-time by mid-March, he immediately faced an unforeseen headache: wartime shortages. As Shesta later recalled in his memoir, the modest collection of equipment contributed jointly by Pierce and himself was insufficient for RMI's initial operations. Lawrence quickly applied for a priority application "from the proper government agency" for extra lathes and machine tools, but was told that "such allocations were granted only to shops operating three-shifts on a twenty-four hour basis."[2] "There was no way we could operate on this basis," Shesta continued. "We therefore had to 'make do.'" It was only some time later and many trips to Washington by Lawrence "that our difficulties were finally resolved and we received the needed equipment."[3]

AEROJET

It so happened that in the same month the RMI/Navy contract was sealed, some 3,000 miles away in Azusa, California, the Aerojet Engineering Company was founded (on 19 March) by Dr. Frank J. Malina and others. This made Aerojet the second rocket-engine company formed in America. It was a spinoff from the GALCIT (Guggenheim Aeronautical Laboratory of the California Institute of Technology) Rocket Research Project started in 1936 by the young predoctoral student Malina, under the direction of Dr. Theodore von Kármán, the eminent Hungary-born aerodynamicist. RMI therefore had a competitor very early in its history. Aerojet was similarly set to develop JATOs of both liquid- and solid-propellant types, although RMI concerned itself only with liquid-fuel JATOs. Nevertheless, RMI was spared early competition from Aerojet—for now. The simple reason was that Aerojet's principal customer for this type of product was the U.S. Army Air Corps, the predecessor of the U.S. Air Force that was formed in 1947. At that time, and for years thereafter, RMI's main and only customer was the U.S. Navy.

RMI'S EARLIEST MOTOR DESIGNATIONS

Because Jimmy's final version of his rocket motor, M-15, led to RMI's formation, and their first Navy contract called for a continuation of the development of this motor up to the 1,000-lb-thrust level, Lawrence and his small team logically agreed to simplify matters and name all the RMI follow-on motors the M15-G series. The "G" probably meant these motors burned LOX and gasoline rather than LOX and alcohol, as did the original M-15. From RMI's initial M15-G model, there evolved a succession of improved and more powerful variants that included the M15-G1, the M16-G, and on up through the M18-G1, the M18-G2, the M19-G1, and the M19-G2 motors. This scheme very logical and easy to follow.

It is also significant that this rocket-motor designation system was the first in the history of the American rocket-motor industry and was unique to RMI. Yet their system was operational only during the bulk of war years. By the mid-1940s, as Chapter 4 will show, RMI's original Wyld motor-based designations had quickly become outmoded and they gave way to still another system. Upon the rapid and dynamic growth of the rocket industry by late 1940s, RMI adopted the new U.S. military standard with XLR ("Experimental Liquid Rocket") designation, which was then in industrywide use.

RMI'S EARLIEST ENGINE DEVELOPMENTS

RMI's very first motor was a near duplicate of Jimmy's M-15, but it operated on LOX and gasoline and became known as the Wyld motor, Serial No. 2. It is now in the collection of the National Air and Space Museum and currently on exhibit. Close by in the same exhibit case stands a facsimile of Wyld's original M-15 (or Serial No. 1), mounted on ARS Test Stand No. 2. Serial No. 2 does look almost exactly like Serial No. 1, except that the name "Reaction Motors, Inc." can be seen stamped around the periphery of an internal retainer ring, followed by the designation number "M15G-1-1." But the "1-1" part of this designation appears suspicious, as it may be expected that Serial No. 2 would simply be designated as "M15G-1." This specimen may not truly be Serial No. 2, but a follow-on; NASM records now refer to it as a "variant" of Wyld Serial No. 2.

Many years later, in a letter to Warren P. Turner, RMI's applications engineering and contracts head, Lawrence helped solve this mystery in his side remark: "Of interest—two weeks before the final acceptance test [of Serial No. 2]—this motor melted and we re-designed it and [it] passed without [Benjamin F.] Coffman [Jr., then the Navy's civilian administrator of its rocket program] finding out until later."[4] Shesta corroborated this story when he said that Serial No. 2 "quickly burned out…and a complete re-design was necessary."[5] The redesign entailed substituting the aluminum nozzle for a copper one. This explains what happened to the real Serial No. 2 and why M15G-1-1 was the "survivor" motor. These accounts also dramatically show that the true history of Serial No. 2—RMI's very first motor—was a bit more complicated

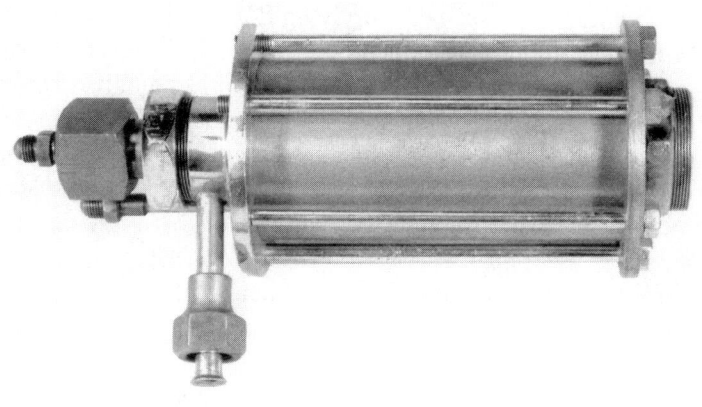

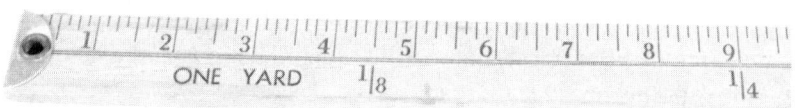

Fig. 3.2   Wyld Serial No. 1 motor in the National Air and Space Museum. The RMI designation "M15-G-1-1" stamped around the retainer ring is not visible to museum visitors. (Courtesy National Air and Space Museum.)

than first appears, and that its most untimely meltdown nearly cost RMI their first contract.

Quick thinking and action on Lawrence's part saved the day, and RMI fulfilled the second stipulation in the contract by developing and successfully demonstrating the 100-lb-thrust motor comparable to Jimmy's prototype model, although it ran on gasoline and LOX.

The surviving variant motor must have experienced its own considerable history. According to Aerojet pioneer William C. House, once this motor was turned over to the Navy, it was fired as many as 50 times for a few seconds to more than a minute in duration in tests at their Naval Engineering Experiment Station in Annapolis, Maryland, to determine the best propellant-mixture ratios for optimum performance.[6] The motor's now-missing fuel inlet line seems to confirm this early, heavy usage.

Although the history of Serial No. 2 is fairly well covered, precise details about the majority of RMI's other earliest hardware are lacking, despite the fact that some excellent examples of this hardware are also in the collection of the NASM. The underlying problem is that almost all these pieces are very poorly documented.[7]

Probably early in 1967, in preparation for the donation of this hardware by the Reaction Motors Division of the Thiokol Chemical Corporation to the museum that year, Thiokol only furnished NASM with an eight-page typed list of some 80 objects from which the museum could make its final selections. This list was drawn up by one or more surviving RMI veterans, although the name of the principal compiler is not included. The list itself is most frustrating in that it offers only a brief description of each object and its respective significance—or in some cases, just a speculative remark. This is the museum's single documentary source on the final selection of early RMI objects accepted for donation from Thiokol. Incidentally, NASM wound up selecting about 60 objects on the list, although the majority of them remain poorly identified.[8]

The situation is further confounded by the fact that RMI's earliest developmental and testing files no longer exist, although there is the partly completed RMI *Log Book* dating from March 1943 that reports on tests. One surprising source that does enlighten us is a small, separate collection within the NASM archives of early vintage RMI photographs, simply titled "Reaction Motors, Inc. Collection" and was most likely originally collected by Wyld; he was also likely the original photographer. As a bonus, a number of these photographs contain Jimmy's brief, clear, handwritten identifications and dates. In some cases, the captions reveal significant RMI milestones.

One photograph in particular—actually, one of a small series of similar images—is especially valuable because it shows an array of more than half a dozen motors developed by RMI from about May or June 1942 to May 1943. The photograph presents a group shot of several of RMI's key M15- variant motors and their respective identifications. The specific motors are Wyld's original regen model, which he dubbed the M-15; RMI's first duplicate of this motor, known as the M15-G1; the M16-G; the M16-G-1; the M17-G-2; the M18-G; and the M-19-G1. (Variations of this picture include the same view of Lawrence with the motors and another with him and the workers with the motors.) Although some intermediary models are missing, this array represents RMI's first family of rocket motors and the remarkable and steady technological evolution of more than the year of work that went into them. The thrusts they represent range from the 100-lb-thrust M-15 and M15-G1 to the 1,000-lb-thrust M19-G1.

A cursory examination of this photograph, as well as comparisons with existing motor specimens in the museum's collection, demonstrates that almost all the early RMI motors had a decided "RMI look." They are usually long, straight cylinders, and their internal cooling jackets (or "liners," as they were called) are fitted over the motors proper. Both the photograph and specimens revel that the internal de Laval exhaust nozzles remain hidden from public view. While these motors were successively scaled up, one model, the M18-G1, stands out. It is somewhat shorter and narrower—and therefore

Fig. 3.3 Lovell Lawrence Jr. with an array of RMI's earliest motors. Left to right, the motors are Wyld's original regen model, the M15-G1, the M19-G1, the M16-F1, the M16-G, the M18-G, and the M17-G-2. (Courtesy Smithsonian Institution, 78-12180.)

more compact—and it may have retained an approximate thrust rating similar to the larger M19-G1. RMI's technical progress toward their 1,000-lb-thrust goal was a painstaking evolutionary process, with at least these half-dozen variants produced up to late spring of 1943. Shesta corroborated that RMI gradually "scaled-up" Wyld's original design, as well as making modifications along the way.[9] It is evident that this considerable body of work must have presented an extremely steep learning curve for RMI's four founders and their all too few additional employees.

It is reasonably certain that the "G" in these early designations referred to the switch to gasoline. Thus, the key technical question that needs to be asked at this point is: Why did the Navy insist upon gasoline in lieu of alcohol for the follow-on Navy versions of Wyld's motor?

In using gasoline (especially because it had to be aviation gasoline), it is logical to surmise that the Navy brass were being practical in considering the possibility of 1) somehow bleeding off the aircraft's own fuel (a small portion of it) for JATO utilization when needed, thereby avoiding extra weight and keeping space taken up by an extra tank for rocket fuel; 2) a cost savings of gasoline over alcohol; and 3) making the JATO system far more compact and efficient, as well as cheaper in the long run.

RMI'S INITIAL ROCKET TESTING

For all their regular rocket testing, the small RMI group not only regularly used the borrowed ARS Test Stand No. 2, but there is photographic and other evidence that they modified or somewhat customized the ARS equipment for their own purposes. Among the changes was the installation of a peculiar drum-shaped device at the top center to perform throttling tests that were part of the contractual requirements. Another alteration was the removal of the original thrust dial and its replacement with much larger and bolder Roman numerals for easier visibility in filming the readings. Other structural changes were made, and calibrations of the stand were also conducted inside one of RMI's facilities. The stand, which still exists and is now on exhibit in the National Air and Space Museum, differs in appearance in several respects from when it was first used by the ARS during their own experimental days in the late 1930s and early 40s, as pointed out in a special study on ARS Test Stand No. 2 by the author.[10]

Early RMI photographs show that Lt. Fischer occasionally showed up to check up on RMI's progress—and perhaps witnessed a test or two at Haskell. One photograph shows a dated sign next to ARS Test Stand No. 2, indicating that he visited them in mid-May 1942. Notations in the RMI *Log Book* document that Fischer was also on hand for a number of tests in 1943, even though he had been succeeded at the Navy's rocket desk in December 1942 by Lt. John S. Warfel. This proves that Fischer was always very keen on the

Fig. 3.4   ARS Test Stand No. 2 on the RMI truck. Man in overcoat, back to camera, is Lawrence. To his right is Lt. C. Fink Fischer. Date unknown. (Courtesy Smithsonian Institution, 78-12176.)

progress of RMI and persistently encouraged them. The early period of RMI's history was definitely one of very intensive activity that was highly challenging and exciting to all those involved.

Rocketry, of course, can be an extremely dangerous business, and its overall history, especially in the early pioneering years, is replete with accounts of explosions and other mishaps. RMI was no exception. Lou Arata remembered that typically "We would test a rocket cylinder in the morning and spend the rest of the afternoon trying to extinguish the fire we started in the surrounding woods. Everyone from the President [Lawrence] on down joined the bucket brigade."[11] Arata does not say when this happened, but this type of accident was frequent throughout RMI's early history. Existing film footage of early RMI tests show the most common problem to be leaks that led to breakouts of fire.

Apart from designing and testing the basic engines, Lawrence and his team sought to further refine Wyld's method of cooling. In addition, there were myriad other technical details to work out, ranging from devising reliable igniters and injectors to fabricating special valves and valve systems. Not to be overlooked is the design, fabrication, and testing of bulkier propellant and pressurizing gas tanks than they had ever previously used.

Igniters were small and relatively simple things that were absolutely crucial in the operations of rocket motors; for the sake of convenience, according to Arata, they were tested "about fifty feet from the main street [Wanaque

Avenue] of Pompton Lakes."[12] "As well as you can imagine," he went on, "this would cause quite a disturbance among passerbys [sic.], so half the [RMI] personnel ran out and joined the crowd and tried to distract their attention, by starting a fight if necessary, while the other half tried to hide the equipment."[13] (Ignitor tests may also have been conducted a little farther away, on the bank of the Wanaque River behind the building.)

ESTABLISHING THE FRANKLIN LAKES TEST SITE

One of the earliest orders of business for the infant RMI company was to procure a permanent test site and create their own static, or fixed, test stand at the site, although these tasks took longer than expected. As later recounted by Bob Lawrence:

> It was difficult to locate a suitable site within the immediate vicinity. [Finally,] Mr. Lovell Lawrence approached the owner of the Franklin Lakes Airport who agreed to sub-let a small area located within the boundaries of the airport property. This was of the essence in constructing the stand so the formalities of a written lease and the securing of township approval to run tests was dispensed with. The stand was completed in a very short time [and] preparations were made for the first test. I believe the engine tested was the newly developed 300 pound [thrust] rocket [although this may have actually been the 350-lb thrust M16-G motor].[14]

This occurred in September 1942. This new test facility at Franklin Lakes in nearby Bergen County, New Jersey, was placed "in back of the air field," according to the book *Franklin Lakes: Its History and Heritage* by Maria S. Braun of the Franklin Lakes Chamber of Commerce.[15] It appears to have been near a hangar that stored light planes used by the airport, although the general location was then in a secluded area. "The rockets were secured and were stationary," she added.[16] The test site was situated about six miles from the Pompton Lakes shop on Wanaque Avenue.

As for the owner of the small, rural Franklin Lakes Airport, whom Lawrence approached about building his site, he may have been the aviation enthusiast and entrepreneur Don Nelson. He had come to Franklin Lakes in the late 1920s, purchased a large tract of land, cleared it, and set up an airport that became locally acclaimed for its air circuses and stunt flying. Because "Nelson's Airport" was cofounded by H.E. Merchant, a transplanted Texas barnstormer who had a pilot's license and was the airport's flying instructor, Lawrence may have dealt with this man instead. Technically, from 1932 to 1943, this flying facility was successively known as Nelson's, North Jersey, Oakland, and finally, the Franklin Lakes Airport.[17]

Upon the opening of the brand new Franklin Lakes testing facility, the old-but-modified ARS Test Stand No. 2 was put aside and placed in storage, or

Fig. 3.5 Franklin Lakes test site with 300-lb-thrust test stand in center, September 1942. (Courtesy Smithsonian Institution, 90407588.)

rather, in their "museum" as they later called it. Again, the very first motor tested at Franklin Lakes was likely the 350-lb-thrust M16-G.

> Bob Lawrence later vividly recalled their initial run there. The test, he said, was run under conditions that produced the loudest and most inspiring sounds ever heard in that area. Calls were received by the State Police reporting the crash of B-17 airplanes in the vicinity, and the occupants of nearby houses reported earth tremors. It took a considerable amount of time to convince everyone that an engine that you could hold in one hand was responsible for the noise. The [Franklin Lakes] town council held a special meeting and ordered the Company to desist from further testing. However, after all the facts concerning the work were disclosed, the Company was allowed to continue their operations.[18]

This initial noise complaint was a foretaste of what was to come.

Thanks to the diligent research of local Franklin Lakes historian Jack Goudsward, we can now pin down the exact date and other circumstances of this first run. A second-page story of the event appeared in the nearest local newspaper, the *Wyckoff News* (Wyckoff, NJ), for Thursday, 24 September 1942, headed "Test Causes Excitement at Airport":

> The peace and quiet of the Franklin Lakes section of the Borough was broken last Saturday night [19 September 1942] by a loud roar and flash that caused many residents of the neighborhood to have visions of another army plane crackup. The racket came from the Franklin Lakes airport where an experimental laboratory was recently set up by a Pompton Lakes man. Neighbors hearing the roar of a motor and seeing the flash of

fire, decided an aircraft had come to grief and called the State Police and
the Oakland Fire Department to the scene. By the time they arrived the
noise had subsided and the Pompton Lakes man emerged from a small
cement building to explain it was only an experiment. The erection of the
small building was investigated recently by the Borough zoning authorities when it was found it had been erected without benefit of a permit.
After explaining the use of the building the permit was granted.[19]

Note the conspicuous absence of the word "rockets" in this news report.

Assuming this is a more authentic account of what happened, it is impossible to determine exactly how Lawrence ("the man from Pompton Lakes")
explained the use of the buildings at the time to the satisfaction of the local
authorities. For now, it seems that whatever artful ruse Lawrence used succeeded; the true purpose of his cement building was kept secret.

Lawrence, with his years of experience at IBM in reading and laying out
intricate electric circuitry schematics, played a major role in designing the
electrical systems for the new test site and stand at Franklin Lakes, This is
amply evidenced by his signed drawing dated 30 July 1942 of a "Proving
Stand Control Panel."[20] There are an impressive number of similar highly
detailed drawings in the Lovell Lawrence Jr. Papers. An earlier drawing, dated
11 March 1942, was considered such an advanced idea that Lawrence later
had the sketch notarized. There are also sketches of test stand layouts by
Shesta and evidence that Jimmy Wyld also significantly contributed to the
design of the Franklin Lakes test facility.

Another, undated document, in Shesta's own hand, is part of a lengthy written textual description that is followed by a drawing showing a very large
rocket motor. It shows the nozzle facing downward and blasting away into a
bowl-like concrete ditch to deflect the billows of exhaust smoke safely upward,
much like modern stand arrangements. Shesta wrote: "New test stand about
$3,000," although there is no evidence that this more advanced test stand plan
was ever adopted by RMI during this early period.[21]

LAYOUT OF THE FRANKLIN LAKES SITE

Previous knowledge of the Franklin Lakes test site was limited strictly to a
small, so-called blockhouse. This very modest structure first came to national
attention in a long *New York Times* article of 19 August 1977, which reported
the efforts of Edward J. Lenik, the director of the archaeology laboratory of
the Van Riper-Hopper House museum in Wayne, New Jersey, to try to place
the former RMI site in the National Register of Historic Places. The 1786
house, one of the oldest in New Jersey, was situated just five or six miles
northeast of the blockhouse. But it was not until 1975 that the RMI structure
first came to Lenik's attention "from an archeological standpoint...when a

Fig. 3.6   Shesta by the RMI blockhouse at Franklin Lakes, circa September 1942 when the site opened. (Courtesy Smithsonian Institution, 83-2855.)

friend of his [now identified as fellow member of the North Jersey Highlands Historical Society William Mead Stapler], who once worked for Reaction Motors…mentioned an early tests site of the company in Franklin Lakes."[22] Lenik later toured the area and found the site. He and his volunteers made a survey and succeeded in having the building placed on the National Register. More details of this development —and the aftermath of the site's story—are related in Chapter 11.

As determined by Lenik and his assistant, the 13 × 13-ft blockhouse was made of coarse, grayish-brown, reinforced cinder blocks. The side facing the test stand (the east wall) featured two horizontal rows of narrow, rectangular, horizontal viewing ports, each fitted with shatterproof glass; there were six such ports or windows above, and five below. These viewing ports faced eastward, while the curved, snakelike bank of the Lower Lake of Franklin Lakes was situated many feet away, marking the northern boundary of this part of the overall property. Lenik had also found that the blockhouse rested on a large concrete pad, although "now [in 1977] covered with brush and undergrowth," according to an illustrated story in the *New York Times* about his survey of the small building, toward his aim of making it a national historical landmark.[23]

Some 40 years later it is possible to examine rare and marvelous black-and-white silent film footage from NASM's collection of Franklin Lakes tests from 1942 to 1943. The film provides an extraordinary closer peek and the opportunity to learn much more about the formerly hidden Franklin Lakes site.[24] The brief footage is accompanied by typed subtitles for each of the tests.

The film shows that the little structure with the portholes was not the blockhouse. Instead it is labeled the "control room."[25] Indeed, the term "blockhouse" does not appear to have been part of the RMI rocketeers' jargon, although it persists in present-day usage. By some definitions, a blockhouse is part of a launch complex, although the mission of the Franklin Lakes facility was entirely devoted to monitoring static tests, not launches (flights). From this point on in the current book, the correct RMI terminology of "control room" will be adopted.

The control room was but one of the structures on the site, albeit the main one. The film footage does not give an overall view of the complex, primarily because of the facility's tight quarters and the cameraman's very selective shooting. An adjacent concrete wall faced the porthole of the control room; it served as both the mount and barrier for the test stand proper. (Fig. 3.5) A few feet east of the control room stood an ingeniously constructed, moveable, corrugated-metal storage shed that was wheeled over the stand on tracks when testing was completed. The shed, with the stand within, was locked at night. The footage also hints at other work/storage sheds, perhaps as many as four, that might have been used to house tools, equipment, and propellants. (All these structures, including the control room, were in a row a few feet from each other and faced eastward from what is now Kent Place and Lake Drive.)

Directly facing the viewing-port side of the little control room, just a few feet away from it, stood an L-shaped, concrete, wall-like structure. Its shape provided greater strength and protection should there be an explosion.

Picture a drawing in which the small rectangular control room is on top and the long side of the L faces the control room, while the smaller leg of the L faces downward, to the right. Along the opposite side of the concrete wall were placed the propellant and pressurizing gas (nitrogen) tanks, with piping connected to the stand proper. The stand appears to have been mounted on a step section or block placed against the front of the leg of the L and jutting out slightly to the right. Viewers within the control room could thus clearly witness—and film—each test, including readings from the small array of instruments, including gauges and a clock. The plume of the rocket motor being tested would have faced northward, toward the direction of the Lower Lake River. Sandbags piled around the test stand offered additional protection. Still another component of this very modest test complex was at least one storage and/or work shed, placed farther east of the back wall of the inverted L-shaped structure. Overall, the entire RMI site—that included a large cement floor area extending beyond the blockhouse—was recorded by Lenik to have measured about 20×26 feet. Lenik had no knowledge of the previous additional sheds or other structures that had once occupied the entire test complex. These other buildings may have either filled up the additional cement area, or one or more of these other sheds may have had their own separate pads. In any case, the actual test area was exceedingly small and cramped by present-day standards, although it was considered ideal at the time.

Within the interior of the little control room, as recollected by Les Collins about a decade after the formation of RMI, "There were none of the elaborate set ups we have today [in 1951] such as recording instruments, etc. We [simply] had regular pressure gauges and manometers which were photographed by movie cameras during runs. (Occasionally Mr. Wyld, [who also served as] our photographer, would forget to push the button!)."[26]

Significantly, one of the cameras was electric—perhaps high tech for its day—and was employed outside the blockhouse. It came with a very long extension cord that Wyld reeled in after a test. Electricity for the site, including the camera, may have been supplied from a nearby Nelson Airport hangar, although it is likely they turned on the camera from inside the control room. The camera did not show the gauges, just a different angle of the running motor. There are, indeed, indications that direct electric current was available from an adjacent airport hangar for other instruments and purposes—mainly for the ignition-system controls that would have been inside the control room. An outlet for basic lighting may have been included. The existing film footage shows the control room operating at night. According to another article on the Franklin Lakes site in the *Patterson News* (Patterson, NJ) for 21 March 1980, "Cobwebs have replaced the instrument panel that used to stand on the south wall of the blockhouse, leaving only two cast iron pipes that formerly covered ignition [and perhaps other] cables running to the engines."[27]

As for the stand proper, film footage does not show it in closeup, and details of its workings remain unclear. Again, it is evident that several, if not all, the instruments from the ARS Test Stand No. 2 were borrowed and affixed to the newer RMI stand. Other components, such as tankage, valves, and piping, may have likewise been cannibalized for adaptation onto RMI's first test-stand model that was very much patterned after ARS Test Stand No. 2. Later, when RMI graduated to a larger stand and no longer needed some of the components and other parts from Test Stand No. 2, this hardware was properly returned and reinstalled onto that stand, which was then kept in a display and/or storage area that RMI termed its "museum." But some components were still retained for the work. It seems that the original ARS Test Stand No. 2 clock, as well as the vertical glass propellant-volume gauges were still considered useful because they surface again and are found on one or more of RMI's larger and more advanced stands.

The extant film footage also reveals that after their more successful tests, the small group of hearty RMI experimenters toasted each other with apple-jack (fermented apple juice)—the "official" drink of rocketeers, as they phrased it.

Thrust capacities of all the iterations of RMI's earliest stands remain unknown. But the caption on a partly faded photograph of their very first moveable stand (it stood on four casters and was secured to a barrier in use) indicates that that model was rated at 300-lb capacity (Fig. 3.5). It soon may have been upgraded to accommodate the increased thrust of the M16-G-class motor at the time. Upon the opening of the Franklin Lakes site in September 1942, all RMI stands were fixed, heavy-duty structures. Although closeups of the new stand are lacking in the film footage, surviving photographs show the follow-on "large test stand" of December 1942 to now be a very imposing structure set in a massive concrete block with some six dials on its face, in addition to utilizing the borrowed clock from ARS Test Stand No. 2 and a pair of vertical propellant-volume or capacity gauges.

Les Collins described post-test procedures. He wrote, "After the tests, all the equipment [although evidently excluding the stand itself] had to be man-handled back into the test house or loaded into the truck for the return to the shop [at 280 Wanaque Avenue in Pompton Lakes]. We only had one set of tools then, so could not leave any behind as they might be needed in the shop."[28]

MOVE TO POMPTON PLAINS

RMI's newer, more-powerful stands reflected their considerable progress. They also necessitated a drastic need for an expanded new facility and head-quarters. Les Collins explained:

> In January 1943, we moved into our new plant at Pompton Plains, a three-story frame building. The siding was such that the owners had stucco coated [to] the whole outside. This building had started life as a silverware factory in the late 1800s; after the death of its owner, it had remained empty for years, and then it was re-opened as a night club (The Silver Circle). When RMI took possession, parts of the bar and kitchen were still in place…The move went off very smoothly and all hands were in their pitching. There were the President, Treasurer, Secretary, and Foreman [Collins], and all the other employees (three [of them], wiring up machines, moving equipment into place, drilling holes in concrete, setting up the engineering department and stock room. By Monday, January 17, 1943, the new plant was all ready to go.[29]

RMI's facilities thereby jumped from about 900 square feet at 280 Wanaque Avenue, Pompton Lakes, to 8,900 square feet at Pompton Plains.

Fig. 3.7 RMI's new headquarters and shop in the former "silver factory" at Pompton Plains, NJ. (Courtesy National Archives.)

Collins's often-used description of the building as a former "silverware factory" is not quite accurate and requires further definition. Others claim that the building was later turned into a Prohibition-era speakeasy known as the Silver Circle. That, it turns out, is close to the truth and is a shocking story in and of itself.

As for its original use as a silverware factory, the 85 x 38-ft building technically started off in the 1890s as a silver electroplating company run by the Britain-born Thomas Shaw. Shaw, who had emigrated to America just after the Civil War, became very prosperous in the electroplating business, first in Rhode Island and later in Newark, New Jersey. He retired to a country house in Pompton Plains, where he became a prominent real estate developer and also ran his own electroplating concern on the side. The Shaw plant at Pompton Plains produced fine cutlery for a number of exclusive clients, including Tiffany and Co., Thomas Brown and Sons, and Relich & Company. Then, about 1924, Shaw's widow sold the plant building to Emma B. Dunn, and it was Mrs. Dunn who leased what was then called the "Dunn barn" to RMI.

There is great credence to the claim that the barn was the former Silver Circle speakeasy. It seems that during those turbulent and violent times, one Frankie Dunn of Union City, New Jersey, owned a major illegal brewery that supplied beer to bootleggers all over Manhattan and northern New Jersey. He rose quickly to become a notorious beer-baron gangster, dubbed the "beer king." Then, on 7 March 1930, Dunn was ruthlessly gunned down in Hoboken, along with James "Bugs" Donovan, by the rival New York City bootlegger known as Waxey Gordon. The multiple assassination was big news at the time and made all the New Jersey newspapers as well as the *New York Times* and other papers across the country.[30]

These stories show that Frankie Dunn owned property in Pompton Plains, and from the census records we additionally learn that his widow, Emma, and her two children continued to live there after his untimely demise.[31] (The place was called either Pompton Plains or by its original Indian name of Pequannock; Pompton Plains is part of Pequannock Township.) Emma had moved to this location in 1924 and may have been the original purchaser of the property. Furthermore, the Dunn barn was on this land. Consequently, it is highly likely the barn really had been one of Frankie's Prohibition speakeasies. When Lawrence rented the building early in late 1942 or early 1943 from Frankie's widow, he must have known, or came to know, the shady past of the barn. But for obvious reasons the gangster connection never became part of the official history of RMI. For that matter, it is not known how Lawrence came to learn of the available barn.

From Emma's side, as revealed in Lovell's later *Daily Log*, we know that she must have known the true nature of RMI's business, and is not known to have ever made complaints about it; she was even afforded a tour of the premises after the war, on 26 September 1946, when their business was no longer a secret. Emma's house was on the same property and a short walking distance north of the barn where she had kept her three or four horses prior to the rental to RMI. Emma died in 1967 at age 87 at what had become known as the "Dunn place" in Pompton Plains. The street still exists, although her house and the nearby Dunn barn, where much aerospace history by RMI was made, have long since disappeared.[32]

When the barn was rented to RMI, Emma was still using the top level for storage. Just the first two floors were utilized and partitioned off for RMI's dual administration/drafting room, laboratory, and factory. The barn was short of space, but it was a far cry from their previous shop quarters at 280 Wanaque Avenue in Pompton Lakes. Technically speaking, RMI's address at the time, when it was situated in the Dunn barn—and then the overall combined RMI headquarters and plant—was simply given as: "Reaction Motors, Inc., State Highway, Route 23, Pompton Plains, N.J." Presently at the former RMI location, at NJ State Highway Route 23, Boulevard, and West Parkway, there is an historic marker off the Parkway that actually honors the RMI test site, a sand pit that was in the back of the Dunn barn. The site is particularly known for its role in the first tests of the 6000C-4 engine of the famous Bell X-1, which is covered in Chapter 4.

THE 1,000-LB-THRUST MILESTONE IS REACHED

Progress at RMI continued rapidly upon their move into the Dunn barn. Already on 20 March 1943, the second major RMI milestone of 1,000-lb thrust was reached with the successful firing of their almost 20-in.-long M19-G1 motor. At about this time, the RMI team amounted to eight men. The 1943 RMI staff now included, in addition to the four founders, foreman Les Collins, machinist Lou Arata, and newer men Edward "Ed" Cahill and Harry Smith. Apart from their new plant, RMI retained their Franklin Lakes test site, although the previous stand had given way to one of more than 1,000-lb capacity and was anchored to a concrete slab as a fixed stand. Lawrence also astutely took advantage of the 3.5 acres around the Dunn barn to erect three additional and more powerful stands with thrust ratings as high as 1,500 lb. Again, Lawrence himself was largely responsible for all the increasingly complex electrical circuitry layouts for their designs. He may undoubtedly be considered an early pioneer in American rocket test-stand technology.

The important development of the 1,000-lb M19 motor could be credited to all members of the small RMI team. For instance, Les Collins later recalled:

> I well remember two of my first assignments at RMI—one was making an injector for the 1,000 pound unit out of a piece of about four inch stainless [steel] on an Atlas lathe...What a job trying to drill a 1 ¼ inch hole through; it just would not pull. The other job was welding up the head of a 1,000 pound unit. Although I had plenty of experience before on electric welders, I had never run across such a specimen as we had to use...wow—how that gadget could throw weld splatter around everywhere except the joint. Finally, we managed to get it made and all the [previous] leaks stopped.[33]

Lawrence had not only concerned himself with the electrical circuitry in this motor development, but also drew flow diagrams for the safety-valve system that could be remotely operated by gear work and small solenoid electric motors. He and Wyld applied for a patent on 6 July 1943 for such a throttling system, although this was not granted until 23 August 1949 as U.S. Patent No. 2,479,888 for a "Controlling System for Reaction Motors." This patent therefore traced its roots back to the development of the 1,000-lb-thrust M19 motor.

PIERCE'S ROLE IN RMI

As for Pierce, no sketches or the like by him have been found within the Lovell Lawrence papers. However, the RMI *Log Book* (starting in March 1943) does show he was present at every test, with continued work on the M19-G1 and successor motors. He also contributed to writing several of the *Log Book* reports and handled the equipment; he must have done the same throughout RMI's first operating year of 1942, and from there on. During the JATO test, Pierce was very much in evidence. His value to RMI is well proven in an existing Navy Bureau of Aeronautics (BuAer) document within the

National Archives that shows that in mid-April 1943, when Lawrence reported that both Pierce and Arata's draft status was under review, Lawrence immediately submitted affidavits for both men to have their occupational classifications retained so they would not be considered for the draft. "Since the registrant [Pierce] is an authority on jet reaction motors," Lawrence wrote on the form, "it will be impossible to replace him."[34] Pierce's important roles in the live JATO flight tests covered later in this chapter were also very much in evidence.

It is relevant to note that despite Pierce's lack of academic and technical credentials, Lawrence treated him as a fully equal partner in RMI and officially referred to him in the affidavit and other documents as an "engineer." From the start, all four of the founders had also received exactly the same salary, in addition to 125 stock shares in the company.

THE 3,000-LB-THRUST MOTOR

The successful completion of the 1,000-lb M19-G1 immediately led BuAer to issue a second contract to RMI. "By this time," Shesta wrote in his memoir paper, the Navy had already begun

> to realize that they needed a much larger motor than a 1,000 lb unit for what they had in mind, namely, the Jet-Assisted-Take Off (JATO) for their large sea planes. Thus, our second contract called for a 3,000 lb motor. It was to be installed in a PBM [Patrol Bomber of the Glenn L. Martin Company], and test flown in accordance with detailed specifications. One of the qualifications called for a member of the firm [RMI] being on board for the test flights[35].

"My contribution to the design," Shesta went on, "was the development of a so-called multi-nozzle injector," but he obviously did far more than this.[36] Already, on 6 May 1943, the 3,000-lb thrust was reached, although with an earlier model motor: the M18-G1 fitted with Shesta's multinozzle feature. In truth, the reaction (thrust) recorder registered 3,180 lb for about 24 s, according the *Log Book*. The *Log Book* adds other interesting comments as recorded in Wyld's handwriting:

> Prelim.[inary] for [exhaust] velocity [about] 6,000 ft/sec. Flame straight, not remarkably long, but intensely hot—started a number of small fires in woods and scorched ground. Also burned corner off some nearby sandbags, but did not damage nearby glass view mirror appreciably, although paint burn off frame and rubber mounting gaskets melted. Noise not much greater than 1,000-lb type [motor]…Motor appears to be in excellent shape after run—no burnouts. Combustion was quite steady…Dead [tree] stump about 80–90 ft from motor was set afire during [the] test.[37]

Testing continued into the fall. This involved water injection mixed with gasoline to further help cool the motor; it worked very well, but cut down on the brilliancy of the flame, although clear Mach waves were still observed. "Our greatest problem during those years," Arata once told the author, "was heat transfer. Of course, regeneratively cooled systems were the only way to go; but this did not mean that all the cooling problems were solved, especially for long-duration runs."[38]

There were also efforts to improved the propellant, or propellant flow, but one of these measures backfired. According to Shesta, "Tergital" was supplied by the Linde Company "to emulsify the gasoline and make it mix with water." But this presented its own complications because "in cold weather," he added, it "would form a pasty slush which would choke the flow in valves and metering orifices. Eventually, it [gasoline] was abandoned in favor of alcohol."[39]

Lawrence also strove to perfect the ignition system, which was initiated by a spark plug and entailed an ingenious arrangement of pressure switches and relays to make the control automatic. The *Log Book* reveals that the ignition trials did not always go smoothly, what with valves often failing, overloads resulting in loud bangs and roars of escaping gas, and so forth. Nonetheless,

Fig. 3.8   Wyld, at left, with probably Fischer's replacement, Lt. Cmdr. John S. Warfel, and Lawrence next to the 3,000-lb-thrust motor, perhaps before the JATO tests. (Courtesy Smithsonian, 89-1847.)

in the same period another milestone was reached. On 14 May 1943, the longest run yet was attained with the M19-G2 motor when it ran for "probably over one minute."[40] Previous tests characteristically lasted only seconds, or half a minute or so at most.

THE JATO PHASE

The test of 28 May was a demonstration for both Navy and Glenn L. Martin Company representatives from their plant in Baltimore. The guests included Lt. William L. "Bill" Gore of the Marine Corps, who was designated to serve as the pilot for the later "live" JATO flight trials on a Martin PBM Mariner patrol bomber seaplane. Gore was another early believer in JATO and had authored a paper in 1938 on his own version of the concept. Later in his career he became president of the ARS and a vice president of Aerojet.

On Friday, 1 October 1943, the first of the PBM-3 mockup trials were started at the Franklin Lakes test facility. BuAer had shipped a full-scale mockup of the boat tail of the plane to RMI so that Lawrence and his team could not only familiarize themselves with this part of the aircraft, but work out details on fitting their JATO, as well as making practice runs of the 3,000-lb JATO motor. The mockup was crudely fashioned from plywood covered with corrugated galvanized iron. "It did not look very much like a plane," Shesta later commented, "but it served the purpose."[41] Flight tankage (the propellant and pressurizing tanks) were added, and RMI made test runs with it "from time to time."[42]

Undoubtedly due to the long installation and preparation time, the initial test with the motor on the mockup started late in the day, at 6:30 p.m., when it was almost dark. According to the *Log Book*, with all four RMI founders and others present and the cameras "open full," the ignition was started but failed.[43] Wyld's write-up continues: "After several [other] attempts to fire, it was decided to shut down as it was 7:00 p.m. and practically dark. Test was therefore stopped."[44] The "trouble," Wyld concluded in his report, "was eventually traced to a defective spark plug."[45]

Hence, it was not until Monday, 4 October of that year, with the entire shop crew present, in addition to Mr. Henry C. Kornemann of the Linde Air Products Company that supplied LOX to RMI, that the first true trial with the 3,000-lb motor was again tried on the PBY-3 boat tail mockup.[46] Once more, ignition problems surfaced and tests were stopped. And so it went. The boat tail runs were nearly always troublesome and never entirely satisfactory. RMI and the Navy still persisted with the JATO project until the major difficulties were essentially corrected by the end of November, and Wyld was sent to Annapolis to begin supervising the installation of the 3,000-lb-thrust motor onto a real PBY.

GODDARD AND RMI

In those rugged pioneering days of American rocketry, setbacks were far more routine with the new technology than nowadays; even the highly experienced Dr. Goddard, then also stationed at Annapolis working on his own JATOs, faced his full share of technological frustrations as well. But characteristic of all these true rocket pioneers—the RMI team included—he remained steadfastly optimistic in the great promise of this technology.

Were there interactions during these years between RMI and Goddard? There were, and they were always cordial, although both sides were respectively competitive with each other. Goddard noted in a diary entry for 15 September 1942: "Went with Lt. And Mrs. [C. Fink] Fischer to New Jersey, and saw [for the first time] the Reaction Motors setup. Had dinner with them [the Fischers] and Lawrence and Shesta."[47] Arata also remembered occasional visits by Goddard, as well as by Truax. Toward the end of the war, interactions between Goddard and RMI became even closer, and Lawrence asked Goddard if he could join forces with the company, but this never transpired.

PREPARING FOR THE LIVE JATO FLIGHTS

Numerous sketches within the Lovell Lawrence papers offer tantalizing clues of constant brainstorming and an incredible level and accelerated pace of activity during this time that were the hallmarks of all these pioneers. For instance, there is a drawing by Lawrence, dated 26 January 1943, of a LOX flow-meter design; a "Proposed [PBM] Mock-Up Layout" of 7 July 1943; his sketch of 8 July 1943 for a "Proposed Electrical Pressure Switch Control Circuit for U.S. Navy Mockup—PBM-3"; a long textual description, with a drawing by Wyld, dated 5 September 1943, of a "Rocket Meteograph"; undated working notes, also with a unique drawing perhaps by Lawrence, simply titled "Navy DU-2" (that is, a droppable JATO unit) that includes lists and exact dimensions of materials needed for this unit; a very complex electrical schematic, also probably by Lawrence, undated and titled "DU-500 Test Stand Electric Control Circuit"; and another undated drawing, probably by Lawrence, of the entire internal arrangement of a "3000# [3,000-lb-thrust] Drop Unit."[48]

Although not described in Shesta's memoir paper or other sources, some of the above items now present rich, new details on the nature of RMI's own planned JATO. Like the Navy and Aerojet's JATO efforts (and probably Goddard's as well), the RMI model was similarly designed to be droppable. This meant it was encased in a cigar-shaped container more than 8 ft long, with a rounded head that contained a folded parachute. Following its deployment as a JATO by a PBY taking off from a Pacific island runway, the unit was to be ejected from the seaplane, then descend into the water where it could be retrieved by a boat stationed nearby. After, it would be throughly cleaned and refueled, including adding new pressurizing gas, for reuse.

The 3,000-lb-thrust RMI unit was pressure fed with nitrogen and it utilized 55 gallons of LOX, as well as 28 gallons of gasoline fuel that was cooled by 29 gallons of water. In this case, as gasoline was definitely the fuel and is shown in the sketch in Lawrence's papers as water cooled, it is certain this particular design of a "droppable unit"—a term within the JATO community that had evidently come into general use—was RMI's.

THE JATO FLIGHTS

"Shortly before Thanksgiving that year [before 26 November 1943]," wrote Shesta of the actual live JATO test preparations, "F. Pierce, Frank [T.] Muth [RMI's new head draftsman] and I loaded the old RMI truck [with the JATO-ready motor] and started for Annapolis."[49] Lawrence, Wyld, and Les Collins joined them later. "At Annapolis," Shesta continued, "we found the [Navy] plane crew rather skeptical and uneasy about the forthcoming tests, especially so because of a rather minor accident that happened shortly before to another plane with a rocket that was built by another agency [in another JATO test at Annapolis, perhaps by Aerojet]…Pierce stepped into the breech nobly and pointed out that our equipment was eminently safe and in fact thought nothing at all of flying with it personally. This went a good way toward restoring their confidence."[50] "The work of installation [in the real PBM]," Shesta went on, "was long and arduous, because we all had to do all the piping ourselves and at the same time there were always fifty sailors in the plane work on brackets and bulkheads."[51]

The first live JATO flights started on 12 January 1944. Lt. Gore of the Marine Corps was the pilot. According to the contract, one RMI member had to be aboard to monitor matters, but as Shesta put it more bluntly, this was done "just to make sure nothing would fall apart."[52] In his other account of this phase of the RMI story in a memoir paper, Shesta wrote: "Lawrence and I took alternate flights. Occasionally, Wyld or Pierce would also fly along. The test program called for flights with gradually increased aircraft loading up to, I believe, 60,000 lbs."[53] This was done with sandbags hauled aboard the plane by sailors. "The tests went on for about two weeks," Shesta continued. "Toward the end, there were so many sandbags in the passageway that one could hardly get through. Besides that, extra tanks were installed in the bomb bays. Because of the cold weather and danger of freezing, we also loaded up on aviation gas[oline]. As Lt. Gore would say, we had enough fuel on hand to fly non-stop to Honolulu."[54] Shesta concluded:

> Eventually, the tests were over and done with. We felt relieved. There had been no trouble at all. The only thing that remained to be done was to run a demonstrated static test for a group of admirals and other dignitaries. In

Fig. 3.9   PBM seaplane at Annapolis, MD, with RMI JATO seen at an angle under the belly of the plane. (Courtesy Smithsonian, 92-145.)

due time the guests arrived. The plane was wheeled out on the apron, the engines reved up and the rocket turned on. Nothing happened. Only huge billowy gusts of raw gasoline spewed out of the nozzle. No oxygen. We tried again and again without success. Finally, the tests were called off and the dignitaries departed in disgust. We were all embarrassed and mortified by this fiasco. The next day I dismantled the assembly and found out what happened. The special safety valve at the base of the oxygen tank had been exposed to salt air....Corrosion had set in the aluminum parts.[55]

To this he added: "After completing the PBM flights we had definitely given up gasoline as a propellant and switched [back] to alcohol which we felt much better suited for rockets than gasoline."[56] As Wyld already knew, alcohol mixed with water in the right ratios better helped cool the motors.

Further details on the overall PBM flight tests are found in a letter sent by Lawrence to Fink Fischer, written many months later, on 8 June 1944. Unquestionably, Fink had played pivotal roles in the early history of RMI and Lovell had thus befriended him. Therefore, after Lt. Fischer was transferred to other duties by November 1942 and his position as the Navy liaison with RMI taken over by Lt. John S. Warfel, Lawrence lost contact with him. Lawrence's letter thus begins:

Dear Fink, I have been trying to get a letter off to you for these past twelve months, but it seems that every time I sat down to do it I could not locate your [new] address. This is a poor excuse I know, but I will now makes amends. The BuAer has kept us pretty busy. We have completed the

Fig. 3.10   RMI and PBM crews, probably after the JATO tests at Annapolis, 1944. (Courtesy Smithsonian Institution, 78-12186.)

PBM3 [sic.] flight tests and are four months underway in an acid-aniline, oxygen-gasoline parameter study program. The PBM3 flight tests were run off without a hitch—largest gross load 57,000 pounds. It was necessary to run 75 seconds before getting clear of the water, with the jet [rocket JATO] 28 seconds. We had originally desired to go to 60,000 pounds; however, the plane was so cumbersome that we decided that this was a good point to stop the tests.[57]

THE AFTERMATH

Despite all their long and tough pioneering work to arrive at this point, it turned out the Navy never did adopt RMI's JATOs—nor Aerojet's, nor Goddard's, nor other liquid-propellant models either. What had happened?

While not spelled out by Shesta, part of the answer lay in the Navy's insistence on gasoline as the fuel. Aerojet and Truax had favored *hypergolic* combinations, such as redfuming nitric acid (RFNA) and aniline, that ignite on contact with the fuel and oxidizer, thereby eliminating the need for an ignition system and greatly simplifying the rocket, although RFNA presented its own headaches in corrosiveness and toxicity if not handled properly. Gasoline was cheaper as well, although very problematical as a fuel. Compared with alcohol and other fuels, gasoline also has a lower *specific impulse,* which is a measurement of the efficiency of propellants and therefore has a lower *exhaust velocity.* The higher the specific impulse and corresponding exhaust velocity of a propellant and rocket motor, the more powerful the performance. Therefore, after all RMI's frustrating experiences with it, gasoline became the least favored propellant for potential JATOs. Overall, the Navy found the hypergolic (acid/aniline) units to be more convenient in the long run and more powerful, and for a short while the Coast Guard was furnished with some of Aerojet's units.

Upon greater reflection and inhouse comparisons of all the various experimental JATOs by 1944, the Navy brass came to realize that the basic liquid-propellant units—as compared to Aerojet's far simpler and infinitely less temperamental solid-propellant types—could not be easily serviced in the field. Liquid-propellant types were also more complex and expensive. The Navy and Army Air Corps therefore opted for the solid units instead, and these are the ones that soon became operational. The once highly competitive era of experimental liquid-propellant JATOs was more or less over before the war's end.

These are all valid general reasons for the apparent demise of the liquid-propellant programs from the Navy side. Another recently discovered document in the Lawrence papers, anonymously authored and undated, appears to paint a far bleaker and bigger and mysterious picture. In all likelihood, the author was Lawrence, writing hastily on hotel stationary after a top secret meeting with Navy brass in Washington, D.C. The document lists a number of points and comments, as follows:

> 1.—Liquid oxygen not in fleet for two or three years—[and] tankage training not avail.[able]; 2.—Absolutely no chance for us [RMI] to mfture [sic., manufacture] any equip.[ment] even if needed—financial structure [for this inadequate]; 3.—Aerojet to get [$]3,000,000 worth in aniline, acid—drop units...powder [solid-propellant JATO types]; 4.— [I] found that [Aerojet's Andrew G.] Haley would offer us to effect a merger with Aerojet; and 5.—[I] will try to get our present contract straightened out; 6.—[The Navy] would like us to carry on the research & development but felt that it would be under the same conditions as as at present.[58]

The conclusion of RMI's JATO program—and its prospects—seems to have evaporated mysteriously overnight for a variety of far more complex and, in some respects, dire issues than is generally realized. Shesta's memoir paper and other sources on the early history of JATOs for the Navy are silent on these developments. Therefore, the above-cited document presents a series of provocative historical questions involving interrelated logistical, budgetary, and perhaps internal Navy political issues that await deeper research into BuAer and other records. The full extent of these multiple problems and the true reasons for the demise of RMI's venture into JATOs, which paradoxically had helped them to get started, may never be known. Years later in the postwar period, proposed RMI JATO projects occasionally did surface, although nothing came of these either.

One thing was certain, however. As seen by Shesta's remark, after the live JATO tests, RMI gave up on gasoline as a propellant and reverted back to Wyld's original choice of alcohol as a fuel. Under another Navy contract, RMI had also begun investigating the possibilities of nitric acid and aniline. Within the Lawrence Jr. papers is another of his most interesting and sophisticated drawings, dated 29 February 1944, titled "Proposed Dual Test Stand – Manual (Aniline-Acid—Water—Gasoline)."[59] There is also his note of 25 October 1942, headed "Priority Sequence," in which Lawrence lists as Priority No. 1 an M17 G-2 1,200-lb-thrust "acid version" of the motor, with a stainless steel liner and nozzle—stainless steel to prevent severe corrosion caused by the acid.[60] Among other priorities on the same list are RMI's fabrication of DU-1 tanks, "two more M17G-2" motors, four M15G-5 types, a test of M17G "with oxygen, and two M15G-2 motors."[61]

Despite the fact that their venture into the development of JATOs never fully materializied, multiple miscellaneous BuAer contracts continued to flow into RMI and kept them thriving. BuAer was now keeping them busily engaged in the relatively newer field of missiles and soon thereafter, of rocket planes.

RMI ENTERS THE MISSILE FIELD—THE GORGONS

The first missile project that RMI became engaged in appears to have been BuAer's radiocontrolled Gorgon air-to-air type, although the available records on this project are scant. The Gorgon was really a family of missiles with diverse modes of propulsion. As more fully discussed in Chapter 6, the origin of the Gorgon goes back to as early as 1937. At that time, however, it was merely the rudimentary concept of an unmanned piston engine-powered aircraft that could be suitable as either a target drone or "aerial weapon." It was not until the summer of 1943 that the missile program itself started and became known as the Gorgon. Later that year, a rocket power plant was adapted to the missile. But the rocket was not the only mode of propulsion associated with this weapon. Throughout the subsequent long, complex history of the Gorgon family, which lasted until 1953, there evolved more than half a dozen models that included: single and dual-powered rocket versions, a pulsejet-powered model, a ramjet-powered model, and eventually one with a turbjojet. Here, only the rocket versions are reviewed.

First, there was the Gorgon 2A model missile, 14.5 ft long and 11 ft wide, with a 25-mi range. It originated in 1943 and was designed to employ a 350-lb/130-s-duration-thrust rocket motor burning monoethylene and mixed sulfuric and nitric acids. Claimed as the first U.S. liquid-propellant guided missile, the Gorgon 2A was intended by the Navy to be deployed against Japanese naval and merchant shipping; the missile is also referred to as the Gorgon II-A pilotless aircraft air-to-air or air-to-ship missile. The Navy's team of Lt. William Schubert and Ens. Robertson Youngquist was largely responsible for the design and initial development of the motor, but RMI was called in to start making preproduction models for further developmental purposes.[62] Consequently, from 1944 to 1945 they built approximately 50 of the missile's motor, based on engineering drawings furnished by the U.S. Naval Experiment Station, even though in 1946 the missile was cancelled. An existing, beautifully restored example of a Gorgon 2A, with an original RMI manufacturer's label attached to its internal motor, is now on exhibit at NASM's Udvar-Hazy Center near Dulles International Airport in Chantilly, Virginia, about 30 miles from Washington, D.C.

Second, RMI went on to build the CML-2N motor—as it was now designated— of the same thrust and duration for the Gorgon 3A, or III-A, that started as another air-to-air model; after the war, it became reassigned as a "control test vehicle" (CTV), or strictly experimental vehicle. Third, RMI built the twin motors (two in tandem of the Gorgon 2A type) for the Gorgon 3C, or III-C, also known as the KU3N-2. Only a dozen of these unique missiles were built and were mainly deployed by NACA for early postwar high-speed research and by the Navy for performance and stability tests. A quite rare example of this early experimental American missile (doubtless, the only surviving example)—fitted with its original twin RMI motors—is also found in NASM's collection, although it is not presently restored and remains in storage.

RMI's Gorgon experience directly led the company to become involved in another early U.S. missile program by the end of the war: the surface-to-air Lark. That activity began in 1945 and is also discussed in Chapter 6, along with their postwar missile-motor developments.

RMI'S EARLY WORK WITH PULSEJETS

There was one other missile project that *did* involve RMI during the war years—the infamous German pulsejet-powered V-1, popularly dubbed the "Buzz Bomb," due to the monotonous pulsating noise it made before it struck its target. Technically not a rocket, the air-breathing V-1 power plant still counted as a reaction motor and therefore logically came under the bailiwick of RMI. Shesta related what happened in July 1944:

> Our engineers were summoned to the Bureau of Aeronautics in Washington for consultation in connection with the German V-1. At the same time, certain sketchy details of the construction of the power plant were relayed to us and we were requested to construct and test a duplicate thereof. Work on this project was started in the middle of July and was completed on or about the first week in August 1944. These tests continued to about the middle of September 1944, at which time we took the motor to the Philadelphia Naval Aircraft Factory…for further testing.[63]

Elsewhere, Shesta informed the author that the project started when the Navy had gathered up the remains of some pieces of German Buzz Bombs that were bombarding London and forwarded these on to RMI.

RMI came to adopt the name "resojet" for the pulsejet, in addition to the more curious and puzzling nickname of "Peenemunde 16 [sic.]," after Germany's massive Army Research Center, in Peenemünde-East, where the true rocket-powered V-2 missile had been created. The V-1 had been developed in Peenemünde-West, at the Luftwaffe, or German Air Force side; whether Lawrence's team were aware of these fine historical distinctions at the time is unknown. Wyld, incidentally, executed his own detailed drawing, dated 27 July 1944, of the "Feed-System of 'Peeenemunde 16' Robot Plane."[64] The origin of the nickname "Peenemunde 16" remains a mystery.

RMI had set up a dedicated pulsejet test stand at their Pompton Plains facility, and Shesta also became heavily occupied with this novel and seemingly promising technology. Their resojet research continued well into the postwar years, although these efforts never led to any operational U.S. version of the V-1. There was an American V-1 counterpart named the Loon, with a pulsejet motor developed and produced by the Ford Motor Company, not RMI.

It is interesting that the Navy's requested work from RMI on resojet development was simply tacked on to their contracts with RMI as other "tasks," along with their continued assigned rocket work. The inside story of how RMI did *not* become part of the important development of the Loon, also called the JB-2, (Jet Bomb 2) has yet to be told and requires more research. It may simply be that although RMI had successfully modified the V-1's "aeroresonator," extending the theoretical valve life to about four times that of the German valve, tiny RMI was no match for the giant Ford Motor Company in generating a viable program toward potentially mass-producing a whole new missile technology.

As matters turned out, however, that once-exciting pulsejet trend withered into a relatively short-lived phenomenon. The fundamental reason is that the subsonic Loon, and other pulsejet-powered missiles (including the Gorgon 2C model), were limited strictly to subsonic velocities and performances. They became quickly outmoded, in sharp contrast to the infinitely more powerful and highly promising turbojets, rockets, and ramjet vehicles then coming on the scene—and all within the supersonic realm. The Loon, which had progressed to become test flown in 1947 from the submarines USS *Carbonero* and USS *Cusk*, was cancelled as early as 1950.[65]

RMI STARTS EXPLORING ROCKET AIRCRAFT

The rocket plane was a natural extension of RMI's work with JATOs. Jimmy Wyld had already combined his dual passions for aviation and rockets to arrive at ideas along these lines. About 1940, before RMI was formed, he had authored a 20-page, still-unpublished manuscript titled, "The Application of Regenerative Liquid-Fuel Rocket Motors to Military Aircraft."[66] In it, he conceived two modes of application. One was a rocket assist, or JATO arrangement, although he did not use that term. The other was the rocket as a *primary* source of propulsion for aircraft. It is striking that for the latter he suggested applying a regen motor of 1,000-lb thrust of Shesta's design and proposed that it be built *into* the aircraft.

Nothing germinated from either of these ideas and it cannot necessarily be assumed that his plans for a JATO became the germ for the Navy's request for a JATO development by RMI upon its formation the next year. By that time the basic concept of what became known as the JATO already had a long history, and solid-propellant types had even been tried successfully on a Junkers seaplane in Germany by the late 1920s. Much later, the U.S. Navy's own Robert Truax championed the development of liquid-propellant types that led the Navy to establish its U.S. Naval Engineering Experimental Station at Annapolis, Maryland, in July 1941 for the express purpose of developing a "rocket unit to assist the take-off of large flying boats"—five months before RMI was started.[67]

We also know that by 1944 the U.S. Army Air Corps independently began to consider developing a true rocket aircraft, as opposed to a plane with auxiliary JATO. In 1939, Ezra Kotcher of the Air Corps Engineering School at Wright Field, Ohio, submitted a report that advocated either gas turbines or rocket propulsion for an extensive transonic airplane research program. In 1942, Aerojet—the second U.S. commercial liquid-propellant rocket company, with close ties to the Army Air Corps—initiated its own development of a 2,000-lb-thrust rocket engine for a proposed XP-79 flying-wing interceptor. This became Aerojet's XCALR-2000-A, otherwise called the Aerotojet, that used the hypergolic combination of red-fuming nitric acid and aniline.

A more advanced Aerojet engine design, the XLR-7-AJ-1, that operated with the same propellants, was to deliver 6,000 lb of thrust and became incorporated into Project MX-12, another proposed superperformance, flying-wing-type aircraft of the Army Air Force Materiel Command. The XP-79 never left the drawing board, nor did Project MX-121, although Aerojet's advanced 6,000-lb-engine project concept remained viable—for the time being. By 1944, some of the seeds were sewn that led to the genesis of the rocket-powered Bell X-1 aircraft—the world's first plane to break the sound barrier in 1947—in which RMI came to play a leading role. This story is taken up more fully in Chapter 4.

In the meantime, as the war wound down, Lawrence's very meticulously crafted color sketch of 10 January 1945 of the design of a "Proposed Interceptor Control System," utilizing a system of four throttleable, 750-lb, LOX/alcohol motors emerged as another of those possible seeds of the world's first supersonic aircraft.[68]

ENDNOTES

1. Letter from Walter S. Lehmmon, Radiotype Division, International Business Machines Corporation, to Lovell Lawrence Jr., 5 March 1942, in Lovell Lawrence Jr. Papers, NASM, box 12, folder 3; Letter from Lovell Lawrence Jr. to A.L. Holt, Radiotype Division, IBM, 2 March 1942, Lovell Lawrence Jr. Papers, box 12, folder 7.

   An additional revelation is discovered in Lawrence's letter of 2 March 1942 to his boss, A.L. Holt. It shows that Lawrence had originally merely requested a leave of absence to work in the undisclosed "defense research position," although this was turned down because it was not IBM's policy to grant leaves of absence to employees transferring to such positions. Lawrence therefore felt he had no other option but to submit his resignation. In other words, Lovell well knew that his embarking upon running RMI would be an entirely risky gamble in his career, and as late as early March he was still not fully confident that the Navy contract would come through, nor that RMI would be a success. This is why, up until that time, he had been reluctant to fully leave his IBM job. It is also interesting to note Lawrence's very cautious use of the term "National Defense research work" to IBM. This underscores the fact that the formation of RMI was to be kept as secret possible.

   Lawrence's continued employment with IBM up to mid-March also explains why these critical first four months of RMI's existence are the least documented in the company's overall history. To further complicate matters, for years the Navy has exercised a policy of destroying contractual records once a contract is completed. Consequently, contracts and their accompanying progress reports relative to RMI projects are not found in the National Archives. Fortunately, we are in much better shape for the 1945 to the early 1950s period of RMI's history due to the existence of the Lovell Lawrence Jr. Papers in the NASM Archives, especially Lawrence's invaluable "Daily Logs." The most significant gaps of original RMI documentation pertain to the war years. Nonetheless, Lawrence's collection also contains World War II-era drawings by Lawrence and others, some occasional notes from this period, rare photographs with captions, as well as some short but priceless film loops of their 1942–1943 tests. Shesta's memoir paper and earlier published recollections also provide significant help. Hence, the war-years phase of RMI's history have represented the greatest historical challenges in documenting RMI's history.

2. John Shesta, "Reaction Motors, Inc.: A Memoir," in Frank H. Winter and Frederick I. Ordway III, *Pioneering American Rocketry: The Reaction Motors, Inc. (RMI) Story, 1941–1972*, AAS History Series, Vol. 44, Univelt, Inc., San Diego, 2015, p. 63.

3. Ibid.

4. Letter from Lovell Lawrence Jr. to Warren P. Turner, 1 March 1967, Lovell Lawrence Jr. Papers, NASM Archives, box 13, folder 27.

5. Winter and Ordway III, *Pioneering American Rocketry*, p. 107.

6. For an account of how the probable Wyld Serial No. 2 motor variant came to be saved and eventually wound up in the collections of the National Air and Space Museum as donated by Aerojet through William C. House, see Winter and Ordway III, *Pioneering American Rocketry*, pp. 52, 58.

7. For other accounts of the history of Wyld Serial No. 2 motor, consult Winter and Ordway III, *Pioneering American Rocketry*, pp. 49, 107–108, 317, 335–336, 353, 385.

8. For an annotated listing of RMI's wartime motors, including the Wyld Serial No. 2 in NASM's collection, see Winter and Ordway, III, *Pioneering American Rocketry*, pp. 317–319. Note, however, that some of these artifacts are German and of World War II-vintage and were collected by Lawrence during his fact-gathering trip to Germany, shortly after the war in 1945.

9. Shesta, "Reaction Motors," p. 63.

10. Frank H. Winter, "Rocket History through an Artifact: American Rocket Society (ARS) Test Stand No. 2 (1938–1942)," in Kerrie Dougherty, ed., *History of Rocketry and Astronautics*, AAS History Series, Vol. 41, Univelt, Inc., San Diego, 2014, pp. 259–262.

11. Arata, "How Many Remember?" *The RMI Rocket*, Vol. 2, Dec. 1951, p. 7.

12. Ibid.

13. Ibid.

14. Robert Lawrence, "Excursion into Rocketry," *The RMI Rocket*, Vol. 2, Dec. 1951, pp. 5–6.

15. Maria S. Braun, *Franklin Lakes: Its History and Heritage,* Phillips-Campbell Publishing Co., Cedar Grove, NJ, 2nd printing, 1976, p. 82.

    Other statements in Braun's brief mention of RMI are inaccurate: calling Lawrence "Buddy Lawrence," for instance, and saying that RMI was making antitank rockets.

17. Braun, *Franklin Lakes,* p. 81.

    Braun says the airport was established on the "old Fletcher estate" between Colonial Road and Kent Place off Franklin Lakes Road. Later, in 1943, it was shut down due to noise and other complaints. Today, there is no sign at all of the existence of the old small airport, and there are affluent residential homes there that face just south of the Lower Lake of Franklin Lakes. Today, the roads that run through the former airport are the same, except that the former blockhouse site was located near the curb of what is now Dogwood Trail.

    For more on the Nelson Airport that operated from 1932 to 1943, see online, "A Short History of the Franklin Lakes Airport" by Jack Goudsward.

18. Lawrence, "Excursion," p. 5–6.

19. "Test Causes Excitement at Airport," *Wyckoff News* (Wyckoff, NJ), 24 Sept. 1942, p. 2.

20. Lawrence Jr., drawings, Lovell Lawrence Jr. Papers, NASM, box 8, folders 13 and 15.

21. Ibid.

22. Telephone interview with Edward J. Lenik by Frank H. Winter, 24 Jan. 2017.

23. "Landmark Status Asked for Rocket Site of the 1940s," *New York Times*, 19 Aug. 1977, p. 45;

24. Film in NASM film collection, "RMI Tests," FB 00370. A similar reel exists that turns out to be a duplicate. This footage, running 11 minutes and 23 seconds, shows tests from 29 Dec. 1942 to 2 March 1943 and includes the following motors: M-17-G1, M-19-G1, and M-17-G3.

    In addition to this footage in NASM, the author is indebted to Ken Montanye of Butler, New Jersey, an avid collector of all things RMI, for the opportunity to study his own similar footage. Mr. Montanye likewise owns the only known remaining brick of the control room, or blockhouse. Again, more on the fate of the blockhouse is found in Chapter 11.

25. Ibid.

26. Leslie W. Collins, "No. 280 Main Street," *The RMI Rocket*, Vol. 2, Dec. 1951, p. 6.

27. Ron Duhl, "Early rocket experiments conducted in Bergen blockhouse," *Patterson News* (Patterson, NJ), 21 March 1980. p. unknown.

28. Collins, "No. 280 Main Street," p. 6.

29. Ibid.

30. For the *New York Times* coverage, see, "Four Kill Beer Chief, Turn Machine Guns on Hoboken Police," *New York Times*, 8 March 1930, pp. 1, 4.

31. One or more the papers reporting on the Dunn assassination mistakenly claimed he owned property at Pompton Lakes that was not far from Pompton Plains.

32. Emma Dunn's obituary can be found in *Suburban Trends* (Butler, NJ), 18 Oct. 1967, Section 4, p. 3.

33. Collins, "No. 280 Main Street," p. 6.

34. "Affidavit—Occupational Classification (Industrial)" for H. Franklin Pierce, in RG 72 (BuAer), [U.S. Navy], General Correspondence, 1943–1945, National Archives, box 4943, file QM (8586) (Reaction Motors, Inc.), 1943.

35. Shesta, "Reaction Motors, Inc.," p. 66.
36. Ibid., p. 67.
37. James H. Wyld, in Reaction Motors, Inc., *Log Book — Motors Tests R.M.I. — Book 1*, in "Reaction Motors, Inc." file, NASM, entry for 6 May 1943.
38. Winter and Ordway III, *Pioneering American Rocketry*, p. 109.
39. Letter from John Shesta to Frank H. Winter, 12 Feb. 1979, in "John Shesta"file, NASM.
40. Wyld, Reaction Motors, Inc., *Log Book*, entry for 14 May 1943.
41. John Shesta, "Pioneering in Rockets," Part 1, *The RMI Rocket*, Vol. 2, June 1950, pp. 1–2, and quoted at length in Winter and Ordway III, *Pioneering American Rocketry*, p. 355.
42. Ibid.
43. Wyld, Reaction Motors, Inc., *Log Book*, entry for 1 October 1943.
44. Ibid.
45. Ibid.
46. Wyld, Reaction Motors, Inc., *Log Book*, entry for 4 October 1943.
47. Esther C. Goddard and G. Edward Pendray, eds., *The Papers of Robert H. Goddard*, 3 Vols., McGraw-Hill, New York, 1970, Vol. 3, p. 1478.
48. Lovell Lawrence Jr., drawings, Lovell Lawrence Jr. Papers, NASM.
    It is not known for certain whether the DU-2 sketch in the Lawrence papers was the Navy's own Droppable Unit-type JATO, or RMI's version. RMI may have been subcontracted by the Navy to them to work on it, or part of it.
49. Shesta, "Pioneering in Rockets," p. 2; also in Winter and Ordway III, *Pioneering American Rocketry*, p. 355.
50. Winter and Ordway III, *Pioneering American Rocketry*, p. 355.
51. Ibid.
52. Shesta, "Pioneering in Rockets," p. 2.
53. Shesta, "Reaction Motors, Inc.," p. 72.
54. Ibid.
55. Ibid.
56. Ibid; also found in John Shesta, "The Chesapeake Fiasco," *The RMI Rocket*, Vol. 2, Dec. 1951, p. 5.
    Incidentally, the terms "rocket motors" and "rocket engines" are often used interchangeably. Generally, they mean the same thing, although "engine" usually denotes a power machine with moving parts. For this reason, "motor" is more applicable to solid-propellant rocket units that have no moving parts, like pumps. Motor is also preferred for smaller-sized units. Therefore, in later chapters of this book, when we talk about larger-scale rocket power plants, the term engine is preferred and is used more frequently there.
57. Letter from Lovell Lawrence Jr. to Lt. Cmdr. C.F. Fischer, Lovell Lawrence Jr. Papers, NASM, box 1, folder 13.
58. List, handwritten, probably by Lovell Lawrence Jr., undated and untitled, except for the initials "OP-51" on top, in Lovell Lawrence Jr. Papers, box 8, folder 17.
59. Lovell Lawrence Jr. drawing, "Proposed Dual Test Stand...," 29 Feb. 1944, Lovell Lawrence, Jr. Papers, box 8, folder 10.
60. Lovell Lawrence Jr. note, "Priority Sequence," 25 Oct. 1942, Lovell Lawrence Jr. Papers, Box 8.
    In the same box is a Lawrence drawing of 23 June 1944 of a "diaphragm valve and injector (concentric) for aniline." Obviously, those new propellants required their own new safety-handling technology and means of measuring their performances, and RMI therefore contributed toward those other pioneering roles, as did Aerojet.
61. Ibid.

62. Youngquist, a rocket pioneer in his own right, had also been an active ARS experimenter. After the war he joined RMI, most likely directly through his connections with Lawrence. He eventually became RMI's chief development engineer.

63. Winter and Ordway III, *Pioneering American Rocketry*, p. 128.

64. James H. Wyld, drawing, "Feed-System of 'Peeenemunde [sic.] 16' Robot Plane," 27 July 1944, Lovell Lawrence Jr. Papers, NASM, box 8, folder 13.

65. For a history of the Loon (JB-2), consult Kenneth P. Werrell, *The Evolution of the Cruise Missile,* Air University Press, Maxwell Air Force Base, Montgomery, AL, 1985, pp. 67–68, 113, 224.

66. James H. Wyld, "The Application of Regenerative Liquid-Fuel Rocket Motors to Military Aircraft," handwritten, unpublished, undated, in "James H. Wyld" biographical file, NASM.

67. Winter and Ordway III, *Pioneering American Rocketry*, p. 351.

68. Lovell Lawrence Jr., drawing, "Proposed Interceptor Control System," 10 Jan. 1945, in "Lovell Lawrence" file, NASM.

Chapter 4

RMI's "Black Betsy" in the Bell X-1

"Short prayers, etc., and fire!"

–James FitzGerald, contributing RMI engineer to
6000C-4 development.

THE FINANCIAL WOES

Toward the close of World War II, the tiny firm of RMI operating out a converted barn in a small town in northern New Jersey was barely prospering financially. They were constantly creating the most unconventional products—rocket motors that could be applied to almost anything imaginable. But their insatiable need for new equipment to create these technological marvels was very costly and prevented them from accumulating working capital. Their president, Lovell Lawrence, all too often met RMI's requirements for the 1942–45 period by demanding bank loans and advance payments on Navy rocket contracts. This is why, by the end of 1945, RMI held a net worth of $30,168.05, yet had purchased $44,246.09 in capital equipment.

THE NOISE COMPLAINTS

On top of this precarious financial situation, and in spite of careful security measures, only oblivious residents of Pompton Plains could have been unaware that RMI was up to something in the course of developing and testing one of its new products. Indeed, RMI's Dunn barn plant was located in close proximity to a state highway—at the intersection of Boulevard and New Jersey Route 23—as well as residential areas. Their closest test area consisted of a sand lot on an extension to the parking lot behind the barn. And on occasion, the shrapnel of unsuccessful rocket tests showered on the plant-barn. In one instance, an explosion caused by a stuck oxygen valve nearly resulted in disaster. But even successful test runs could be earsplitting. One former Franklin Lakes police chief, Arthur Pickering, remembered that "It sounded like real blasting, like dynamite. You could hear it for miles."[1]

RMI had thus become increasingly non grata to the residents of Pompton Plains and its notoriety, reflected in numerous stories in the local and statewide papers throughout 1945 in particular, culminated in damage suits. In one, a local farmer charged that RMI's excessive noises kept his chickens from laying eggs; in this case RMI reimbursed the farmer for his losses. In addition to the incessant noise, heavy vibrations from the tests caused cracking of plaster and masonry in nearby homes. Health complaints were also lodged. (John Shesta spoke of the "noxious fumes from our nitric acid motors which we also tested from time to time."[2] The seriousness of the complaints reached the congressional level.

In the beginning, residents of the two adjacent municipalities—Riverdale Borough and Pequannock Township—joined forces and circulated petitions to "stop Reaction at once."[3] The signers acknowledged that "jet propulsion devices may be the conqueror of time and space," but nevertheless, "it's a headache to us."[4] The locals knew the general nature of RMI's business, which could hardly be hidden, but not the technical details. Nonetheless, the situation escalated further, and Pompton Plains police chief Walter Sweetman was

Fig. 4.1 Montage of 1945 Pompton Plains, NJ, newspaper stories on the noise and damage complaints against RMI during their testing of what became the Bell X-1 engine. (Courtesy Frank H. Winter collection.)

called in; he notified RMI that its renewal of a permit to store dangerous materials had been denied. Lawrence often could not be reached for comment when he was sought out by the papers or, if contacted, was defiant in attitude. "Go ahead and stop us from operating—if you can," he challenged at one point.[5] At another time he responded to a local who complained about noise that there would be "bigger ones" that day.[6] He also made it known that compressed oxygen, nitrogen, and gasoline could not be construed as explosives.

An injunction initiated by the citizens of Pompton Plains against RMI was successfully fought off by Lawrence, although this prompted irate citizens to bring the matter to certain members of Congress. In turn, the congressmen took action and the secretary of the Navy was requested to stop all testing at Pompton Plains. The secretary pointed out that "if the testing…did not continue, the entire development of the [Navy's] rocket program would be seriously hampered."[7] At this time, there was a general sense of alarm within the aviation community—military, civil, industrial, academic—that the United States was very far behind Germany's wartime progress in jet and rocket-propelled aircraft. The Cold War with the U.S.S.R. was not yet on, and would not be until 1948 (starting with the Berlin Airlift), but the status in fall 1945 of U.S. general progress in advanced aviation was not good. In the end, an agreement was reached by the secretary and the members of Congress that "the Navy would be allowed a period of two months from 25 February 1946 to 26 April…to relocate the development and testing work in connection with the rocket motor program being conducted by Reaction Motors, Inc., in Pompton Plains, N.J."[8]

Move to Lake Denmark

It was decided to move the testing and development work to a strip of land owned by the Army and adjacent to the U.S. Naval Ammunition Depot at Lake Denmark, New Jersey, not far from the city of Dover, or more than 30 miles west of New York City. Because the rocket-development program was also of vital importance to the Army, that branch transferred 430 acres adjacent to the Lake Denmark Navy property. Most fortuitously, the new land to be occupied by RMI was ideal and exceptionally suited to their requirements; it was acoustically shielded by wooded hills set within a secluded natural bowl, yet within easy access of major population centers and their ample supply of skilled workers and test fuels and other supplies. The severe noise problem was thus solved, and the initial difficulties turned out to be an incredible blessing in disguise: RMI gained a far superior testing environment along with more spacious and professional facilities, in addition to abundant room and opportunities for growth. The move was completed in June 1946 and was termed a "milestone" in RMI's history by Shesta.[9] "Almost overnight," he added, "we had become a large concern—at least in our estimation."[10]

RMI's new home to house their rocket-motor design, construction, and production departments consisted of two large modern buildings, converted warehouses in an old ordnance battalion area of Lake Denmark. Smaller nearby structures served as their administration and supporting-services departments. Many more employees were also added, the number jumping from 55 on 31 December 1945 to 120 by mid-1946. Officially, they were now at Dover—although this was strictly their mailing address.

The testing took place in a more remote area of Lake Denmark, at the Naval Ammunition Depot adjoining Picatinny Arsenal, already a well-used testing site for standard ordnance. Eventually, the entire area came under the Picatinny Arsenal. But in contrast to what they had faced at Pompton Plains, the arsenal community welcomed their new guests. Notably, *The Picatinny Arsenal* base newspaper for 1 March 1946 announced RMI would be joining them and added: "In addition to the noises that its own testing makes, Picatinny will soon enjoy the louder testing of experimentation in rocket bombs and jet propulsion."[11] Of course, RMI did not make "rocket bombs." But perhaps this reporting error was due to the secret nature of their main project. (The official announcement of the impending move, by the way, had been made a week before, in late February.)

THE RESOJET WORK CONTINUES

What, then, had RMI really been up to that had caused all the noise at Pompton Plains in the first place? There are really two answers. One was surprisingly their wartime pulsejet, or "resojet" work, as briefly related in Chapter 3, to initially improve upon the German V-1 power plant. Their involvement with pulsejets, which are extremely loud objects by themselves, continued for some time after the war. In late May 1945, Lawrence signed a contract with Elman Borst Myers, a tireless private inventor who had been developing this type of power plant even before the war, to temporarily work at RMI "as a consulting engineer…[in] the designing, developing, and testing of the…Resonating Jet Propulsion Motor [sic.]" in the interest of the Navy.[12] Myers seems to have later become a regular RMI employee.

We also know that in June 1945 it was reported that among RMI's three test stands, one was dedicated to testing resojets and was evidently placed in the sandpit in back of the Dunn barn. It featured a $50,000 government-provided wind tunnel "equipped to furnish ram pressure" and operated by a large Pratt & Whitney 1830-88 aircraft engine that powered a four-bladed propeller. The propeller rammed in air at 175 mph through an attached cylindrical testing chamber.[13] This setup alone, not to mention the resojet motor within the chamber that also blasted away, must have always been run with a prolonged and deafening roar.

Later, in 1947, RMI received contracts for Project Squid, an ambitious postwar defense effort by the Office of Naval Research (ONR) to investigate and improve upon the dynamics of pulsejets, and later rockets as well. The Squid program was aptly named after the sea animal that propels itself by a form of natural jet propulsion. Consequently, RMI undertook Project Squid pulsejet testing at Lake Denmark, on top of its heavy workload in rocket development.

BACKGROUND OF X-1 ENGINE DEVELOPMENT

The main culprit, of course, for all the noise from 1945 was the rocket engine that was destined to become the power plant for the Bell X-1. At first, it started as a single 1,500-lb-thrust unit, but underwent rapid and intense development throughout the spring of 1945 especially, to soon become a four-barrel, 6,000-lb-thrust arrangement that blasted away for up to four minutes. The initial firings took place just yards from the Dunn barn RMI plant, also in the adjacent sandpit testing area. Just how far away the rocket test stand was situated from their resojet—or pulsejet stand—is not known.

The original RMI inhouse designation of the rocket engine was the 1500N-4C, meaning 1,500 lb from four chambers, and its propellants (LOX and alcohol) were pressure fed into each chamber by nitrogen, hence, the letter "N." RMI's service representative, Frank Iwanowsky, and engineer Harry Burdett Jr. were jointly credited for this new designation. Later, by March

Fig. 4.2   The primary cause of the noise complaints, RMI's four barrel 6000C-4 engine firing during a typical test run. (Smithsonian Institution, 89-20570, courtesy Orbital ATK.)

1946, the engine was redesignated the 6000C-4, evidently to signify it as the production model required by Bell Aircraft, but interesting nicknames came to be adopted as well. But from here, it is necessary to look backward to briefly trace the origins of the rocket plane itself.

EARLIER BEGINNINGS OF THE ROCKET PLANE
IN THE U.S. ARMY AIR FORCES

Earlier, from 1944, the U.S. Army Air Forces had begun to seriously consider the development of a true rocket plane. At the same time, Major Ezra Kotcher of the Army Air Forces Engineering School at Wright Field, Ohio, was also thinking of a transonic research aircraft.

On 30 November 1944, Kotcher was paid an unexpected visit by Robert J. Woods, a prominent aircraft designer from the Bell Aircraft Corporation. During their chat, Kotcher took the opportunity to inquire of Woods whether the Bell Corporation would be interested in advancing the idea of a transonic, or near-supersonic-speed plane. For maximum performance, Kotcher suggested it might be rocket propelled. The risks for such a venture by Bell were many, as all this technology had yet to be developed in the United States. Unexpectedly, though, Woods agreed to Kotcher's casual query, but it was not until 16 March 1945 that a Bell–Army Air Forces contract was signed for the development of an aircraft to be designated the Bell XS-1 (Bell Supersonic Experimental 1), later shortened to the Bell X-1. From the start, then, the Army Air Forces was more focused upon entering the supersonic realm.

Woods really preferred a turbojet rather than a rocket engine. His preliminary sketches incorporated the General Electric I-16 turbojet (later designated as the J-31) as the power plant of more than 1,600-lb thrust that had been intended as a more powerful replacement engine for the Bell P-59 Airacomet fighter aircraft, the first U. S. turbojet airplane. Production of the I-16 had started in 1943—the first jet engine to be mass-produced in the United States—and it was fully proven and available by 1945.

Reluctantly, then, Woods agreed to evaluate Aerojet's 6,000-lb-thrust, more advanced version of their Rotojet—also known as the Aerotojet—under development for the proposed Northrop XP-79 flying-wing rocket-propelled interceptor. The Aerojet motor was to use red-fuming nitric acid and aniline—a hypergolic (self-igniting) combination. To first determine the nature of hypergolics, Woods witnessed a basic experiment by Bell engineers in which they placed the liquids in two separate bottles, taped the bottles together, and then threw them against a rock close to the Bell plant in Buffalo. The bottles shattered at once, and the contents erupted in flames. Understandably, Bell immediately dropped Aerojet from their plans for the new plane.

Very fortunately, Benson Hamlin of Bell's XS-1 design team had learned of the existence of RMI at a Navy (BuAer) meeting on power-plant possibilities on 5 March 1945, in which Lt. Cmdr. John S. Warfel, who had recently replaced Fink Fischer as the Navy's liaison with RMI, was present. Benson's existing notes on the conference rightly say that RMI was a "small outfit" then working on a priority (JATO) project for the Navy.[14] Hamlin followed up this lead with a phone call to RMI, apparently intending to later make a personal visit. In any case, Lawrence, along with Shesta and John A. Pethick, then the general manager of RMI, took the initiative, and all went to Bell in Buffalo to explore the possibility of RMI's contribution to the XS-1 project. Hamlin and other leading members of the XS-1 design team were present.

Unfortunately, there are no transcripts of this meeting, which must have been a most significant one because it firmly established the RMI–Bell connection that culminated in an agreement in which RMI became the developer of the aircraft's power plant. There was a followup visit by Robert M. Stanley and Roy Sandstrom of the Bell XS-1 team to RMI at Pompton Plains on 30 March "for further discussion on the problem of furnishing them with a 6,000 lb thrust unit."[15] Thus, by this time the general requirements for the XS-1 engine had already been set. Meanwhile, there was a concurrent competitor research rocket-airplane program, though as it turned out, a complementary one to the XS-1; in addition, there were other U.S. rocket-aircraft projects during the closing months of the war.

EARLIER BEGINNINGS OF THE ROCKET PLANE IN THE U.S. NAVY

Earlier, by late 1944, it so happened that the Navy and one of its prime contractors for naval aircraft, the Douglas Aircraft Company of El Segundo, California, had also moved toward developing a transonic aircraft, although it was to incorporate a turbojet power plant. That was the start of the joint NACA/U.S. Navy research D-558 program that soon also encompassed a supersonic phase. Early in the program, Douglas applied their own designation to it: the Douglas Model 558 High Speed Test Airplane. The supersonic aspect evolved in stages. Already, in early March 1945, NACA recommended that the Navy procure the Douglas D-558 proposal in both turbojet and combination turbojet/rocket forms.

Then on 13 April, Douglas submitted a three-phase proposal to the Navy's BuAer. The phase-one aircraft, hence known as the D-558-1, was to be jet powered and have straight wings for investigating aerodynamic data up to Mach 0.89 and became known as the D-558-1 Skystreak. The phase-two aircraft, the D-558-2, was conceived as a rocket and jet-powered supersonic research aircraft with swept-back wings; it was later fittingly called the Douglas D-558-2 Skyrocket. As for phase three, according to aviation historian Richard P. Hallion, Douglas was to submit engineering proposals and construct a mockup of a combat aircraft, based upon the research results of the original D-558 aircraft. But soon after, on 26 April, Hallion asserts that because rocket technology applied to aircraft was still a relatively new and complicated field for the Americans, the Navy's Engineering Division suggested that "the proposed rocket-propelled phase be delayed pending further information on the structural details of the rocket installation."[16]

In early May, BuAer and the assistant secretary of the Navy approved the Douglas program. A little more than three months later, on the morning after V-J Day—15 August, the day of Japan's unconditional surrender that marked the final end of World War II—plans for Phase 2 firmed up in the start of a meeting at the Douglas El Segundo plant between Douglas, Navy, and NACA officials. This is the accepted historical start of what became known as the supersonic, swept-wing Douglas D-558-2 Skyrocket. BuAer was to maintain close contact with the Army Air Corps to ensure it did not duplicate their XS-1 program, while the Skystreak represented a more conventional design than the XS-1, notably its straight wings.

It is most curious, however, that while both the D-558 and the XS-1 programs had started in late 1944, and it appears that RMI was involved first with the Army Air Forces XS-1 project (see earlier section on the Beginnings of the Rocket Plane in the U.S. Army Air Forces), there is no question that RMI really worked first on their 6000C-4 engine for use in the Navy's Douglas D-558-2. After all, the engine was later transferred from the Navy to the Air Force for the XS-1. Unfortunately, though, documents are lacking in this critically important phase of the history of the 6000C-4 that involved the D-558-2. Matters are further complicated in that the D-558-2 development really began later in 1945. For the same reasons, we also do not know the exact chronology of the introduction of the 6000C-4 into the D-558-2 program.

We can therefore only surmise that because the Navy was always RMI's number one customer, the Navy may well have either submitted a contract—or a Memo of Understanding or other arrangement for the D-558 program—to RMI very early, prior to RMI being awarded Bell's subcontract for the power plant of the XS-1 on behalf of the Army Air Forces. By the same token, the initial rocket engine concept for the D-558 may have been altogether different from what became the 6000C-4.

For certain, says Hallion, "From the start, Douglas, the Navy, and the NACA planned to incorporate rocket propulsion in the plane, since they felt there was little point in testing the swept wings on a turbojet airplane inasmuch as existing turbojets lacked the thrust capability to propel the plane to sufficiently high Mach numbers where the swept wings could be full evaluated."[17]

Hallion significantly adds that the NACA originally stipulated that the aircraft's rocket engine deliver 4,000-lb thrust for a 2-min powered operational flight, although later they increased this requirement to 6,000 lb for 100 s, or 1.6 min.

THE NAVY'S ATTEMPTED WARTIME ROCKET-PROPELLED INTERCEPTOR

In addition to these advanced research aircraft, by late 1944 the Navy's BuAer had also pursued the design and development of a secret high-performance, strictly rocket-propelled interceptor for the war effort. Their Ships Installations Branch was instructed to seek a suitable power plant for it. The initial requirements for the interceptor was that it was to operate on a series of four 700-lb, pump-fed LOX/alcohol motors, although they later changed this requirement to LOX/gasoline. They turned to RMI, who were to be responsible for the chambers and controls. Surely this was the true origin of Lawrence's 10 January 1945 sketch of his design of a "Proposed Interceptor Control System" utilizing a system of four throttleable 750-lb LOX/alcohol motors for a total thrust of 3,000 lb. The main difference is that Lawrence had obviously slightly up-rated the motors from the Navy's concept.

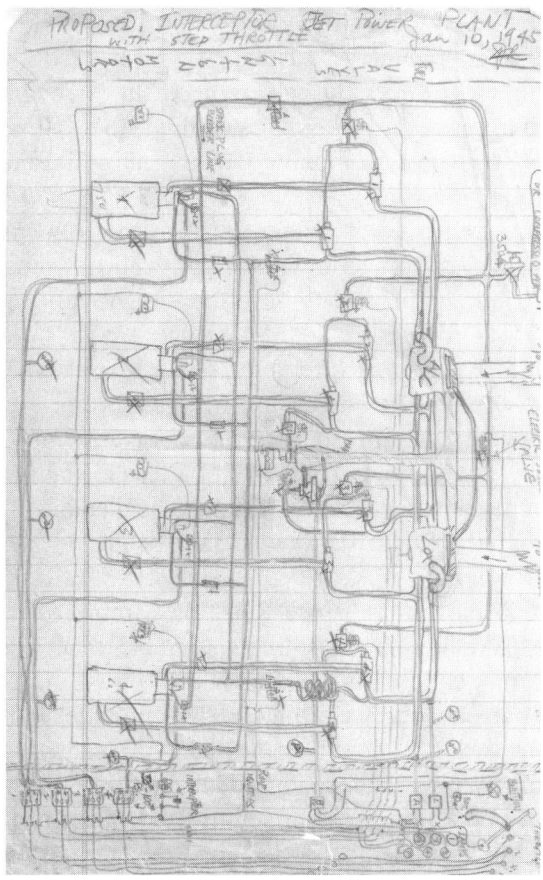

Fig. 4.3   Drawing by Lovell Lawrence Jr., dated 10 January 1945, of a "Proposed Interceptor," using four 750-lb-thrust rocket motors that seems to have also played a role in the 6000C-4 story. (Courtesy Smithsonian Institution, 87-13582).

Ironically then, although RMI's great efforts toward the development of JATOs had come to naught, the invaluable experience they gained from developing these large and powerful liquid-propellant units—in addition to their associated hardware development and tooling—fully enabled the small company to embark upon a brand new phase in its short history: the development of rocket motors as main power plants for aircraft.

GODDARD'S INVOLVEMENT

Goddard as well was involved in the Navy's interceptor project. His diary entry for 30 December 1944 reads: "Had conference with Reaction Motors Incorporated and Navy men on interception with jets [i.e., the rocket- or "jet-propelled" interceptor.] We [he and his assistants] are to work on pumps, turbines, and gas generator—they to work on chambers and controls and perhaps valves."[18] The Navy wisely chose to enlist the services of Goddard because he clearly had more experience with rocket pumps than RMI. "The Navy," he wrote in a letter on 3 January to Robert L. Earle of the Curtiss-Wright Corporation,with whom he also had an arrangement, "intends to have us then work together to improve the power plant, with the idea of eventually using a turbine in the blasts to pump a monopropellant [nitromethane, although this never transpired]."[19]

Goddard's diary entry for 1 January records the following: "Had large pumps and gas generator set up in morning. Lovell Lawrence came; talked over pumps etc.; and he offered to have me merge with Reaction Motors, Inc."[20] This surprise proposed Goddard–RMI merger was turned down.

Lawrence attempted another offer later, on 27 March, when he suggested that Goddard have a one-fifth interest in RMI—to become a co-equal partner with the four founders. Again, Goddard refused, and by mid-April he made arrangements anyway to join Curtiss-Wright after the war to help them start their own rocket development section, although he died on 10 August.

In his letter to Earle, Goddard provided more details on his joint work with RMI on the interceptor that was to be conducted under BuAer contract NOa(s) 192. He informed Earle that "since Reaction Motors…has already made chambers using oxygen with a gasoline-water mix [that is, the JATOs], the Navy is having this concern make four 700-lb thrust motors of the same design, using an alcohol and water mix."[21] This provides indisputable proof that RMI's JATO work was directly converted to their newly assigned task of developing similar motors as the main propulsion for an aircraft. But it is not the only proof. Goddard further confided to Earle that "It so happens…that [by July 1944] we developed a 700-lb thrust oxygen-gasoline motor, with tangential water cooling."[22] The Navy took advantage of this situation and to greatly hasten matters they had Goddard's pump and generator shipped directly to RMI to adapt it to their motor. Yet from then on, documentation on

RMI's own progress on the interceptor and follow-on project, is disappointingly sparse.

Inevitably, the Navy's rocket interceptor project soon fell by the wayside and never did materialize due to several factors. One was the Navy's rapid progress towards their transonic, and later supersonic, programs, which emerged as the Douglas D-558-2 Skyrocket. Another reason for the disappearance of the rocket interceptor was the cessation of the war—there was simply no longer any wartime need for this aircraft. On the other hand, there is evidence that RMI's modest four-chambered 750-lb thrust power plant originally conceived for the interceptor helped prepare them for its considerably more powerful, though similar, four-chambered 6,000-lb-thrust power-plant arrangement for the Douglas D-558-2 project, although here, too, the exact timing is unknown. It is known that in turn, the same 6000-lb-thrust engine was next transferred from the Navy's program to the Army Air Forces for their Bell XS-1.

TRANSFER OF ROCKET ENGINE FROM THE NAVY TO THE ARMY AIR FORCES

These events happened rather quickly. First, a surviving *Final Progress Report* for BuAer contract NOa(s) 4525 says that work was started on an RMI rocket motor "for high-speed aircraft" in February 1945, although the specific aircraft project is not mentioned.[23] The same report notes: "Some preliminary tests were made using a 350-lb [thrust] liquid-fuel motor; [while] a more complete set of tests were made in April and May 1945 using a 750 lb thrust [motor]. The fuels used included gasoline, methyl alcohol, and ethyl alcohol."[24] Water was mixed with the alcohol to assist in cooling the motor.

There also exists an RMI *Activities Report* covering the period 20–31 March 1945 that picks up on this project with reports of a test with the 750-lb-thrust motors and then says: "Actual construction of the [full] 1,500 lb thrust motors for the multi unit aircraft installation will not be started until we have completed this series of tests."[25] Bell's contract with RMI for the XS-1 was then let in April. But soon after this contract was signed, as recounted in an article in the *Newark Evening News,* Lawrence received Navy permission to "take some designs previously made for the [Navy's] Bureau of Aeronautics and incorporate them in the engine for the Bell plane."[26]

This account is backed up by a historically significant document, a letter dated 12 April 1945 from RMI's general manager John Pethick to the chief of BuAer, concerning "Confidential Contract Noa(s) 4525."[27] Pethick wrote that under the provisions of this Navy contract, RMI

> hereby requests the written consent of the Bureau of Aeronautics permitting us to enter into a contract with the Bell Aircraft Corporation of

Buffalo, New York, for the furnishing to that corporation of jet propulsion equipment similar to that presently being developed for the United States Navy. Said equipment [the motor] to be used in the development of high speed aircraft for the Army Air Forces,....Upon receiving the written consent of the Bureau of Aeronautics and upon signing a contract with the Bell Aircraft Corporation, an amendment to the contract...will be negotiated with the Bureau of Aeronautics.[28]

Pethick's letter thus paved the way for the Navy-to-Army transfer of the motor. The actual transfer was made shortly thereafter.

RAPID DEVELOPMENT OF THE X-1 ENGINE

The motor itself underwent rapid development at RMI. According to an 18 September letter from Pethick to Robert M. Stanley, chief engineer of Bell Aircraft, that reported on the general progress of RMI's Model 1500N-4C, the first unit, "which is a prototype only, was completely assembled on 30 August 1945."[29.] Pethick further mentioned that the unit (that is, the overall engine) weighed 225 lb and was "fabricated with mild steel jackets on the individual cylinders."[30] He also reported that "The eight propellant valves on this unit were overweight by two pounds each" and other technical details.[31] "This unit," Pethick added, "was first tested on 31 August 1945."[32] All four cylinders were fired separately and "No effort was made to operate the four cylinders simultaneously."[33] This was because "when the third cylinder was turned on the pressure in the oxygen manifold dropped."[34]

From 1 September this motor was "tested daily."[35] In fact, multiple firings were conducted, including for individual cylinders. Existing test reports show that by 18 October, RMI had reached Test No. 360, although those tests saw frequent blowouts, among other problems. Other difficulties experienced by RMI's team during this heavy period of testing included the usual propellant leakages, ignition failures, stuck valves, etc. Naturally, this intense activity provoked the noise complaints from the local residents.

NICKNAMED "BLACK BETSY"

Shesta later explained that the "mild steel" (or carbon steel) was chosen for reasons of economy, although this material was prone to rusting. Hence, when Bell's Robert Stanley showed up to inspect the prototype motor, it did not look presentable. This alarmed Shesta, a man of few words, who succinctly remarked: "I don't like that. It's got to be stainless steel."[36] The change was made and the finished motor painted black. Harry W. Burdett Jr., a key RMI engineer on the motor's development, later recalled that due to its new appearance and the horrendous noise the four-barrel motor made when fully firing, the men in the shop affectionately, if indelicately, dubbed it the

"Belching Black Bastard," although this name "had to be cleaned up for the press."[37] The motor was consequently rechristened "Black Betsy," after Lawrence's young daughter, Betsy Ellen, born on Christmas Day in 1943. Subsequently, the same motor was officially designated by the Air Force as the XLR-11.

Incidentally, it is claimed that Burdett was RMI's first bona fide engineer employee and had joined the company in August 1945; soon after, on 30 August, he witnessed the first full assembly of the four-chambered 6000C-4. But it was not until 6 September 1945, when all four chambers were operated in succession, that all four fired at once.

There are other versions of the naming of the motor as Black Betsy. One is that this was done when the first operational model was shipped to Bell Aircraft and *Life* magazine wanted to do a story, although the less colorful nickname does not appear in the story that appeared in *Life* for 13 May 1946.

THE MOUSE STORY

Black Betsy—aka the 1500N-4C, aka 6000C-4, aka XLR-11—had other company lore connected with its history. There is the famous mouse story, for example.

It seems that for a prototype of the 1500-N4C for a demonstration for the *Life* magazine story (before its was designated the 6000C-4), RMI test engineer Walter H. "Walt" Myers was assigned to ignite it. Walt, uncustomarily attired in a suit rather than his normal blue-collar shop outfit, stood on a box and lectured to the small audience on the virtues of the motor. Everything went smoothly until he flipped the switches. "We could hear the crackle of the igniters," recalled RMI's Laurence P. Heath,

> but there was no cylinder started. No. 1 didn't fire….The engineers began frantically to hunt for the trouble. They hunted in the fuel line and in the liquid oxygen line. And there, sure enough, was the trouble: a mouse had fallen into the liquid oxygen tank and had been frozen as solid as brick and was...up against the screen in the liquid oxygen line![38]

The mouse was extracted, and the motor reassembled and ignited again. All four cylinders now performed steadily. Shesta, with his characteristic dry humor, watched and paused for a moment, then quietly commented: "I think it works better without the mouse."[39] After this, the mouse incident became a favorite company in-joke.

FURTHER DEVELOPMENT OF THE ENGINE AND THE ENGINE TEAM

The four barrels of the motor permitted a kind of throttling. A pilot of the aircraft fitted with the motor had the option of choosing one cylinder for 1,500 lb of thrust, two for 3,000 lb, and so on, or all four at one time for maximum thrust. Accordingly, the cockpit in the Bell X-1 had four ignition toggle switches—No. 1 to No. 4—in addition to various rocket motor pressure gauges, etc. There was no true throttling, however, and in the contractual language of the day, the overall assembly was more colorfully called a "four-step rocket engine."[40]

Because RMI's overall staff was still quite small, the numbers involved in the 6000C-4 development were also surprisingly very few, considering the later historical significance of their major technological achievement. All four of RMI's founders were on the team, as were RMI's earliest workers like Collins and Arata and newer (circa 1945) people like Burdett and William P. Munger, who came aboard that March.

Lawrence was largely responsible for the electrical circuitry and controls, Shesta with solving vexatious injector problems, while Wyld collaborated with them both on the overall structural and control system. Jimmy is also known to have earlier resorted to a hand-cranked Monroe calculator to work initial out-heat transfer and coolant-velocity calculations for the basic chamber and nozzle of the building-block 1,500-lb-thrust cylinder. Pierce was mainly helpful on the construction side and may have helped in the overall design. Arata greatly contributed toward igniter improvements. Notably, when Lawrence was sent on a three-month naval technical mission to Europe during the summer of 1945 to help the U.S. government investigate prior companies and educational institutions that had developed missile and other rocket power plants in Germany during the war, Arata was busily and independently engaged in devising an upgraded spark-plug-type igniter. When he returned home, Lawrence was quite excited over its superior reliability in tests, compared with the previous version, Arata later recalled, and the igniter was subsequently adopted and further refined for the 6000C-4 motor.[41]

The Marotta Engineering Co.—presently, Marotta Controls, Inc.—then of Boonton, New Jersey, and under Patrick T. Marotta, had undertaken a lot of the valve work for RMI since their 1943 founding, and there is strong evidence from frequent mention in the Lawrence papers that Marotta, too, heavily contributed to the development of the 6000C-4.

PUMP HEADACHES

By far, the greatest technical challenge faced by the team was working out the pump system. Munger was placed in charge of finding the right kind of pump that could push the LOX and fuel at the high rates then required, as well as the most efficient ways to drive the pumps. Numerous pump models were tried, primarily furnished under a subcontract for a fixed fee of $30,000 by the firm of James Coolidge Carter of Pasadena, California, although the technical difficulties lingered. Scheduling became seriously affected, resulting in Bell's

Robert Stanley notifying the Army Air Corps on 3 May that they would now be forced to accept a pressure-fed version of the motor; later XS-1 flights could be made with pump-fed models.

A delay on pump development had been anticipated by Bell, and this decision was thus carried forward and necessitated drastic design and operational changes. The aircraft now had to be fitted with additional, heavy-walled spherical tanks for the gaseous nitrogen for the first two XS-1s; the third plane, with a turbopump, did not appear until 1951. The extra nitrogen tanks imposed heavier weight on the initial planes, and powered durations were also shorter than expected. The end result was that the XS-1 was initially air-launched for its first flights instead of ground-launched, as previously planned for a fully developed pump-fed aircraft, though consequently, it now also had slightly diminished performance capabilities.

EMERGENCE OF THE PRESSURE-FED VERSION OF THE ENGINE

Pump problems notwithstanding, work surged ahead on the pressure-fed version of the motor and milestones were reached. A second unit, completely fabricated of stainless steel, was completed by early October with a reduced weight and greater compactness, along with overall refinements, such as a "new junction box for the ignition and pressure control equipment that has been fabricated from aluminum" and other aluminum parts.[42]

By December, RMI was confident enough to produce their 1500N-4C *Preliminary Operating Instructions* that ran only nine pages, including a flow diagram and pictures. This brevity underscored the simplicity of the motor, although this was the basic pressure-fed model after all. According to Burdett, it also marked the first time such operating instructions were prepared, and later issued, for a man-rated rocket engine. Then, in late March 1946, RMI delivered its first 1500N-4C to Bell Aircraft; it now was known under its new official designation of 60004-C. On 26 April, the motor was introduced to the public by way a much-publicized firing demonstration at Pompton Plains before approximately 100 newsmen, newsreel men, and top Navy officials.

Press reports the following day proclaimed that the 1500N-4C, as it was still called, weighed only 210 lb, or one-fifth the weight of a turbojet, although it produced about twice the thrust. The *New York Times* and other papers claimed this thrust equated to 8,000 HP, although Navy officials remained closed-mouthed about the motor's intended application.[43]

The public firing of the 6000C-4 on 26 April also marked the last test at Pompton Plains. Thereafter, this engine and others were tested at Lake Denmark, although due to the RMI move there and site preparations, there was a hiatus until 10 June 1946, when testing resumed with the firing of a single Black Betsy chamber.

TESTING AND DEVELOPMENT AT LAKE DENMARK

More testing and improvements were carried out at Lake Denmark, especially because the motor had to be "man-rated"—a term not then in use—prior to its installation in the XS-1. The critically important job of bringing the motor up to flight standards for use in a manned aircraft was largely turned over to James W. FitzGerald, who had joined RMI that June. He described the original unit merely as a "test quality device."[44]

Rocket testing methods at that time FitzGerald came aboard were still frighteningly primitive by modern standards, as he recalled years later and were "hellishly dangerous."[45] For one, the level of required precautions was excessively risky. In one notable example, RMI test engineer Walt Myers, who had to prepare the stands for firing, often came far too close in the event adjustments had to be made, even as a rocket blasted away. Ridiculously, he barely protected himself with only a leather football helmet, although the flames of a motor could leap some 20 feet to the rear. FitzGerald also remembered that the firing room for the full engine test stand was likewise about 20 feet away and was merely an unprotected Quonset hut.

Henry A. Jatczak, another new member of the 6000C-4 development team, described the motor test stand itself as "just a lump of concrete to which the

Fig. 4.4   Static firing of a single chamber of RMI's 1500-4C motor, 30 August 1945. The man at right is probably test engineer Walter H. "Walt" Myers. (Smithsonian Institution, 89-14180. Courtesy Orbital ATK.)

four-chambered motor was mounted…That's the way we tested engines in those days, the *Buck Rogers* days."[46] More accurately, this stand was simply a heavy, concrete square or rectangular base. Firings began with a horn blown 30 seconds prior to a test, FitzGerald added, then a small countdown, a red light turned on, observers in place, followed by "short prayers, etc., and fire!"[47]

Despite all the shortcomings of the low level of technology at the time, FitzGerald was extremely competent and innovative in meeting his own challenge of man-rating the 6000C-4. "I brought it up to a higher quality," he explained, "with attention to material, welding standards, location of [aircraft] airframe mounting pads and parallelism of thrust vector."[48] Beyond this, he later designed a pivoting static-firing boom named the Large Rotary Beam test stand for test firing the motor at various angles, because Bell Aircraft was concerned about stressful pressure buildups as the aircraft climbed or dove. This unique and invaluable testing device was built by March 1947 by a local firm in nearby Dover, evidently the McKiernan-Terrry Corporation, and was installed at Lake Denmark. The Beam itself was a 40- or 50-ft-long steel grating platform pivoted on two steel "A" towers set on a concrete base and it stood 30 ft high. A "bull" gear was attached on one side of the platform, and a small electric motor linked to this gear, including a pinion gear, enabled the platform to be elevated at any pitch attitude, or rotated from −45° to +45° while the attached rocket motor was firing, "making for a great spectacle," according to FitzGerald.[49]

The setup also came with an inverted V-shaped exhaust flume within the base to deflect exhaust gases and was cooled by large water hoses. "At first," FitzGerald continued, "there was a little trepidation, but no real problems developed and after the initial firing there was cause for a celebration."[50]

Interestingly, the Large Rotary Beam test stand became Project "A" of BuAer contract NOa(s) 8239 to RMI, and was specified for use in the development of engines for such aircraft "as…the Douglas D-558."[51] It turned out that the same basic 6000C-4 served as the power plant for both the Army Air Forces' Bell XS-1 and the Navy's Douglas D-558-2, with slight modifications for the latter. Normally, the Kiernan-Terry Corp. built pile drivers and were quite capable of constructing this very large, rugged though intricate test structure.

Gradually, other important refinements were integrated into the technology. On the propellant side, the fuel for the 6000C-4 was hardly a generic alcohol, but was classed as a "synthetic alcohol" made according to a specific formula. In late March 1946, RMI proposed to BuAer that further improvements be made and that "a research and testing program be instigated to prove the efficiency and reliability of such fuels and determine the proper amount of water to be mixed with each."[52]

**Fig. 4.5 A 6000C-4 engine being fired on RMI's Large Rotary Beam test stand, 1946.
(Courtesy Smithsonian Institution, 87-17050.)**

PUMP PROBLEMS SOLVED

Meanwhile, William Munger labored on pump development, with assistance from Albert G. Thatcher and also later from Henry Jatczak and others. "There was nothing unique in the hydraulic design of the 6000C-4's pumps," Munger explained, but in the case of the 6000C-4 inside the extremely cramped quarters of the XS-1, exceptionally light weight for all components was an absolutely crucial requirement, in addition to "reasonably close toler ances."[53] Apart from these very demanding considerations, the gas generator for the pump was itself like a small rocket engine. "But the problem was we never were able to get the [normal] gas generator sufficiently reliable...[It] could have run ten times beautifully and the next time you would start, it would blow up. That also held up the program."[54] "The explosions were powerful," he added, and "when one of those things blew up, all the connected lines, valves, and things were damaged. A lot of these things [pump components] were...one of a kind so you went back to the shop to make another one. All that took time."[55] On top of this, the pump explosions could not have happened at a worse time—when the noise complaints surfaced.

"Eventually," Munger says, "I was able to develop a [gas generator] system using hydrogen peroxide."[56] When subjected to a catalyst "it developed hot gas which would run the turbine very nicely. So the final design in the [later Bell XS-1 # 3 and other] aircraft...forced the peroxide into the gas generator which ran the turbine which ran the pumps which forced the propellants into the combustion [chamber]. And when we got that all together, including the thrust control mechanism, we finally had something."[57]

Pump-development-problem delays were so severe that Bell was forced to also subcontract with General Electric to help out, but GE encountered difficulties as well, in addition to contending with a workers strike. But in the end, Munger and his team did succeed.

During World War II, the Germans used a hydrogen-peroxide gas generator in the power plant of their Me-163 rocket fighter, but Munger was adamant that this wartime technique was not "borrowed," and that RMI tried different systems on its own until they found that a hydrogen peroxide generator was the most efficient. Also, Jatczak emphasized that "the German and American systems were quite different."[58] The major difference was that the Germans based their system on 70%-strength hydrogen peroxide, whereas RMI went with a far higher and more potent 90% solution.

For RMI's system, Munger and his team sought out the Buffalo Electrochemical Company (BECCO) for advice on the use of high-strength hydrogen peroxide; mainly from BECCO they learned about handling the hydrogen and storing it in special tanks. The turbines and bearings came from quite an unexpected source—surplus wartime B-29 bomber gas superchargers. Frank A. Coss, another early RMI veteran, commented that RMI bought all the surplus superchargers they could and they remained in the company's inventory for many years.[59]

This is not to say that RMI was not aware of the Me-163 hydrogen peroxide-fueled power plant—far from it. Lawrence had first examined one of these engines during his Nav Tech mission in Germany in the summer of 1945. In addition, Task No. 7 of BuAer Noa(s) 7866 contract to RMI stipulated that they additionally study and test Germany's wartime operational Hs-293 air-to-surface missile motor that employed hydrogen peroxide activated by "Z stoff," a permanganate in aqueous solution.

An even more bizarre source of materials for the 6000C-4 was that for the catalyst for the hydrogen peroxide. It contained 90% pure manganese dioxide that came from the Caucasus Mountains in Russia. When Lawrence learned this, he immediately ordered general manager John Pethick to "Buy two tons of the damned stuff."[60] This was done, and he used the same batch for twenty years, although RMI afterward developed a better catalyst based upon silver-plated screening. Eventually, after their completed development and design refinements, James Coolidge Carter of Pasadena, California, furnished the operational propellant pumps to RMI's specifications.

Fig. 4.6   Early developmental pump of the 6000C-4 engine. (Courtesy Smithsonian Institution, 89-1857.)

THE 6000C-4 INSTALLED IN THE X-1 AND FIRST FLIGHTS

The first operational model of the pressure-fed version of the 6000C-4 was installed in Bell XS-1 #2 at Bell Aircraft's plant near Buffalo, New York. The engine was not ready when XS-1 #1 was rolled out on 27 December 1945. (Minus its power plant, XS-1 #1 underwent glide-flight tests, with ballast in place of the engine.) But it was not until 9 December 1946 that Bell XS-2 #2 logged its first powered flight, with Bell company test pilot Chalmers H. "Slick" Goodlin at the controls. Ten seconds after the release of XS-1 #2 from its carrier plane, Goodlin pressed one of the rocket switches. The burn was very smooth. Minutes later, he tried all four engine chambers, and the plane shot forward with "terrific acceleration."[61] Very shortly after, a fire-warning light appeared on the dashboard and remained on until Goodlin safely landed. An RMI confidential report reveals that loose igniter nuts had caused a leak of fuel, which ignited. FitzGerald said the RMI engineers soon fixed the problem, incorporating a harder-copper igniter gasket, rewiring the engine, and taking other measures.

Bell contractor-required flight testing—20 powered flights—was continued with this aircraft until mid-1947, followed by Air Force flight trials for maximum performance, based at Muroc Army Air Field (now Edwards Air Force

Base) in California. By this point, the Air Force assigned its own designation for the engine—the XLR-11 (Experimental Liquid Rocket-11). The original production version was the XLR-11-RM-1; those installed in the first two flight planes were XLR-11-RM-3 models.

YEAGER'S FLIGHTS

Capt. Charles E. "Chuck" Yeager was one of the three volunteer Air Force pilots for the XS-1's operational flights. These men were sent to Bell's Niagara Falls facility, where they were introduced to the plane and learned how to fly it and operate its rocket engine, starting with ground tests that included static firings. These tests also helped verify the power-plant performance. Interestingly, the aircraft was tied down only by a solitary cable running to an attachment point under the fuselage, even when all four chambers of the 6000C-4 were fired.

Fig. 4.7   Full view of the 6000C-4 engine, pressurized and first operational version. (Courtesy Smithsonian Institution, 87-17051.)

Then, on 6 August, Yeager's XS-1was carried for the first time by a B-29 and dropped over Muroc in a "familiarization" flight; the first powered flight after ejection from the carrier plane was successfully completed on 29 August. Yeager made several other powered flights after this, each one at successively higher speeds, the eighth attaining Mach 0.997—a new world's speed record and the closest any pilot had come toward breaking the mysterious and dreaded "sound barrier."

For years, the sound barrier (a popular term for the sudden increase in aerodynamic drag and other effects upon aircraft as it approached the speed of sound) invoked fear. Namely, it was greatly feared that this might be the uppermost limit of flight. Flight beyond this point would be extremely difficult, if not impossible, with all those forces at play; hence, the term's negative second word—barrier. Indeed, upon reaching the barrier speed, the aircraft might simply be torn asunder.

It was on his eighth transonic flight with XS-1 #1 on 14 October 1947 (the 600C-4 had finally been installed in it in March) that Chuck Yeager attempted the supersonic run about six weeks after he first flew the aircraft, which he had nicknamed *Glamorous Glennis*, in honor of his wife. Now, on this bright October morning, he prepared for his ninth powered flight in the orange, needle-nosed plane. After Yeager climbed into its open cockpit from the B-29 carrier aircraft and the cockpit was shut, the XS-1 #1 was dropped at a launch altitude of 20,000 ft. Seconds later, Yeager ignited two of Betsy's rocket chambers to attain initial acceleration and climb up to 40,000 ft, where he leveled

Fig. 4.8   The Bell X-1 *Glamorous Glennis* on its first supersonic flight, 14 October 1947. (Courtesy Smithsonian Institution 97-17485.)

off. He next flipped the third rocket switch, then stared at the needle of his Mach-meter dial. The needle quickly shot to Mach 1, and continued to Mach 1.06 (about 700 mph) before it halted. Other than a slight turbulence, the magnificent goal had smoothly been reached.

Yeager's historic first supersonic flight was kept secret by the Air Force and all others concerned for two months, until *Aviation Week* for 22 December leaked the news, which quickly grabbed international attention. Henceforth, Yeager's flight entered the history books as ushering in the era of supersonic flight. RMI's Black Betsy (now called XLR-11) had made this flight possible and would go on to achieve many more triumphs in the history of world aviation.

ENDNOTES

1. Ronald J. Dupont Jr., "Power for Progress: A Brief History of Reaction Motors, Incorporated 1941–1972," from author's unpublished manuscript, p. 7; available also online on GardenStateLegacy.com, Issue 12, June 2011.

   As mentioned in Chapter 3, there is now a historic marker near where the old Dunn barn site used to be. It was put up by the Pequannock (NJ) Historic District Commission and others. The marker sits as close as possible to the exact location of the former building, near the Chilton Medical Center, at 242 West Parkway, Pompton Plains, NJ. The three intersecting roads at this point are NJ Route 23, Boulevard, and West Parkway. A smaller street that runs along the edge of the former Dunn sandpit where the 1500N-4C was first tested is now called Dunn Place.

2. Shesta, "Reaction Motors, Inc.," in Frank H. Winter and Frederick I. Ordway III, *Pioneering American Rocketry: The Reaction Motors, Inc. (RMI) Story, 1941–1972*, AAS History Series, Vol. 44, Univelt, Inc., San Diego, 2015, p.74.

3. Winter and Ordway III, *Pioneering American Rocketry*, p. 217.

4. Ibid.

5. Ibid.

6. Ibid.

   Typical of the numerous articles on the noise complaints are "Residents Cite Damage Done By Tests of Jet Propulsion," *Newark Evening News*, 1 Oct. 1945, p. unknown; and "Township Revokes License For Testing Here," *Patterson Morning Call* (Patterson, NJ), 25 Oct. 1945, p. unknown.

7. U.S. Navy Contract NO(s) 7866, Amendment No. 1, 5 April 1946, p. 2, in National Archives, RD 72, U.S. Navy, Records of the Bureau of Aeronautics, Records of Divisions and Offices, Contract Records, 1940–1960, box 1854, file "NOa(s) 7866."

8. Ibid.

9. Shesta, "Reaction Motors, Inc.," p. 74.

10. Ibid.

11. "Jet Propulsion Research Firm To Be at Navy, *The Picatinny News*, Vol. 4, 1 March 1946, p. 2. The author is indebted to Dr. Patrick J. Owens, historian, ARDEC (Armament Research, Development and Engineering Center) Historical Office and Museum, Picatinny Arsenal, for this interesting find.

12. Untitled, (Contract between Myers and Lawrence), 22 May 1945, in National Archives, RG 72, U.S. Navy, BuAer, General Correspondence, 1942–1945, box 4172, file "QM (8586)."

13. Report on RMI manpower and facilities, 11 June 1945, in National Archives, RG 72, BuAer General Correspondence, 1943–1945, box 4172, file "QM (8584)."
14. Telephone interview with Benson Hamlin by Frank H. Winter, 16 Aug. 1989; letter from Benson Hamlin to Frank H. Winter, 4 Nov. 1989.
15. Reaction Motors, Inc., "Activities Report for the Period from March 30, 1945 to and including March 31, 1945," in National Archives, RG 72, U.S. Navy, BuAer, General Correspondence, box 4172, file "QM 858."
16. Richard P. Hallion, *Supersonic Flight: The Story of the Bell X-1 and Douglas D-558,* Macmillan Company, New York, 1972, p. 62.
17. Ibid, pp. 71–72.
18. Esther C. Goddard and G. Edward Pendray, eds., *The Papers of Robert H. Goddard,* 3 Vols., McGraw-Hill, New York, 1970, Vol 3., p. 1558.
19. Ibid., pp. 1558–1559.
20. Ibid., p. 1558.
21. Ibid., p. 1559.
22. Ibid.
   Consult the same reference for Goddard's own five progress reports on the rocket interceptor project, the second and third of which allude to RMI's participation, as follows: Vol. 3, (Report No. 1) pp. 1568–1570; (2) 1573–1577; (3) 1584–1587; (4) 1593-1594; and (5)1599–1602.
   Additional proof that RMI's JATO work led directly to the full power for aircraft include statements by Shesta and former RMI engineer Hartmann J. Kircher. Shesta, for instance, says in his memoir: "The type of motor we finally decided upon [for the Bell XS-1, but which also had roots in the Navy rocket interceptor project] was basically similar to the one used in PBM tests, though they were smaller [more compact] and improved units. Shesta, "Reaction Motors," pp. 72, 213–214;
   Kircher agrees that the 6000C-4 motor in the Bell XS-1 "had a similar type thrust chamber and…the same type of ignition system" as the type used in the PBM JATO tests.
   Telephone interview with Hartmann J. Kircher, 11 June 1987, notes in "Reaction Motors, Inc. Gen." file, NASM.
23. Reaction Motors, Inc., *Final Progress Report [for] Contract NOa (s) 4525, July 1, 1945,* p. 1, copy in "Reaction Motor, Inc." file, NASM.
24. Ibid.
25. Reaction Motors, Inc., "Activities Report for the Period from March 30, 1945… ."
26. Albert M. Skea, "Pioneer Rocket Engine Exhibited," *Newark Evening News* (Newark, NJ), 29 Aug. 1950, p. unknown.
27. Letter from John A. Pethick to Chief, BuAer, U.S. Navy, 12 April 1945, copy in "Bell X-1" file, NASM.
28. Ibid.
29. Letter from John A. Pethick to Robert M. Stanley, 18 Sept. 1945, in National Archives, RG 72, U.S. Navy, General Correspondence, BuAer, 1943–1945, box 4172, file "1945 QM (8586)."
30. Ibid.
31. Ibid.
32. Ibid.
33. Ibid.
34. Ibid.
35. Ibid.
36. Interview with Shesta by Frank H. Winter, 10 June 1982.
37. Telephone interview with Harry W. Burdett Jr. by Frank H. Winter, 25 June 1982.

38. Interview with Laurence P. Heath by Frederick I. Ordway III, 3–6 July 1982.
39. Ibid.
40. Memo from BuAer to RMI, re Contract NOa(s) 7866—Assignment of Tasks, 9 Jan. 1946, in National Archives, RD 72, U.S. Navy, Records of BuAer, Record of Divisions and Offices, Contract Records, 1940–1960, box 1854, file "NOa(s) 7866."
41. Telephone interview with Lewis F. "Lou" Arata by Frank H. Winter, 9 June 1982, notes in "Louis F. Arata" file, NASM.
    For the names of several other members of the 6000C-4 developmental team, consult Winter and Ordway III, *Pioneering American Rocketry*, pp. 213, 215–216, 232, 418–419, 423–425.
42. Letter from Pethick to Stanley.
43. Telephone interview with Harry W. Burdett Jr. by Frank H. Winter, 25 June 1982, notes in "Reaction Motors, Inc." files, NASM.
    For the *Times* coverage, see, "Rocket Air Engine Revealed By Navy," *New York Times*, 27 April 1946, pp. 1, 12.
44. Telephone interview with James W. FitzGerald, 11 June 1987, notes in "Reaction Motors, Inc." file, NASM; Letter from James W. FitzGerald to Frank H. Winter, 24 Aug. 1989, in "Reaction Motors, Inc." file, NASM.
45. FitzGerald interview.
46. Interview with Henry A. Jatczak by Frank H. Winter, 18 June 1987; "Rocket Air Engine...," p. 1.
47. FitzGerald interview.
48. FitzGerald interview; Letter from FitzGerald.
49. Ibid.
50. Ibid.
51. William C. House, *United States Navy Project Squid Field Survey Report—Liquid Propellant Rockets*, Princeton University, Princeton, NJ, 30 June 1947, Vol. 2, pp. 35–36, online.
52. Letter from John A. Pethick to BuAer, "Proposal for the Issuance of a Task Letter Under Contract NOa(s) 7866...", 25 March 1946, in National Archives, RG 72, U.S. Navy, BuAer, Records of Divisions and Offices, Contract Records, 1940–1960, box 1854, file "NOa(s) 7866."
53. Interview with William P. Munger by Frank H. Winter, 18 June 1987.
54. Ibid.
55. Ibid.
56. Ibid.
57. Ibid.
58. Jatczak interview.
59. Consult Frank A. Coss, "XLR-11-RM-13 Notes," handwritten copy in "Reaction Motors, Inc.—6000C-4" file, NASM.
60. Ibid.
61. Hallion, *Supersonic Flight,* p. 94.

Chapter 5

Betsy's Legacies: Other Supersonic Planes to Lifting Bodies

"While the Skyrocket bored steadily toward the heavens, I prayed silently
to God. 'Don't let me goof this one!'"

–A. Scott Crossfield on his Mach 2 flight, *Always Another Dawn:
The Story of a Rocket Test Pilot*, World Publishing Co.,
Cleveland, 1960, pp. 176–177.

X-1's Milestones Beyond the First Mach 1 Flight

The bright orange, bullet-shaped X-1 that opened aviation's Supersonic
Age with Chuck Yeager's flight is now fittingly exhibited, suspended from the
ceiling in the Milestones of Flight gallery of the National Air and Space
Museum (NASM). The Bell XS-1 itself—renamed the X-1 later in 1947, following that flight—reached additional milestones. The plane attained a new
top speed record of Mach 1.45 (957 mph) on 26 March 1948; took off from the
ground under its own power for the first time on 5 January 1945; and attained
a maximum altitude of 71,902 ft on 8 August 1949. The last flight of the aircraft (X-1 #2) was made on 23 August 1951, although its XLR-11 engine cut
out following two ignition attempts, and the propellants had to be jettisoned.
The mission thus had to be completed as a glide flight. Altogether, Bell X-1 #1
and #2 made 135 powered flights, excluding several aborted flights, but those
did not interfere with the program.

Bell X-1 #3—also known as the X-1-3—was not as lucky. Because of the
pump-development delay by RMI, which was further aggravated by funding
problems for the aircraft's development on the Air Force side, this project fell
three years behind schedule. It was not until April 1951 that the completed
plane was finally delivered by Bell to the Air Force's Muroc Dry Lake Air
Field in California. This aircraft had vastly increased capabilities over the X-1
#1 and #2. Its new engine, designated the XLR-11-RM-5, weighed more than
the preceding pressure-fed models, at 345 lb vs. about 210 lb for the pressure-
fed XLR-11-RM-1. The new engine also required a propellant increase of 437
gallons of LOX and 293 gallons of diluted ethyl alcohol, compared to 311
gallons of LOX. The ethyl alcohol amount remained the same. On the other
hand, the duration of the new power plant was considerably longer at 4.2 min
vs. 2.5 min for the pressure-fed model of the engine. The gain in duration

Fig. 5.1   The Bell X-1 at NASM. (Courtesy Frank H. Winter.)

doubled the performance of the #3 aircraft. Its rate of climb was now 45,000 ft/min, compared to 28,000 ft/min for the earlier version of the plane. This meant the X-1 #3 was theoretically capable of Mach 2.4 flights, although this was not fated to happen.

WEIGHT AND DIMENSION DISCREPANCIES OF THE X-1 ENGINE IN THE LITERATURE

Prior to relating the story of the attempted flight of X-1 #3, and in order to get a better picture in regard to all rocket-propelled research aircraft with RMI power plants in them, it is first necessary to step back and explain the wide discrepancies in dimensions and weights of XLR-11 (or 6000C-4, or Black Betsy motors) that have appeared over the years in the literature. These discrepancies may be attributed to a number of factors.

First, there were slight technical changes among the different variants within both the original nitrogen pressure-fed family of motor models and those within the more advanced pump-fed series, even though both types and their respective variants overtly look remarkably alike, and their respective thrust ratings always remained the same at 6,000 lb.

Other reasons for discrepancies may have been due to occasional rounding-off of dimensions or to typographical or copyist errors. Most surprisingly, it appears that neither exact dimensional nor weight data for the engine for Yeager's first supersonic flight of X-1 #1 are available. Chapter 4 showed, however, that the press were told by RMI at the first public demonstration of the motor on 26 April 1947 that this power plant weighed 210 lb. Later, evidently in 1948, Shesta wrote a news release on the "Model 6000C4 Rocket Engine" that still cited the 210-lb weight.

On 14 January 2014, the author of this book inspected a circa 1946–1947 version of the same type of engine—that is, a pressure-fed type—in the collection of NASM (Cat. #1953-0051), measured it, and also had it weighed. It was found to weigh 205 lb, although this particular engine may lack some standard components. It is thus reasonable to deduce that the weight of the engine for Yeager's first supersonic flight really was "about 210 lbs." Interestingly, the accession records on this artifact that were examined identifies it as a "proto-type" of the one used in Yeager's first supersonic flight. But unfortunately, no further details are offered in the available paperwork on this object.

The overall length of this motor measured 55 in. (or 56.5 in., including a protruding cable); the maximum width measured 18.75 in., and the minimum width, 13.75 in. The term "width" better applies here rather than diameter, because the four chambers are arranged in a diamond-shaped pattern consistent with the Navy's version of the engine (designated XLR-8), such as that installed in the Douglas D-552-2 Skyrocket. The length of each chamber of the NASM specimen inspected is 21.5 in.; the outside diameter of each chamber, 7 in.; and inside diameter of each chamber, 5 in.

The Bell X-1 was taken down from the ceiling of NASM in June 2015; it had hung suspended since the building opened in 1976. It was placed on the floor in preparation for the museum's revitalization, a gallery-by-gallery refurbishment. Flashlight in hand, the author took this golden opportunity to examine the innards of the Bell X-1 as best he could to see what more could be learned about its engine. The remarkable discovery of a green Air Force paper tag on the engine, marked Edwards Air Force Base and dated 15 May 1950, revealed that it was "XLR-11-RM-1, S/N 7," that is, serial number 7 of that model engine. Subsequent research indicated that both X-1 #1 and #2 exclusively used RM-1-variant engines from 1946 to 1949.

On the other hand, it should not be assumed that this XLR-11-RM-1, Serial No.7 model was the same one that powered the aircraft's world-famous first supersonic flight. The author's careful earlier examination of the *Log Book* of the museum's D-558-2 #2 Skyrocket—the first plane to reach Mach 2— reflects a history of its operational life from 1953 to 1955. It shows some 21 instances of removals and reinstallations of either the same engine, or the installation of another engine with different serial numbers. Some removals were for inspections, others were for overhauls. In fact, this plane started with LR-8-RM-6, Serial No. 16 and wound up fitted with LR-8-MR-6, Serial No. 15. The same or perhaps a more complicated scenario surely must have happened with Yeager's Bell X-1 #1.

That aircraft had a long post-first-supersonic-flight history that included a couple of flights with engine fires and one with an engine explosion on 2 May 1949. These incidents alone would have necessitated engine changes or major overhauls. The plane made its final flight (No. 59, also by Yeager) for the 1950 Howard Hughes film "Jet Pilot." Even that flight experienced a small flash fire in the engine bay from a slow-closing alcohol valve, but the flames were quickly doused by an automatic extinguisher. The aircraft was retired after this flight and subsequently donated to the Smithsonian Institution; it is presently suspended from the ceiling of NASM's Milestones of Flight gallery. It appears safe to say that Bell X-1 #1 was always fitted with an RMI XLR-RM-1 engine, but the serial number of the engine that produced the first Mach 1 flight remains undetermined. However, during the author's close examination of this most historic plane, it was not possible to actually view the engine within the deep and largely closed-off interior of the aircraft, other than to inspect and photograph the four nozzles imbedded in the tail of the aircraft. For the same reason, no direct measurements could be taken of the aircraft's engine—if indeed, it is still there in place behind the nozzles.[1]

X-1 #3 AIRCRAFT AND CONCLUSION OF BELL X-1 PROGRAM

Following a test glide flight of X-1 #3, the aircraft began a captive test on 9 November 1951 to try out the propellant-jettison system while the plane was still mated to its carrier aircraft, a converted B-50 Superfortress. Distilled water took the place of hydrogen peroxide, simulating the volume and weight of the peroxide. Normally, this X-1 carried a load of the alcohol and LOX propellants. During the brief test flight, the distilled water was successfully jettisoned. This was to be followed by a jettisoning of the propellants by gaseous nitrogen, but the pilot, Joseph Cannon, discovered that there was insufficient pressure to carry out this second task. Consequently, the test crew was forced to cancel the procedure while airborne and to land the B-50 mothership with X-1 #3 still mated to it.

The carrier plane then taxied to the propellant-loading area and acquired sufficient pressurized nitrogen in preparation for a propellant jettisoning on the ground. A tractor was hooked up to the carrier plane and it towed the plane to a suitable area on the runway; a ground crew cleared the area as fire trucks moved in, and stood by as a precautionary measure. When pilot Joe Cannon was given the go-ahead signal, he initiated the pressurization of the LOX. A moment later, there was an explosion deep within the rocket plane. Fearing it would escalate, Cannon quickly exited the cockpit, but he was knocked down on the ground by a secondary explosion that also set the plane afire. As Cannon crawled away, Bell personnel rushed toward the fallen pilot, pulled him away to safety, and then lifted him into a Bell pickup truck that then sped away to the base hospital. Cannon was badly burned, although did he recover in time. However, the explosions and fire had destroyed both the X-1 #3 and its mother ship.

Later, the X-1 #3 accident-investigation board judged that the explosions were due to a failure in the high-pressure nitrogen system. In his account, aviation historian Richard P. Hallion wrote that the specific cause was the use of a tube-bundle assembly made of 410 stainless steel, a material that had extremely low tolerance at low temperatures, even though the steel's manufacturers did not recommend its use at low temperatures. Hallion concluded: "The accident board theorized that the cold liquid oxygen lowered the temperature of the surrounding structure, so…when Cannon pressurized the system, the nitrogen tubing shattered." However, another aviation historian, Jay Miller, in his book *The X-Planes*, found that "The explosion eventually was determined to have been caused by the incompatibility of Ulmer leather gasket material and liquid oxygen." Despite these unfortunate materials problems, Miller concluded: "The X-1 program proved to be one of the most productive and technologically successful in aviation history. It proved unequivocally that manned aircraft could, indeed, fly faster than sonic velocities, and do so safely."[2]

Understandably, the emphasis in aviation histories on this world famous aircraft is upon the aerodynamic design and associated technology that enabled it to break the sound barrier—its primary mission. But there is no question that the development of the X-1's power plant was an incredible achievement in and of itself, especially as rocket technology in America had barely left its infancy. Additionally, this magnificent achievement was undertaken within an extremely short period of time and by a pioneering company of only a couple of dozen men.[3]

This situation is fully corroborated in a 1950 assessment in *Aviation Week* by Richard P. Gompertz, then with the Power Plant Branch, Edwards Air Force Base, in which he characterized the engine of the X-1 as an "orphan child." "Rocket engine evaluation," he said, "has had to take a back seat in the X-1 flight test program because of the interest centering around the aerodynamic performance of the airplane. This orphaned status has meant actual flight data pertaining directly and only to the engine have been very hard to come by."[4]

DOUGLAS D-558-2 SKYROCKET

Before proceeding to the second generation X-aircraft, it is illuminating to turn back to the progress of the Douglas D-558-2 Skyrocket, which had technically led the way to the Bell X-1. Lovell Lawrence, in his letter of 1 March 1967 to RMI's Warren P. Turner, explicitly stated that their original 1,500-lb-thrust chamber was "a preliminary design for the Douglas Skyrocket program which later went into the Bell XS-1."[5] It may be that because the joint Navy–NACA–Douglas Aircraft transonic program was so broad in originally seeking, from the latter part of 1944, "large scale flight test research…to obtain aerodynamic data required for the efficient and safe design of future high-speed aircraft approaching the speed of sound," that this slowed down the overall pace of the program that eventually led to the D-558-2 Skyrocket.[6]

Chapter 4 showed that Phase 1 of this program became the straight-wing D-558-1 Skystreak that utilized a turbojet, whereas Phase 2 became the swept-wing D-558-2. Moreover, the original design of the D-558-2 was a combination jet- and rocket-propelled aircraft. The jet was to operate the aircraft at lower speeds and altitudes and the rocket provided bursts of power for higher speeds and altitudes. Overriding and contrasting with these factors, the singular goal of the XS-1 program was unequivocally simple: the development of an aircraft capable of supersonic flight and utilizing only rocket power. For the same basic reasons, the XS-1 project assumed a primacy from start. Perhaps this is why there seem to have been no obstacles that we know of for the Army Air Forces to request the Navy to transfer the rocket engine meant for the D-558-2 to the far more straightforward Bell XS-1 program.

As it turned out, RMI benefitted doubly in contributing and modifying the same engine for both the Bell and Douglas programs. Not only was the Large Rotary Beam employed in testing the engine for both aircraft, both were to have the identical 6,000-lb-thrust ratings for their respective rocket power plants. In a memo from RMI engineer Bill Munger to chief engineer Mason W. "Whit" Nesbitt dated 18 March 1947, Munger made a comparative-weight study between an oxygen-alcohol and a hydrogen peroxide gas generator installation for the "D-558 airplane," while he was still tackling the overall pump development for the Bell XS-1.[7]

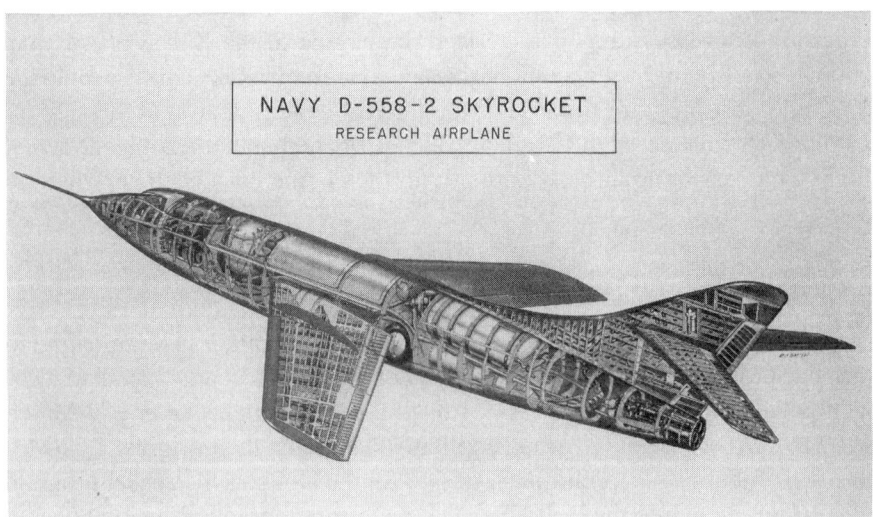

NAVY D-558-2 SKYROCKET
RESEARCH AIRPLANE

Fig. 5.2 Drawing, Douglas D-558-2 Skyrocket, showing engine installation. (Courtesy Smithsonian Institution, 88-17873)

Fig. 5.3   The "diamond pattern" of the RMI engine chambers in the Air Force's Bell X-1. (Courtesy Frank H. Winter.)

In their final forms, the engines in both aircraft looked identical, yet there were differences, primarily in the way the thrust cylinders were arranged to fit different aft contours of the respective planes. In the X-1, this arrangement was in the diamond pattern; in the D-558-2 Skyrocket, the cylinders appeared in the square pattern. That is, the bottom and top cylinders are much farther apart in the X-1, and the two middle cylinders much closer together than in the more even pattern in the Skyrocket.

There were also different official designations for the engines. The official Air Force designation for the X-1—and next-generation X-planes and other vehicles, like the X-15 and lifting bodies—was always the XLR-11. To distinguish their engines from those of the Air Force, the Navy chose the designation XLR-8, or LR-8. One source reported that the Air Force preferred odd-number designations and the Navy favored even numbers. Bottom line, each service had to have its own distinguishing designation.

In addition, the form "LR-8" theoretically denoted that this was an operational model, as opposed to an experimental one. In practice, the "XLR-11" designation on the Air Force side is more common in the literature throughout the history of that engine. The Navy specifically incorporated the XLR-

Fig. 5.4   The "square pattern" of the RMI engine chambers in the Navy's D-558-2
Skyrocket. (Courtesy Frank H. Winter).

8-RM-6 model in their D-558-2 program, although model variants RM-2,
RM-4, and RM-5 are also known.

There were internal differences in the engines as well. Notably, according
to Munger's previously cited memo, both the LOX/alcohol and hydrogen
peroxide systems for generating the turbopump were adapted for the D-558-2
by 1947. Compressed nitrogen was additionally employed to activate the turbopump. In the end, helium was adopted—to save weight—over nitrogen
preferred by the Air Force. The Navy's choice of helium necessitated valve
and other control changes, as well as different types of pressures and tank
configurations. Ignition systems also came in different modifications. James
FitzGerald, who had also worked on the Bell X-1 engine, elaborated that
"little mockup work was done by RMI for the X-1, much more for the D 558
2.[8] The use of aluminum tubing, in particular, worried the Navy, as did the
engine's vibration and "screaming" problems.

The old 6000C-4 often experienced a fierce and baffling howling sound
during its earliest testing that the RMI rocket engineers came to call "groan-
ing" or "screaming." It was later determined that this awful noise was due to
combustion instability. Various complicated theories and fixes were propounded and tried, by both the RMI and Bell people; these went on for some
time. Eventually, one pragmatic approach was found to be very effective:
shutting down immediately and restarting whenever screaming occurred,

FitzGerald later recalled that "this did not do much for a possible occurrence in flight when the scream might not be heard by the pilot!"[9]

In January 1947, according to Lawrence's *Daily Log*, new team member Frank Coss significantly alleviated the problem when he discovered that a single thrust chamber that was running normally could sometimes be induced to scream by simply firing its igniter. Consequently, FitzGerald, with Lawrence's help, quickly revised the control circuitry to activate only the igniter in the chamber(s) being started, instead of firing all four igniters on all four chambers on every individual chamber start. Igniter life was greatly increased, and the incidence of screaming was reduced to a secondary problem.

During the development of the engine for the Douglas D-558-2, Douglas Aircraft had also been concerned "about fatigue failure of the relatively unconstrained tubing on the engine," said FitzGerald.[10] For these reasons, line dampers and "anti-chafers"—a means of reducing the friction of one part of the engine rubbing against another part—were installed.[11] "It [fatigue failure] had given us and Bell no problems," he adds, "but test time was usually in minutes, not hours, so no one knew for sure."[12]

By March 1947, Douglas and the Navy cancelled three Skystreaks from the increasingly expensive program and amended the contracts so that they would be replaced by three Skyrockets. That November, the first D-558-2 was completed, but its power plant—still designated the 6000C-4—was not installed in the aircraft until a full ground test was performed to ensure the engine was flight-worthy. The aircraft's initial flight trials were thus made only with the turbojet. The first jet-only flight was made on 24 May 1949, with Robert A. "Bob" Champine serving as the pilot. The first Skyrocket with rocket engine installed was D-558-2 #3; it made its first ground-takeoff flight using both turbojet and rocket modes on 8 January 1949, with Eugene F. "Gene" May as the pilot. On 25 February 1949, May completed the plane's first rocket flight. The RMI rocket was found to increase the speed of the plane by more than 100 kn (115 mph), compared with the turbojet alone. It also increased its corresponding maximum Mach number from about 0.82 at 20,000 ft to 0.99 at 20,000 ft. But both the Navy and Douglas were still not fully aware of the aircraft's full capabilities with rocket power.

On 24 June of that year, the same aircraft, again flown by May, became the first Skyrocket to fly beyond Mach 1; it used both the jet and rocket power plants. "As soon as the Skyrocket began flying with its rocket engine," says Hallion, "the potential advantage of air launching [with the rocket] became obvious."[13] On the other hand, takeoffs for the Skyrocket, fully loaded with both jet and volatile rocket fuel, were extremely hazardous.

In the jet and rocket modes, the takeoff ground roll began under turbojet-thrust alone and remained that way for a minute—the Skyrocket flying at about 100 mph. At that mark, the pilot switched to rocket power for the actual takeoff; either one or two of the rocket chambers would do it. In those years, another launch technique harnessed even more thrust from a couple of Aerojet solid-propellant JATOs; it was adopted by early 1949. Using JATOs put less strain on the heavily laden aircraft as the spent JATO bottles were ejected after use. The very first flights of the D-558-2 were jet only. Again, the combined jet and rocket flights of the Skyrocket were very difficult, and were further complicated by the extra weight that restricted the aircraft's full potential.

This situation prompted Douglas to a study a modification of one aircraft in which the turbojet was removed and replaced with increased rocket propellant. The study proved invaluable, as it demonstrated that, theoretically, an air-launch version of the D-558-2 from a Boeing B-29 could easily attain velocities Mach 1.46 to Mach 1.6. Comparisons made with flight results of the straight-wing Bell X-1 also pointed toward highly favorable results with an all-rocket-powered Skyrocket with swept-back wings. Dr. Hugh L. Dryden, NACA's director of research, fully supported going ahead with the major modification to the Skyrocket and recommended this mode for D-558-2 #2. Another of the two remaining Skyrockets could also be modified.

Dr. Dryden consequently forwarded his recommendations to the Navy on 1 September 1949 and on 25 November this extra task, or amendment, was added to the BuAer contract to Douglas that affected D-558-2 #2 and #3. But it turned out the D-558-2 #3 retained both its turbojet and rocket power plants, and that Douglas simply installed retractable mounts for the air hooks.

By contrast, D-558-2 #2 underwent a complete makeover. Its Westinghouse J-34-WE-40 turbojet and accompanying gasoline tank were completely removed, and in their place technicians substituted LOX and alcohol rocket-propellant tanks. There were now two tanks each for the fuel and oxidizer, for a total capacity of 378 gallons of the alcohol/water fuel as well as 345 gallons for the LOX oxidizer.

The aircraft's jet exhaust outlet was also removed; the air intakes for the jet engine were replaced with flush fuselage panels; and an upgraded RMI LR-8-RM-2 engine, with its accompanying turbopump, was installed. These modifications were completed by August 1950, and the plane returned to Edwards on 8 November. On 17 November, D-558-2 #3 exceeded Mach 1 during a shallow dive using both the turbojet and a single rocket cylinder. From here, a jump to the air-launched flights of the "all-rocket" D-558-2 is warranted.

AIR-LAUNCH FLIGHTS OF THE "ALL-ROCKET" D-558-2

In a typical all-rocket air launch for these rocket-research aircraft, the first rocket cylinder was fired right after the drop of the test plane from its carrier

Fig. 5.5   Douglas D-558-2 in flight. (Courtesy NASA.)

aircraft, followed by the other three cylinders in succession. The Skyrocket then soared upward quickly.

The first such air-launch Skyrocket flight was accomplished on 25 January 1951, piloted by William B. "Bill" Bridgeman, who hit Mach 1.28. Mach numbers would steadily increase—and dramatically—in the next few flights. Mach 1.88 was attained on 17 August. After, A. Scott "Scotty" Crossfield was assigned to the Skyrocket and made a series of lower Mach-level flights as part of the plane's routine longitudinal, lateral stability, and other aerodynamic testing. However, from 1952 and into 1953, NACA shifted its focus to more closely examining the aircraft's supersonic behavior that Richard Hallion characterized as "Scotty's glory days in the Skyrocket."[14]

As an added measure to afford greater protection to the aircraft's rudder from rocket exhaust during supersonic flights, in September 1953 NACA technicians devised and installed a nozzle extension on the LR-8 rocket engine cylinders. The extension also afforded a bonus 6.5% increase in the engine's thrust at 70,000 ft. Scotty made the plane's first nozzle-extension flight on 17 September, reaching Mach 1.85 at 74,000 ft.

A more exceptional performance with the extension occurred on 14 October, when Scotty sped to Mach 1.96. This achievement created a momentous turning point when the High Speed Flight Research Station requested Dr. Dryden's permission to attempt Mach 2. Permission was granted, and a flight plan was worked up by an aeronautical engineer in which Scotty would climb to about 72,000 ft while still under rocket power, execute a "pushover," or transition from climbing to diving flight, and reach Mach 2 in a shallow dive. On 20 November 1953, he carried out the plan and the D-558-2 #2 Skyrocket thus became the world's first aircraft to pass Mach 2 (1,291 mph, when the plane reached 62,000 ft). This singular accomplishment was arguably the highlight of the D-558 program; the record-breaking aircraft is now exhibited near the Milestones of Flight gallery of NASM. Although not always fittingly recognized in the annals of aviation and rocketry, the milestone was another major triumph for RMI—then just a dozen years old. As it happened, the D-558-2 #2 never made another Mach 2 flight.[15]

On 2 June 1954, BuAer's Ben Coffman wrote a letter to NACA informing them of the "development of an improved version of the LR8-RM-6."[16] Coffman had played a prominent role in the Bureau's JATO program during World War II, and witnessed early RMI testing through the narrow viewing ports of their original control room (blockhouse) at Franklin Lakes back in the early 1940s. This upgraded model, he wrote, "can be rated at either 6,000 lbs, or 8,000 lbs. of thrust. This is possible since the new engine components have been stressed designed, proportioned and are being tested at the 8,000 lb. Level."[17] RMI's own internal designation for this significantly upgraded model of the LR-8 was TR114.

Toward the end of the following month, on 23 July 1954, Ira H. Abbott, NACA's assistant director for research, wrote to the chief of BuAer stating that it would, indeed, be "desirable to use the 8,000 lb thrust version of the LR-8 engine in the D-558-II airplanes, since the improvement in specific impulse will increase flight time available at high Mach numbers. The increased thrust will [also] improve the flight performance slightly."[18] It seems that the 8,000-lb version did not become operational, most likely because the improved pumps appeared to have arrived late, although did not become qualified for flight status before the D-558-2 # 2 flights ended on 20 December 1956.

Information provided in Chapter 8 and in the conclusion of this chapter reveals that uprated variants at the more-elevated thrust level, and higher, were certainly employed in the interim engines of the X-15 between 1959 and 1960, as well as in the later lifting body program of the 1960s to mid-70s, respectively.

Altogether, the XLR-8 carried out 159 flights with the D-558-2 aircraft with seven engine malfunctions, for a reliability of 95.2%. A brief look at second generation X-series aircraft, also fitted with the venerable XLR-11, is now warranted.

THE ILL-FATED BELL X-1D

Exactly one month after Chuck Yeager broke the sound barrier, on 14 November 1947 the U.S. Air Force authorized Bell Aircraft to undertake a study toward the development of the second generation supersonic X-aircraft. The aircraft were to be significantly improved and offer greater performances. The results of the study led to the Bell X-1A, the X-1B, the X-1C, the X-1D, and the X-1E, each designed to explore different areas of supersonic flight. Although more advanced in aerodynamic design, all these aircraft were to retain RMI's XLR-11 engine, but in its more advanced turbopump-fed form, then still under development.

The X-1D was the first to materialize. The contract for this plane was signed on 2 April 1948 under Project MX-984, which encompassed several of the advanced aircraft. In less than a year, a full-scale mockup of the plane, then called the Model 58D, was produced. The completed X-1D was rolled out of Bell's plant doors in Buffalo, New York, early in 1951 and then airlifted to Edwards Air Force in July of that year. It was fitted with an XLR-11-RM-5— the same variant used in the ill-fated Bell X-1 #3 destroyed later in the year. The X-1D plane was therefore to utilize the XLR-11's newer low-pressure turbopump fuel system that would increase fuel capacity and permit 4.65 min of full four-chamber burn time for its 6,000-lb-thrust engine.

On 24 July, with Bell test pilot Jean "Skip" Ziegler at the controls, the X-1D was launched from a carrier plane over Rogers Dry Lake into an unpowered successful glide flight, although its nose gear failed upon landing. It took several weeks to make repairs, and it was not until 22 August that the plane was prepared for its first powered flight. Air Force Lt. Col. Frank K. "Pete" Everest was the pilot. Aviation historian Hallion commented that it was intended that this flight would "exceed the speed and altitude marks of Mach 1.88 and 79,494 ft. set by the Douglas D 558-2 #2 Skyrocket in early August 1951."[19] But in using a low-pressure system for activating the rocket engine's turbopump, the plane deployed nitrogen stored at 4,800 psi, normally used to pressurize fuel-tank regulators and other aircraft components.

The B-50 carrier aircraft lifted up the X-1D to 70,000 ft. Everest then climbed into the cockpit of the test plane. Upon checking his instrument panel, he noted an appreciable drop in nitrogen source pressure. He therefore climbed down and returned to the carrier to talk this matter over with Air Force flight-test pilot Jack Ridley and Bell rocket engineer Wendell Moore. All agreed to abort the flight after jettisoning the propellants. Everest climbed back into the test plane. While reopening a LOX valve to bring up its pressure, an explosion erupted, almost knocking back the pilot. Flames were spotted from the chase plane.

"Hey Pete! Drop her, drop her, she's on fire!" yelled the voice over the radio from the chase plane.[20] Everest immediately scrambled back into the B-50, and Ridley pulled the drop handle. The X-1D fell and exploded into flames on the desert floor. After this disaster, the accident board determined that the likely cause of the explosion was a fuel leak resulting in alcohol-and-air mist, combined with an electrical spark from the plane's radio transmitter or other source that ignited the dangerous mixture. Wendell Moore, incidentally, later became Bell's own rocket pioneer and gained renown for his development of their "rocket belt."

THE BELL X-1A

The Bell X-1A was far more fortunate. Upon completion, it was airlifted to Edwards early in January 1953. After a few flights, it was returned to Bell for some modifications, then again sent to Edwards to resume flight operations. This aircraft, too, utilized a low-pressure nitrogen feed to the engine's turbopump. On 8 December, Chuck Yeager took up the X-1A to Mach 1.9 at 60,000 ft. Just four days later, he hit a peak of Mach 2.44 (1,650 mph) at 75,000 ft, thereby becoming the world's second aircraft to fly at twice the speed of sound.[21] After this feat, the Air Force Base prohibited future planes to go beyond that mark for fear of excessive aerodynamic stress and possible further catastrophes. Subsequently, the X-1A would only make high altitude flights. It set other records for altitude and experienced a few aborted attempts, some due to faulty ignition starts and turbine overspeed. In the long run, the Air Force's stricture against Mach 2 flights did not hold in practice.

Sadly, the Bell X-1A also met its end in a dreadful explosion. It occurred on its flight of 23 February 1955; the plane's NACA research pilot, Joseph A. "Joe" Walker, survived. The cause of the accident was traced to the incompatibility of the Ulmer leather gasket to LOX, the same defect that caused the disasters of the X-1 #3 and other aircraft. More specifically, it was the tricresyl phosphate impregnated in the Ulmer leather gaskets. After this, all Ulmer leather gaskets were withdrawn from the Air Force's inventory, and no further catastrophic incidents occurred throughout the rest of the Air Force's supersonic program.

THE X-1B

The sister ship of the X-1A, the X-1B was fitted with another variant of the XLR-11—the XLR-11-RM-9, also internally designated by RMI as the E-6000C-4-1. In this case, the modifications amounted to little more than an electric-spark, low-tension interrupter igniter in lieu of the older high-tension model. The plane itself was built for aerodynamic-heating research at high Mach numbers.

This aircraft made its first powered flight on 8 October 1954. On 2 December 1954, Everest flew it to Mach 2.3 at 65,000 ft, contrary to the Air Force's ruling that no Mach 2+-flights were to be made after Yeager's Mach 2.44 flight back on 12 December 1953 with the X-1A. Interestingly, the young NACA test pilot Neil A. Armstrong also flew this aircraft; his first flight occurred on 15 August 1957 and reached an altitude of 11.4 mi. Armstrong, who later became the first man to walk on the the surface of the moon under Project Apollo, was then assigned to help test the X-1B's hydrogen peroxide reaction-control system. Although rudimentary, this system laid important ground work for a similar one developed for the forthcoming X-15 hypersonic rocket-research aircraft. Today, the same X-1B is on exhibit in the Air Force Museum at Wright-Patterson Air Force Base, Ohio.

THE BELL X-1C

The Bell X-1C was envisioned as a supersonic armaments testbed to test various types of machine guns and cannons firing from the nose of a supersonic plane. However, the emergence of the supersonic North American F-86 and North American F-100 jet fighters saw the X-1C as a redundant effort, and this project was dropped with no hardware ever built other than a mockup.

THE BELL X-1E

Last, there was the Bell X-1E, a project started in late 1951. Plans called for the aircraft to be fitted with a low-pressure nitrogen fuel-feed system, although still using the turbopump of the earlier X-1D that had led to the second generation X-aircraft. NACA engineers at Edwards, likely in consultations with RMI, were involved, and based their ideas on what was learned on the D-558-2, the X-1D, and the X-1 #3. Ironically, despite attention to the design of the turbopump in this aircraft, the initial flights of the X-1E experienced a series of pump headaches. They ran the gamut from turbopump failures to start, turbopump overspeeds, and intermittent pump operation. Difficulties continued to plague the program; in the end, the reliable LR-8-RM-5 engine from the Skyrocket program was adopted for the aircraft.

By late 1957, NACA engineers at the High-Speed Flight Research Center proposed boosting the aircraft's maximum design speed of Mach 2.7 to Mach 3. This would be done by boosting chamber pressure in the LR-8, and replacing the LOX/alcohol propellants with a new "super performance fuel" called Hidyne, or U-deta, consisting of 60 percent unsymmetrical dimethylene dimethylhydrazine and 40 percent diethylene triamine. The propellant switch was used in the plane's last flights, but the plane never reached the anticipated higher Mach numbers. The flight of 6 November 1958 turned out to be the plane's last—and the final one in Bell's supersonic X-aircraft series.

That December, an X-ray inspection revealed dangerous cracks in the aircraft's alcohol tank. That discovery and another serious technical deficiency, as well as the much anticipated arrival of the new hypersonic X-15, led to the program's cancellation; subsequently, the X-1E joined other supersonic X-aircraft in retirement.

Overall, the XLR-11—formerly as the 1500N-4 C, the 6000C-4, and Black Betsy—accomplished 191 flights in the combined Bell X-1 series flown from 1946 to 1958, experiencing 14 engine malfunctions, for an engine reliability of 92.5%. But by no means was this the end of the history of this engine.

Fig. 5.6   The Bell X-1E being static tested. (Courtesy Frederick I. Ordway III Collection, U.S. Space and Rocket Center, Huntsville, AL.)

THE XF-91 THUNDERCEPTOR

While the Bell X-series and Douglas Skyrocket were all experimental, the Air Force's Republic XF-91 Thunderceptor was fitted with a variant of the old 6000C-4 engine during its development, and was one of the earliest U.S. supersonic swept-back combat fighters—although it too was experimental at that point. Based on Long Island in Farmingdale, New York, Republic Aviation was probably well aware of nearby RMI. According to RMI's *Daily Log* for 2 May 1946, a Dr. O'Donnell (in all probability Dr. William J. O'Donnell, Republic's chief development engineer) had an appointment with Lawrence regarding the installation of a "motor in one of their planes."[22] This may or may not have been the start of RMI's connection with this project. For certain, the *Daily Log* for 26 August 1948 records that Lawrence had a "discussion of the XF-91" with Air Force officers during a trip to Maxwell Field in Alabama.[23] An RMI *Monthly Project Status Report Summary* for April 1951 also reveals that by that date, XF-91 engine tests were completed.

Inspired by the World War II rocket-powered German Me-163 and the rocket-boosted turbojet-powered Me-262C interceptor prototype, the XF-91 was originally installed with a 16,000-lb-thrust Curtiss-Wright XLR-27-CW-1 rocket engine. However, there were serious problems with the engine, and by late 1950 Republic opted for the lower-powered but reliable 6000C-4 instead, and the plane was never flown with its XLR-27. As hinted above, Republic seems to have already explored the possibilities of RMI's 6,000-lb-thrust engine far earlier—as well as other potential engines from Aerojet and the M.W. Kellogg Company. The latter was another early New Jersey rocket firm, started in Jersey City—at least the company had begun a rocket department by 1946. For reasons unknown, Republic did not initially accept the RMI 6000C-4 engine, but undoubtedly because of the technical hurdles they encountered with the Aerojet and Kellogg units, they later returned to RMI, and the 6000C-4 was finally adopted for the aircraft.

The XF-91 really had two power plants. The primary one was the GE J-47 turbojet of 5,200-lb static thrust, while the RMI rocket was to serve as a built-in auxiliary booster for accelerated takeoffs, climbs, and interceptions. In its 6000C-4—or rather, a modified XLR-11-RM-9—configuration, the entire four-barreled engine fitted directly below the aircraft's jet power plant in a diamond-shaped pattern within a pod. In September 1952, the XF-91 made its first combination rocket-and-turbojet powered flight and from then on, the rocket engine undertook several several other flights in the aircraft.

With both the jet and rockets running, the XF-91 could attain Mach 1.71. Yet of the plane's known 192 test flights, just 19 were made with the RMI built-in superperformance engine, all accomplished without malfunctions for a reliability of 100%. Some time after Republic's test pilot Russell "Rusty" Roth flew the XF-91 on its first supersonic flight in December 1951, the

Fig. 5.7   Republic XF-91. (Courtesy Frederick I. Ordway III Collection, U.S. Space and Rocket Center, Huntsville, AL.)

company's president, Mundy I. Peale, described the plane as "a combat-ready airplane, not purely a research plane" and "a bridge of the gap between jet and rocket planes."[24]

In their great hopes that this aircraft would become operational, Republic, undoubtedly in consultation with RMI, went as far as to design and construct a special 12.5-t LOX-and-fuel-servicing truck manned by two men wearing protective clothing.[25] Yet only two of the aircraft prototypes were built and it never went into production. Among the problems faced by the Thunderceptor was an extremely short flight time of about 25 minutes, which made it highly impractical for helping protect an area as large as the United States. In addition, the Air Force did not have to wait long for the appearance of far superior jet interceptors—and the XF-91 program soon disappeared.

THE XP-92

Even earlier, by August 1945, the 6000C-4 had been considered as a potential high-power booster or auxiliary power plant for the experimental XP-92 manned delta-wing rocket plane as a supersonic interceptor that could reach 50,000 ft in 4 min. Various ambitious configurations were drawn up, including a booster of six uprated 2,000-lb-thrust chambers from the 6000C-4, or 12,000 lb total. A rocket-ramjet version of the XP-92 was also looked at, but nothing came of that project.

THE LIFTING BODIES

As Chapter 8 will show, when the X-15's main rocket engine—the XLR-99—fell terribly behind schedule by the first half of 1958 due to developmental problems, managers of the X-15 program resorted to pulling out retired XLR-11s to serve in sets as the interim engines for the initial missions of the hypersonic rocket aircraft. The last-known official employment of the engine seems to have been in NASA's lifting bodies.

A lifting body is a fixed-wing aircraft with very little fuselage and the wing providing the lift. All lifting bodies were carried aloft by an aircraft, then released for simulated reentry maneuvering and landing. They were used by NASA to gather critically needed aerodynamic and control data on the reentry of reusable manned spacecraft, and were invaluable tools that contributed toward the development of the Space Shuttle following Project Apollo. Several models were made, some installed with XLR-11 rockets.

Lifting body program manager John G. McTigue is credited with the idea of incorporating the XLR-11s. At first, he thought of utilizing an Agena rocket engine in the normally unpowered craft. But the Agena was rated at 16,000-lb thrust, which was far too powerful and expensive, so he turned to the XLR-11s. There are several versions of the story.

According to McTigue's own account, he scrounged several surplus XLR-11s in the most unlikely places: from an old test plane or two on exhibit at a nearby college at Lancaster, California, that we now know to have been the Antelope Valley College, and from a museum in San Diego. McTigue acquired eight engines altogether from his scrounging missions, then contracted with RMI to help rebuild or refurbish them—all for $50,000. Refurbishments were largely done right at Edwards, but an RMD field representative, William "Bill" Arnold, was also available and did participate in this work. The engines were first tested as mated with the X-24A and X-24B Lifting Bodies, although these initial efforts faltered. In fact, this program did not go smoothly for some years before it finally succeeded.

Not all the details are known, but among some of the drawbacks was that the X-24 Propulsion Test Stand (PSTS) at Edwards was suitable for ordinary ground runs of the engine, but could not simulate the variables (far-lower temperatures, pressures, and so forth) that occur in actual flight conditions. Similarly, another huge problem was that the Martin Marietta Corporation, contractor for the X-24A and X-24B, chose not to fully enclose the engine in the aircraft fuselages. Consequently, the XLR-11s were directly exposed to severe cold temperatures from drop altitudes. Delicate electronics in the engine-control boxes especially did not function very well in such conditions. To combat the temperature problem, the NACA team added a heating blanket to the control box, but even this was not enough. Additional problems occurred.

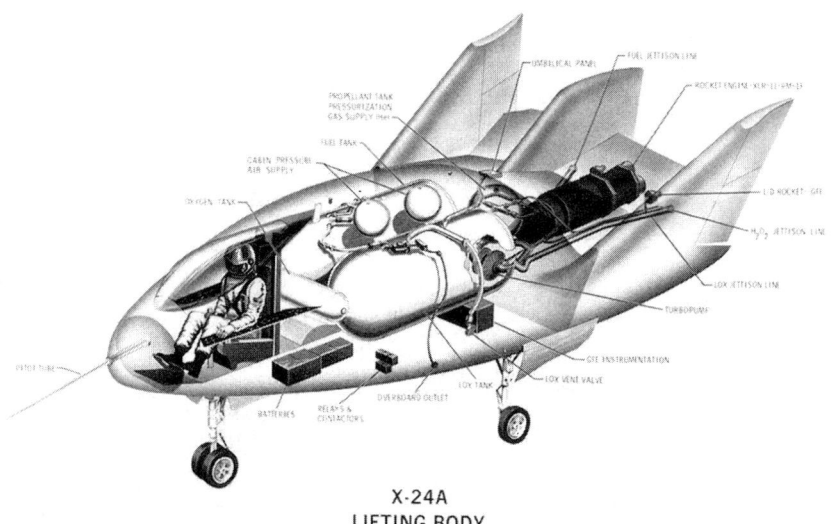

X-24A
LIFTING BODY

Fig. 5.8   Drawing of the X24-A Lifting Body, showing the installation of the XLR-11 rocket engine. (Courtesy NASA.)

One problem was traced to the malfunction of an important pressure switch that helped to open the igniter. An important component in the pressure was the O-ring. This was made of Buna A, a synthetic rubber that was later discovered to shrink under low ambient pressure and was later replaced by a silicone O-ring that proved to work better. Another problem was encountered with igniters that burned out; additional other difficulties presented themselves as well.

In the meantime, other lifting body programs were carried out. Interestingly, in the cases of the M2-F2 and the HL-10 craft, NASA engineer Dale Reed basically resorted to doing the same thing that McTigue had done in procuring XLR-11s—he also scrounged them from a museum, in this instance from the Air Force Museum in Dayton, Ohio.

These hurdles notwithstanding, various lifting body programs with their XLR-11s eventually did succeed. The first operational powered flight in a piloted lifting body was conducted on 23 October 1968 in an HL-10 piloted by Maj. Jerauld R. Gentry after release from a B-52. However, this flight was terminated when one of the two engine barrels failed to ignite. Gentry jettisoned the remaining fuel and landed prematurely. The next month, Pilot John A. Manke won the distinction of achieving the first successful HL-10 powered flight on 13 November, when two XLR-11 chambers were ignited for low-

Fig. 5.9   The M2-F3 on exhibit in the National Air and Space Museum. (Courtesy Frank H. Winter.)

powered planned lifting body maneuvers. Manke reached Mach 0.87 on that flight, but on 9 May 1969 he flew supersonic at Mach 1.13 in an HL-10, although stability and control rather than speed were the uppermost objectives for these craft.

Three Bell Aerosystems 500-lb-thrust/70-s hydrogen peroxide rocket engines replaced the XLR-11s in the HL-10 for the powered approach-and-landing phase of the program, but the more powerful XLR-11 was ideal for the transonic and supersonic flight phases of these vehicles. In all, the HL-10 completed 21 successful flights with this engine. This was matched by 22 XLR-11 flights with the M2-F3 Lifting Body, flown between 1970 and 1972; 18 powered flights with the X-24A from 1970 to 1971; and 24 powered flights with the X-24B between 1973 and 1975.

William H. Dana is credited with the last rocket-powered flight in the X-24B on 23 September 1975, because the vehicle's final six flights were nonpowered glide and landing checkouts. When the NASA/Air Force lifting body program came to a close in that year, the XLR-11 rocket engine had logged in an incredible service life of almost 30 years. That entailed incredible experiences ranging from powering the world's first supersonic flight by the Bell X-1 and the first Mach 2 flight by the Bell X-1A, to serving as an auxiliary power plant for one of the earliest U.S. supersonic swept-back combat fighters (the initial engines for the X-15 hypersonic aircraft) to powering lifting bodies that helped pave the way for the Space Shuttle.

ENDNOTES

1. For examples of the minor technical differences in various XLR-11 models, consult Frank H. Winter and Frederick I. Ordway III, *Pioneering American Rocketry: The Reaction Motors, Inc. (RMI) Story, 1941–1972,* AAS History Series, Vol. 44, Univelt, Inc., San Diego, 2015, p. 422. Consult also John Shesta, RMI press release, "The Reaction Motors, Inc. Model 6000C4 Rocket Engine" (circa 1948), in the Lovell Lawrence Jr. Papers, box 48, folder 1, NASM.

2. Hallion, *Supersonic Flight: The Story of the Bell X-1 and Douglas D-558,* Macmillan Co., New York, 1972, p. 163; Jay Miller, *The X-Planes,* Aerofax, Inc., Arlington, TX, 1988, p. 23.
   See also Miller's Appendix C of that work, pp. 210–212, relating how the Ulmer problem seriously plagued other X-type rocket planes that also used LOX in their propellants.

3. Miller, *The X-Planes*, p. 23

4. Richard P. Gompertz, "Rocket Air Tests: Results of Engine Test Experience with X-1 Reviewed for ASME," *Aviation Week*, Vol. 53, 18 Dec. 1950, p. 34.

5. Letter from Lovell Lawrence Jr. to Warren P. Turner, 1 March 1967, Lawrence Jr. Papers, NASM, box 13, folder 27.

6. E.[dward] H. Heinemann, "The Development of the Navy-Douglas Model of the D-558 Research Project," press release, 17 Nov. 1947, copy in "Douglas -D-558-2" file, NASM.

7. Memorandum from W.P. Munger to M.W. Nesbitt, "Comparative Weight Study...," 18 March 1947, Lawrence Papers, box, unknown, file unknown.

8. Telephone interview with FitzGerald, 11 June 1987, notes in "Reaction Motors, Inc." file, NASM.

9. Ibid.

10. Ibid.

12. Ibid.

13. Hallion, *Supersonic Flight*, p. 153.

14. Richard P. Hallion, in J.D. Hunley, ed., *Toward Mach 2: The Douglas D-558 Program*, NASA History Series, NASA, Washington, DC, 1999, p. 34.

15. For further details on technical preparations for the Mach 2 flight, consult Hallion, *Supersonic Flight*, pp. 172–173.

16. Hunley, ed., *Toward Mach 2*, p. 143.

17. Ibid.

18. Ibid., p. 146.

19. Hallion, *Supersonic Flight*, p. 161.

20. Ibid., p. 162.

21. As for the world's first plane to reach Mach 3, this record was set on 27 September 1956 by the Bell X-2, powered by the Curtiss-Wright two-chamber XLR-25 rocket engine. The aircraft actually flew up to Mach 3.2, although the plane was lost due to inertia control, killing the pilot Milburn G. Apt. Curtiss-Wright (like RMI, also based in New Jersey) was a later competitor in the rocket business and had initiated this phase of activities about 1945. Other than their X-2 power plant, they did not gain any prominence in rocketry thereafter.

22. Lawrence, RMI *Daily Log*, Lovell Lawrence Jr. Papers, NASM, box 5, folder 2.

23. Ibid.

24. "Rockets Push XF-91 Past Mach 1," *Aviation Week*, Vol. 57, 15 Dec. 1952, p. 17.

25. For illustrated coverage on Republic's "completely self-contained" trailer truck for fueling the XF-91, see "Republic Speeds Rocket Servicing," in *Aviation Week*, Vol. 56, 17 March 1952, pp. 34–35.

Chapter 6

RMI's Missile Motors: Gorgons to Bullpups

"I do not want to change the thrust. I do not want leaks. I don't like liquids. I don't want your lousy plumbing."

–Missile contractor's initial comment to RMI on their proposed use of a packaged-propellant liquid motor for the Sparrow III, late 1950s, recollection by RMI's Art Sherman.

THE GORGON FAMILY OF MISSILES

America's first family of missiles, known as the Gorgon series, was aptly named after a sisterly trio of ancient Greek snake-haired monsters. The three sisters were Stheno ("the mighty"), Euryale ("the wide-leaping"), and Medusa ("the queen"). Gorgon itself means "terrible," and each sister had a gaze that could turn mortals into stone. The originators of the Gorgon missile program, which began officially in 1943 (although it had roots going back to 1937), envisioned a variety of missions for these weapons.

There is another reason to apply the name of Gorgon to missiles. Ever since World War II, and especially after that period, it became customary for the U.S. military to name some of their missiles after deities and other beings from Greek, Roman, and other mythologies: Nike-Ajax, Nike-Zeus, Nike-Hercules, Thor, Atlas, Jupiter, and Titan are prime postwar examples; a later example is the Poseidon. The popular practice was also applied to launch vehicles like the Juno and Saturn, and upper rocket stages such as the Centaur. Perhaps, then, the Gorgon was the first U.S. missile that actually started the trend of applying such mythological names to these vehicles.

The Gorgon series, which would come to include America's first liquid-propellant guided missiles, has an interesting backstory. In 1936, while in England attending the London Naval Conference, chief of naval operations Adm. William H. Standley witnessed a British unmanned aerial target, a radiocontrolled Queen Bee spruce-and-plywood biplane in action during a live-fire exercise. Built under contract by the Fairey Corporation and first flown in 1935, the Queen Bee could fly as high as 17,000 ft and travel a maximum distance of 300 mi at over 100 mph. Adm. Standley was so impressed that he decided the U.S. Navy should develop a similar training tool. Lt. Cmdr. Delmar S. Fahrney, a veteran aviator with a master's degree in mechanical engineering, was selected as the leader of this effort; he labeled these craft "drones." Fahrney took on the project wholeheartedly and devoted the remainder of his career to this field, which included missile development. In time, he would be regarded as the "Father of the Guided Missile."

The basic concept of what became the Gorgon sprang up early in Fahrney's investigations on drones. As he later recalled, the basic idea of radiocontrolled aerial weapons occurred when he turned his attention to the missile. Consequently, he included the recommendation in his semi-annual report of July 1937 to the chief of BuAer, recommending that the U.S. Navy's "target drone program be broadened to encompass radiocontrolled aerial weapons that could be carried on combat aircraft for air-to-air and air-to-surface use."[1]

At this early stage, Fahrney did not have a rocket power plant in mind. Rather, he only considered "acceptable and conventional" piston-engine-powered aircraft that were unmanned, radiocontrolled, and carried bombs. Nor was the name Gorgon then applied to his concept. As may be expected, Fahrney's recommendation of the broadening of the drone program—especially in that time of peace and very tight military budgets—was not approved. Undismayed, he continued to "constantly" pursue his ideas on missile development.[2]

On 19 July 1943, the true Gorgon missile program was inaugurated by a classified-secret letter, which included design studies and specifications, from the chief of BuAer to the manager of the Naval Aircraft Factory in Philadelphia. At this time, the code name of Gorgon was also selected. The primary mission of this "aerial torpedo," as it was now generally described, was to "destroy large enemy aircraft" (bombers), while the secondary mission was for use against "light surface craft."[3] The radiocontrolled and target-seeking torpedo could be launched from aircraft or from a ship or land base—hence, it was considered to have multiple mission capabilities.

The next day, the Naval Aircraft Factory was redesignated the Naval Air Material Center. Most of the missile work there was to be carried out by a subordinate command called the Naval Aircraft Modification Unit under now-Capt. Fahrney. The missile program was not rated the highest priority at the time, although it would be raised later.

The main power plant for the initial Gorgon-type missile was slated to be a small Westinghouse 9.5A turbojet engine—military designation J32-WE-2—whose development began in late 1942. The 140-lb engine thrust 260 lb. Due to anticipated delays in the delivery of this mode of propulsion, BuAer then directed the alternate use of a liquid rocket power plant. This was the start of America's first liquid-propellant-powered guided missile.

This development was based upon Lt. Robert Truax's preliminary work on rocket motors at the Naval Engineering Station (EES) in Annapolis. His small team included Robertson Youngquist and a few others. BuAer had also been directed to develop two distinct types of air Gorgon frames—a tail-first, or "canard" type, and a conventional one. In order to have a sufficient number of missiles available for conducting a modest test program, 25 airframes of each type were ordered by October 1943. Fahrney credits the drawing of the rocket-engine installation in the basic Gorgon airframe to Lt. William Schubert.

Undoubtedly due partly to critical wartime shortages of metals, all the Gorgon airframes were to be made of wood. Probably another reason for the wooden fuselages of these missiles is that the Naval Aircraft Factory had already had years of experience in using cost-saving wood for parts of some of their aircraft. Their TDR drone, which became the first U.S. operational drone and first appeared in World War II, is a later instance of this application, using preformed plywood for its wings, fuselage, and other surfaces. But in the case of the Gorgon missiles, planks (boards) were cut into strips and formed into aerodynamically smooth body and wing structures as the missiles were built. This was part of a technique more commonly known in the building of boats. In fact, this technique—the Carvell planking method—had been used by the Naval Aircraft Factory and other aircraft manufacturers to produce flying boats.

The Gorgon rocket motors to power the missiles were to utilize the hypergolic combination of nitric acid and aniline in a very simple arrangement in which compressed nitrogen forced the propellants into the combustion chamber. Despite the simplicity of the Gorgon project, however, one classified document discovered in the National Archives emphatically stated that "Every effort is being made to develop this missile for use in this war."[4]

Another document in the National Archives reveals that on 10 August 1944, following Lovell Lawrence's direction, RMI manager John Pethick submitted to BuAer a "Proposal to amend [the existing] Research and Development Contract [for JATO work] to cover [the] construction of fifty Gorgon IIA jet propulsion [rocket] units of E.E.S. design."[5] The Gorgon IIA—also given as the Gorgon 2A—was the first rocket-propelled missile in this missile family.

The proposal further ordered RMI to "construct…assemble and test all [Gorgon IIA motors] at Pompton Plains, New Jersey, and to deliver…fifty Gorgon IIA Jet Propulsion Units of E.E.S. design….A representative of the E.E.S. shall witness the test of each unit and shall either reject or accept the unit on behalf of the Navy Department."[6] The "total selling price" of these units was $109,014, then not an inconsiderable sum.[7] Of this amount, $20,191.88 was stipulated for "wages and salaries of [RMI personnel]…as defined in the contract."[8] RMI thereby proceeded with the contract, using the Navy's blueprints for the motor and making some improvements in the design. Moreover, according to aerospace writer Lloyd Mallan, "with the personal help of [later] Commander Truax," RMI developed America's first missile motor in the amazingly short span of 45 days, although this motor did not become operational.[9]

The motor—designated the CML 2N by RMI—generated 350 lb of thrust for a duration of 130 s, or a little over two minutes, using monoethylene and mixed sulfuric and nitric acid. As with other RMI motors, it was regeneratively cooled, the nitric acid serving as the coolant.

The Gorgon IIA, 14.5 ft long, with an 11-ft span, was originally designated the KA2N-1. It was intended as a 25-mi-range air-to-air weapon by the Navy, with the Pacific combat theater in mind against Japanese aircraft. But by the time it entered its testing phase, the war was winding down. The missile thus became an air-to-air research test vehicle, occasionally used for air-to-surface experiments. It was nevertheless remarkable that for the combat application, these experiments included the first U.S. attempt to use the very rudimentary development of television as part of the guidance system, probably during the Gorgon's first powered flights from March 1945.

Mainly, the missile was controlled by command radioguidance. The small, onboard TV camera was mounted in Gorgon's clear plastic nose to relay potential target images back to the mother aircraft that had released the missile so that the carrier-aircraft pilot could direct the missile by radio signals to the target. But the immature state of TV transmission produced frustratingly fuzzy images, and the TV application was deemed unsuitable for air-to-air missile guidance, and was apparently dropped.

Fig. 6.1   Closeup of the RMI rocket motor in the Gorgon II-A aircraft missile, specimen in the collection of the National Air and Space Museum. (Courtesy Frank H. Winter.)

Fig. 6.2 Full Gorgon II-A missile on display at NASM's Steven F. Udvar-Hazy Center. (Courtesy Frank H. Winter.)

While the Gorgon's air-dropped launching, propulsion, and aerodynamic flight characteristics were generally found satisfactory, its guidance system failed miserably, primarily because the aiming pilot could not manually steer the sluggish missile towards its interception point. On the other hand, this was the pioneering era of U.S. guided-missile development and the Gorgon IIA continued to serve as a useful test missile, although again, it never become operational.

Gorgon was redesignated in 1946 as KU2N-1, indicating it was a control test vehicle. Early in 1948, this system was simplified and it was redesignated as CTV-4, and after that, CTV-N-4. The intended follow-on—the Gorgon IIB, or Gorgon 2B—was to have been fitted with a turbojet, although a suitable engine could not be found and this project was abandoned.

Despite BuAer's request in 1944 for an initial 50 Gorgon airframes, a far more modest 21 of the II-A models were actually produced. A beautifully restored Gorgon IIA (Cat. #1951-0065) is currently on exhibit at NASM's Udvar-Hazy Center. The canary yellow missile bears the bold black markings "KU2N-1 20" (serial number 20) on its vertical tail fins, making it next to the very last one of this model manufactured.

The author inspected the motor of this specimen prior to the vehicle's suspension from the museum ceiling for exhibit. The motor still has RMI's manufacturer's label attached, showing it was fabricated at Pompton Plains, while the valves were made by the Marotta Engineering Co. of Boonton, New Jersey. Additionally, the motor is fabricated of anodized aluminum, and the missile's propellant tanks and valves remain intact. According to early RMI lore, during one early inspection of RMI facilities by Navy brass, the company's entire array of Gorgon motors on hand—all they had at that time, which were really not many—were laid out to impress the brass.

On the other hand, in a rare early RMI "Shop Weekly Report," dated 21 February 1946, "Gorgon units" are listed as Priority 3 out of nine priorities. Priority 1 was the "Turbine pump" for the Bell XS-1 rocket plane, and Priority 2, the "6000-lb motor for [the] Navy"—the Douglas D-552-2 Skyrocket.[10] "Priority" here, is defined as "All emphasis and manpower is being placed on the first three jobs, while any other available time or men is being allotted to the [others]."[11] This provides an interesting miniperspective on RMI's earliest-known prioritization level of the small amount of missile work underway there at the time.

The remaining priorities in this report are listed as follows: a "gasoline acid" motor (Priority 4); a curious "G.E." [General Electric] 750-lb LOX/alcohol thrust motor for an unknown project—perhaps for an experimental missile or aircraft that was never developed further (Priority 5); the basic pressure-fed 6,000-lb motor itself "for Bell"—the XS-1 motor—that was already well developed (Priority 6); "Resojet [pulsejet] motors" (Priority 7); a rocket motor for an "Autogiro" (Priority 8); and the 1,500-lb motor—the individual chamber for the 1500C4, or 6000C-4 engine, also already well developed (Priority 9).[12] The 6,000-lb "C4 unit" for Douglas (the Skyrocket) and mockup for the same were rated "awaiting priority."[13] The priority of the 2A Gorgon aside, this program was cancelled in favor of the followup Gorgon 2C.

Apart from the Gorgon II-A on display at the Udvar-Hazy Center, there are three other members of the Gorgon family represented there that show two other modes of propulsion deployed by the Gorgons: the pulsejet and the ramjet. The Gorgon 2C, or II-C, ship-to-surface missile (of which 100 were built) utilized a standard pulsejet, similar to those as found on the German V-1 deployed in World War II. It was developed in the U.S. too late for operational use in the war and became a postwar control-test vehicle. There is also the RTV-N-15 Pollux test vehicle, featuring the unusual arrangement of an internally mounted pulsejet in a more streamlined fuselage, in an attempt to increase the missile's overall speed through streamlining to counteract the normally slow performance of the pulsejet. The Pollux's sluggish development led to the project's eventual cancellation, however, and the exhibited specimen is likely the only one extant. A Gorgon 4 ramjet-powered air-to-surface missile, developed from 1946, is also on exhibit. Its ramjet was developed and manufactured by the Marquardt Company, which specialized in ramjets, whereas RMI did not get into this field. As related earlier, RMI was involved with improvements to pulsejets both during and after World War II, although it appears that none were connected with any of the pulsejet-powered Gorgons.

There are additional specimens of Gorgons in the NASM collection. These include two other model IIAs: serial numbers 4 and 9.

There are also three examples of the Gorgon 3A, likewise installed with the identical 350-lb-thrust CML-2N-1 motor. The 3A is very similar to the 2A and has the same airframe. However, it was designed to carry a 257-lb fragmentation warhead for its air-to-air mission, in lieu of the 100-lb special-shaped charge warhead for the air-to-ground 2A. Altogether, 34 3A models were built; serial numbers 14, 17, and 23 are in the NASM collection. After the war, the missile wound up as a control test vehicle.

A more unique and rarer early missile specimen in the NASM collection is the Gorgon 3C (Cat. #1966-0163). This missile has twin rocket units in tandem, both present in the specimen, and each is the same type of motor that propelled the 2A, with a combined thrust of 700 lb. In all probability, the Gorgon 3C (or III-C) had the distinction of being the first U.S. missile with two motors that operated simultaneously. The original rationale for this arrangement was that of the dozen Gorgon 3Cs built, half of these were chosen by NACA for planned high-speed-missile flight research to approach Mach 1 in firing tests. The remainder were to be expended by the Navy in general performance and stability tests. More than likely, the Gorgon 3C in the NASM collection is the only one in existence, and is identified as RTV-4, or "Research

Fig. 6.3 Tandem RMI rocket motors in probably the only existing specimen of the Gorgon III-C missile in the collections of NASM. (Courtesy Frank H. Winter.)

Test Vehicle-4," Serial No. 8. In about May 1946, this missile was redesignated the KU3N-2. Interestingly, like all the other Gorgons in the NASM collection, the body is made of wood, with some metal sheeting over it (such as the protective cowling for the motor section), attached with metal screws and fittings.

THE LARK MISSILE

A direct technological spin-off from the basic Gorgon power plant was RMI's motor for the Lark surface-to-air—or rather, ship-to-air—missile that evolved as a far more substantial missile program. Like the Gorgon, the Lark was also developed for the Navy and had a projected range of almost 40 miles. The Lark was 14 ft long, or 18.5 ft with its twin 2,000-lb-total-thrust Aerojet solid-propellant canted boosters mounted in a peculiar detachable boxlike structure at the aft end of the missile. With booster assembly, the loaded weight of the Lark was 1,200 lb. The warhead weighed 100 lb and was activated by a proximity fuse. The RMI liquid-propellant motor weighed only 20 lb.

RMI's Lark project started slightly later than that of the Gorgon, originating as a U.S. Navy anti-kamikaze program in late 1944. The Lark was therefore conceived as a promising means of destroying Japanese aircraft engaged in kamikaze, or suicide dive-bombing attacks, against American shipping and for the destruction of other enemy planes making low-level antiship attacks or aircraft on reconnaissance missions.

On 6 February 1945, the Navy's newly established Jet Propelled Missiles Board formally approved the preliminary design configuration of the Lark and it was immediately assigned a high-priority rating by the Navy's vice chief of operations. The next month, the Fairchild Aircraft Company was awarded a BuAer contract for 100 of the missiles. Due to Fairchild's slow progress on the project, BuAer was led to award a backup contract in late June to the Consolidated Vultee Aircraft Corporation, later known as Convair. The primary differences between the Fairchild and Consolidated versions of the vehicle involved its controllability and stability; Fairchild chose flaps, and Consolidated went with variable-incidence wings. Fairchild—later, known as the Guided Missiles Division, Fairchild Engine and Airplane Corporation— became the prime contractor.

Meanwhile, BuAer awarded a contract to RMI (number NOa(s) 7070) to develop and furnish the liquid-propellant motor for the Lark. This development was initially easier than it first appeared because it was based upon engineering drawings supplied by the Naval Engineering Experiment Station (EES) of their earlier 350 lb-thrust motor for the Gorgon IIA, in addition to RMI's overall experience on their Gorgon motor developments. RMI particularly benefited from their experience with the then-unusual twin-rocket units for the Gorgon 3C. Indeed, RMI engineers rapidly developed a two-chambered red-fuming nitric acid/monoethylaniline unit for the Lark, consisting of 220-lb and 400-lb thrust chambers mounted together. The former chamber was scaled down from the basic Gorgon motor, the latter was a slightly uprated Gorgon IIA cylinder. The lower-thrust, longer-burning one was to serve as the sustainer for cruising.The larger-thrust, shorter-duration cylinder afforded a boost or boosts for extra velocity as the target approached. Both motors burned together at that point for a maximum thrust of 620 lb. The total firing time for the sustainer was 260 s, or 4.5 min. The 400-lb unit could also be pulsed, or turned on and off, at least eight times per minute. Lark's top speed was Mach 0.85. The missile was radiocontrolled to intercept enemy aircraft up to altitudes of 30,000 ft.

John Shesta's recollections also seem to point to an easy beginning of the Lark power plant when he remarked that "the Gorgon-Lark units were constructed by us [RMI] in accordance with their [the Navy and the main contractors] specifications. All we did was refine some of the manufacturing techniques and nozzle construction."[14] This applied only to the initial configuration of the Lark motor; its development was actually a bit more complicated than this. Robert Truax fortunately provided more details. He said that the Lark engine was

> developed as a cooperative venture between the Navy Bureau of Aeronautics, the Eclipse-Pioneer Division of the Bendix Aviation Corp., and Reaction Motors, Inc. The design was conceived and the general feasibility established by the Naval Jet Propulsion Project [under Truax] at the U.S. Naval Engineering Experiment Station, Annapolis, Maryland. Eclipse Pioneer produced a working prototype engine, using thrust cylinders supplied by Reaction Motors, Inc. Production was a combined effort of Reaction Motors and Eclipse-Pioneer...Injector improvements... increased the performance considerably over that obtained with the Gorgon design. Increased fabrication techniques gave increased cooling ability in the [RMI] regenerative jacket.[15]

What Truax should have added is that Eclipse-Pioneer was mainly responsible for Lark's rocket-driven turbopumps.

From here on, two basic models of the motor eventually evolved—the pressure-fed LR2-RM-12 and the turbopump-fed XLR-6-RM-4—although these were later designations. In the former, the propellants were fed by compressed air or other gas from plastic bags contained inside integral tanks. For the latter, the jet of exhaust gas issuing from the smaller 220-lb-thrust chamber supplied the propellants. This compact system also featured a turbine wheel and fuel and oxidizer impellers, with bearings, in a two-piece cast housing.

But both the conception and development of the Lark had started close to the war's end and it never became operational. Nevertheless, it was a significant missile development for the period with much potential. Follow-on work on the project was carried out that saw at least a dozen different models of the missile and variations of the basic engine.

The original manufacturer's designation for the Lark engine was the A622D2. It also received the Navy designation CML5N, which was essentially the same as that for the Gorgon. Other later Lark engine models include the LR2-RM-8, the LR2-RM-10, and the XLR6-RM-2.

RMI came to build about 500 Lark motors. Naturally, upon RMI's move from Pompton Plains to Lake Denmark in 1946, Lark motor development continued at the latter location, as did the first static-test firings. The Lark missile itself underwent its flight-test phase at the newly established California desert facility of the Naval Ordnance Test Station (NOTS) in Inyokern from 1946. There, multiple modes were used to test the Lark, now designated as KAQ-1. In one, the Lark was adapted and propelled by a "Big Richard" solid-propellant booster down a 450-ft-long elevated ramp until it became airborne. Big Richard was evidently the 14-in.-diam, multinozzle follow-on to the famous Tiny Tim air-to-surface World War II missile. It was most likely the largest air-to-surface unguided rocket ever developed for the U.S. military, although it was never placed in production. Its thrust must have been well over the 3,000 lb for 1 s of the smaller Tiny Tim.

During this period, a total of 62 Lark missiles were fired. In 1948, all Lark testing was shifted to the Naval Air Missile Test Center (NAMTAC) at coastal Point Mugu, also in California, which now permitted ship-launch tests from aboard the USS *Norton Sound* at sea. These tests, utilizing the missile's standard tandem-twin Aerojet solid-propellant boosters, continued until about 1953 or perhaps later. Most of the time, it was designated CTV-N-9, a control test vehicle, although it was additionally employed as an invaluable shipboard launching-crew training missile. Some Larks became training tools for the Air Force and Army Field Forces as well. The Lark then was designated KAQ-1 and finally, XSAM-2. At Point Mugu on 18 December 1950, a Lark scored a direct hit on an F6F drone aircraft—the first direct hit of an aircraft target by a U.S. guided missile. A Lark was also the first missile to launch from the Air Force's Cape Canaveral, Florida, test center. But the Lark soon gave way to more advanced ship-to-air weapons systems like the Terrier. Characteristically, in an attempt to enlarge its market, RMI envisioned other applications of the Lark motor, including a JATO for light liaison and command-type aircraft, but these never came about. A Lark, with its quaint, boxlike booster arrangement, is also exhibited at NASM's Udvar-Hazy Center.

Fig. 6.4   A Lark missile, with booster, on display at NASM's Udvar-Hazy Center. (Courtesy Frank H. Winter.)

MX-774: Predecessor of the Atlas

By the fall of 1945, RMI became involved in another early and very significant program, the MX-774 project that was a forerunner of the Atlas ICBM. It was then that the Army Air Forces Air Technical Service Command very boldly invited the U.S. aviation industry—because a missile industry was virtually nonexistent then—to submit proposals for a long-range intercontinental missile. This occurred in the wake of the recent appearance of Germany's infamous though revolutionary V-2 during the later years of World War II. Consequently, in January 1946, the Consolidated Vultee Aircraft Corp. submitted a proposal that was accepted and led initially to a one-year study that evolved their design for a single-stage test vehicle. By mid-1947, further funding was withdrawn, although Consolidated Vultee was granted permission to utilize unexpended funds to complete the assembly and test launches of three MX-774 vehicles. By June, the Army Air Forces let a contract to RMI for the MX-774 rocket motor, although they had already been working on this design since 1946.

To save money and precious time in this development, RMI cleverly expanded upon their 6000C-4 Black Betsy engine for the power plant for the MX-774, although it incorporated several new features that were technologically revolutionary in themselves. Like the 6000C-4, RMI's XLR-35-RM-1 engine for the MX-774 was configured as four separate chambers, but it could be considered as an advanced version of the 6000C-4. Its most outstanding feature was gimbaling, arranging the cylinders so they could swing back and forth to effect direct control of the missile.

The LOX and ethyl alcohol propellants were fed through two coaxial seals, one on each side of the engine, and also served as bearings. The slight tilting, or movement, of each chamber was controlled by a system of hydraulic cylinders. The smooth coordination of this system, as well as assuring that all four cylinders ignited simultaneously and arrived at the most reliable ignitors, presented enormous interrelated challenges, but they were successfully met.

The gimbaling idea for the XLR-35 seems to have originated jointly by late 1946 with RMI's senior project engineer Peter H. Palen, along with Shesta and Jimmy Wyld. But Palen may have been the lead inventor because he submitted a "Restricted Memorandum" on 1 January 1947 to Mason W. "Whit" Nesbitt, then RMI's chief engineer, as a "Disclosure of Invention—Application for Patent."[16] The inventors all signed their names to the document. This document also relates to the X8000C4 engine, but at this point the engine did not have a proper designation and was known internally at RMI as the X8000C4—meaning 8,000 lbs of thrust total from its four uprated 6000C-4 chambers; the "X" stood for "experimental," and the "C-4," merely meant four chambers. The subject of this memo is subtitled simply "Engine—Controllable Thrust

Fig. 6.5 Swiveling XLR-35 rocket engine for the MX-774 test missile. (Courtesy Frederick I. Ordway III collection, U.S. Space and Rocket Center, Huntsville, AL.)

Angle."[17] In any case, it embodies the gimbaling of what became the XLR-35.

Nesbitt's response was dismissive, however, and ran as follows: "It is known that several patents have [already] been granted on controllable thrust angle rocket engines," although he cited only one to Goddard in 1939.[18] It appears that Palen and his co-inventors were not granted their request to obtain a new patent.

As for Goddard, it is true that from as early as the 1931–1932 period he had independently devised and experimented with moveable tail vanes on his rocket, linked to an onboard gyroscope, to help stabilize rather than steer it. More importantly, he was later granted U.S. patent No. 2,183,311 of 12 December 1939 for a "Means of for Steering Aircraft." Despite the deceiving title for this patent—which was undoubtedly purposely applied by the secretive Goddard to help prevent the idea from falling into the wrong hands—this specification definitely does cover and depict the gimbaling of a liquid-propellant rocket motor. The patent certainly has nothing to do with conventional "aircraft." And there may well have been other rocket pioneers who had arrived at various schemes approximating a gimbaling system toward steering a liquid-fuel rocket. In any case, the XLR-35 appears to have been the earliest known liquid-propellant rocket in the U.S. or elsewhere to incorporate, and successfully flight test, such a rocket. Therefore, Whit Nesbitt may have been too harsh in his judgment.

The basic gimbaling principle as applied to the rocket—or reaction propulsion—can actually be found in several 19th-century concepts of reaction-propelled "flying machines," or aircraft, as a way to steer or control them. But these conceptual craft were usually designed to be propelled by the combustion of solid propellants like gunpowder. Again, the MX-774 may have been the first *liquid-propellant* rocket employing this method of steering in test flights.

As the literature of the history of rocketry demonstrates, the MX-774 vehicle itself also greatly benefitted from the contributions of the outstanding Belgium-born rocket designer then with Consolidated Vultee, Karel J. Bossart. Bossart introduced integral tanks for far greater lightness, while retaining the rocket's structural strength; the separation of the nosecone from the vehicle at peak altitude; and other highly significant design features.

The 31.58-ft-long, 30-in.-diam MX-774 vehicle underwent three flight tests at the White Sands Proving Grounds, New Mexico, on 13 July, 27 September, and 2 December 1948, respectively, with varying results. Its highest altitude, nearly 30 miles, was reached in its last flight. But what is most important about the MX-774 project is that its advances in rocket technology did not remain dormant after the last flight. On the contrary, these features were incorporated years later in the creation of America's first successful Intercontinental Ballistic Missile (ICBM), the Convair Atlas.[19]

Fig. 6.6 **A flight of the MX-774 test missile, predecessor of the Atlas ICBM. (Courtesy Frederick I. Ordway III collection, U.S. Space and Rocket Center, Huntsville, AL.)**

THE RASCAL AND HAWK MISSILES

By the spring of 1951, RMI almost became involved in two other rising and important new missile developments—the air-to-surface Rascal and the surface-to-air Hawk. For the former, they obtained a $300,000 Air Force contract covering the design, fabrication, and testing for "a prototype auxiliary power plant"; for the Hawk, Fairchild sent a letter of intent to RMI to purchase an RMI "Study of a Power Plant for the Hawk Missile."[20] But no details of either of these efforts are known, with one exception. Carold F. Bjork, a propellant chemist at Redstone Arsenal, studied various RMI proposals, including one on the Hawk, and did not feel (by circa 1955) that substituting a liquid-propellant system for that missile offered any significant advantages over the missile's solid fuel. In any case, RMI did not succeed in becoming the main contractors for either missile.

ENTER THE PACKAGED LIQUID-PROPELLANT MOTORS: THE SPARROW III

As the Lark missile program wound down in the early 1950s, RMI created a minor revolution in rocket technology that they called the "packaged liquid rocket engine." This was a pressure-fed liquid-rocket motor in which the storable propellant was hermetically placed (factory-loaded) in a lightweight sealed tank. Thus, this was a ready-made, ready-to-go rocket unit, complete with built-in pressure-feed and igniter systems. RMI's milestone packaged engine successfully produced 5,000 lb of thrust for 2.5 s—a high-thrust, quick-duration performance that was entirely suitable for a short-range missile. From this leap in technology arose a series of propulsion packages for a variety of missiles, some not successful and others, principally the air-to-surface Bullpup series, phenomenally successful.

It is true that during World War II, the German X-4 and Taifun air-to-air liquid-propellant missiles were early forms of packaged liquid-propellant missiles, although RMI and RMD's versions incorporated modern and more advanced materials and methods of mass production.

The earliest of these postwar projects was the Sparrow III air-to-air missile. Normally, the Sparrow was powered by a solid-propellant motor, but BuAer, aware of RMI's progress with experimental packaged liquids, submitted a request for a proposal to explore such a unit for the missile that armed Navy fighters and other aircraft. Among the foreseen advantages were superior performance and better operability in lower-temperature environments. This was pursued by RMI, and led to the development, from late 1956, of their XLR-44-RM-2 (or Guardian 1) motor, as a kind of company trade name. Guardian's development stretched into mid-1958, a few months after RMI had merged with the Thiokol Chemical Corp. to become its Reaction Motors Division (RMD).

According to one RMI/RMD engineer involved, Arthur "Art" Sherman, BuAer favored the packaged-liquid substitution for the solid-fueled Sparrow, whereas the Navy's Bureau of Weapons (BuWeps) did not. Raytheon, the missile's contractor, was similarly against it. This led to a design competition. Among other blunt negative remarks uttered by the Raytheon representative at the opening meeting, were: "I do not want to change the thrust. I do not want leaks. I don't like liquids. I don't want your lousy plumbing. I do not want the center of gravity to shift."[21] The RMI men were "dumbfounded," Sherman noted.[22] "They left the meeting in total despondency."[23]

As a consequence, new competition ground rules were set. The new motor had to be "completely interchangeable" with the old Aerojet one.[24] This led to a very unconventional design, patented by Sherman and some coworkers, and became the basis for both the Sparrow III's Guardian I and follow-on packaged-liquid Bullpup engines. Guardian I's thrust and duration remained the same, but its specific impulse was higher, which meant increased range. In addition, they could now operate at the higher-supersonic level—and therefore, higher operating temperatures—of more advanced planes carrying the missile. By contrast, solid systems were at a marked disadvantage, as supersonic heating could potentially lead to propellant cracks that suddenly increased the burning area, causing possible explosions.

Sparrow III's liquid-engine development was hardly trouble free. One of the development bugs was too much ignition delay. On one harrowing occasion, the pilot of a McDonnell Douglas F4H Phantom jet aircraft carrying the missile during a trial launch pressed the ignition button as an RMD field representative watched from the rear cockpit. But an ignition delay caused the radar-homing missile to suddenly swing around and head toward the Phantom. The Sparrow found its target, but very luckily, the strike against the F4H was not fatal, and the pilot landed as skillfully as he could at the nearest available airstrip. The plane's radar nose was damaged, but otherwise the aircraft was intact. This incident almost cancelled the project, which was set back.

Another headache was the missile's temperamental guidance system. The "kick," or sudden acceleration shock from the missile's liquid-rocket motor, often interfered with the guidance; the solid motor ran a little smoother on takeoff. Rocket engineer George P. Sutton terms Sparrow III's initial ignition-spike phenomenon "a momentary overpressure."[25] RMD toned down the high-pressure surge, although this situation was never completely solved. Meanwhile, a special factory was established by Thiokol near its headquarters in Bristol, Pennsylvania, for the anticipated mass production of the Guardian. Veteran managers from the automobile industry were even hired to ensure that their manufacture would be as efficient and rapid as possible.

In the end, Navy records indicate that the acceleration problem was such that it was determined that the missiles powered by packaged RMD liquid-propellant motors were "Not suitable for fleet use."[26] However, the main problem had been the missile's touchy guidance system. On 30 June 1960, the contract was terminated, even though RMD had already produced some 400 Guardian I engines. The Navy then reverted to an improved Aerojet solid unit. Nevertheless, technologically at least, Guardian I prepared the way to Guardian II to power the Bullpup.

BULLPUP A

Back in September 1958—barely a month after the Sparrow III engine had started up, and just five months after the merger of RMI with Thiokol—BuAer contracted with RMD to develop a packaged unit for the air-to-surface Bullpup A missile. Bullpup was an outgrowth of an urgent Korean War need, then expressed by the Navy, for a precision inexpensive air-to-surface weapon that could be launched by carrier-based aircraft. During that conflict, U.S. airpower faced great difficulty in destroying targets that required precise aiming and were often heavily defended, like large and strategic bridges, as well as pill-boxes, tanks, and truck convoys. But it was not until 1953 that a request for proposal (RFP) was issued to the American aerospace industry for the development of such a weapon. In May (some sources say April) 1954, the Martin Company's Orlando, Florida, division was selected from 14 proposal submissions. Aerojet-General was subsequently awarded the contract for the missile's MK 8 solid-propellant power plant delivering about 25,000 lb of thrust, attached to a simple derivative of a standard 250-lb bomb as warhead. It became the first U.S. mass-produced command-guidance guided missile, operational in April 1959.

The newly designated ASM-N-7 Bullpup was simple and inexpensive for its short range, about 12 miles. In service, it was optically tracked and manually guided by the pilot using a joystick. The pilot in his cockpit aimed the missile by imparting left/right and up/down directions with the joystick. Admittedly, though, this was not easy to accomplish while also flying the aircraft. Tracking was facilitated by pyrotechnic flares in the missile's boat tail. However, the guidance and the missile's short range had their limitations.

From its experiences with the Sparrow III, the Navy evidently was still much taken with the performance advantages of packaged liquids over solids, even if the Sparrow's oversensitive guidance system had led to the cessation of the Sparrow III's packaged-liquids contract. Apart from this, the Bullpup was an altogether different, heavier, and more robust missile, and its guidance was a lot less sophisticated as well as "quirky." The long-term storage and operational superiority in wider latitudes of temperatures of packaged liquids was especially appealing, considering that Bullpups were being deployed in the hot and humid climates of Southeast Asia. Packaged-liquid systems were also less smoky than the Aerojet solid units that could interfere with pilot sighting. For all these reasons and others, the Navy thus surged ahead in requesting a needed redesign of their basic Bullpup to be mated with RMD's promising packaged-liquid technology.

Furthermore, the RMD Sparrow III's Guardian I experience, especially on the manufacturing side—even though the mass manufacturing had been premature—had prepared them well to enter into the quick turnaround time for gearing-up and production of the new Bullpup motors. RMD was fully ready to accommodate the new improved Bullpup variant, the ASM-N-7a, or Bullpup A. This was the version of the missile for which RMD received its prepackaged-liquid contract.

The switch to this motor greatly extended Bullpup's range while still keeping costs relatively low; the model additionally featured a much needed improved guidance system and warhead. Bullpup A entered U.S. Navy fleet

service in 1960 and was immediately deployed in the Vietnam War, following
the example of its solid-fuel predecessor.

In 1962, the packaged-liquid variant of the missile was redesignated AGM-
12B Bullpup-A—or simply, AGM-12, AGM meaning Aerial Guided
Munition. AGM-12 Bullpup armed a variety of patrol and attack aircraft such
as the A-4 Skyhawk, A-6 Intruder, P-3 Orion, F-105 Thunderchief, and F-4
Phantom, among others.

As for Bullpup motors, the first was the LR-58 (also called Guardian II by
RMD) of 12,000-lb thrust for 1.9 s. It was 40.47 in. long, a foot in diameter,
and had a dry weight of 92 lb and a loaded weight of 203 lb. The basic design
elements of the LR-44 (Guardian I) and LR-58 were identical except for size.
Only 14 months after the contract was let to RMD, in November 1959, LR-58
reached the production stage for installation into the missile.

Both the LR-58, and its later, larger, and more powerful improved version,
the LR-62, burned the hypergolic red-fuming nitric acid with mixed amine
(known as MAF-1). The propellant thus differed from that in the Sparrow III.
The exhaust of the burning of a double-base solid-propellant, short-duration
cartridge provided the pressurizing gas to the propellant tanks. This was similar to the German Taifun missile of World War II, although the Bullpup motors
were regeneratively cooled—by the amine fuel that was not found in the
regeneratively cooled Taifun. Another feature unique to the Bullpup motor
(and thus also not found in the Taifun) was the sprayed-on insulation coating
of zirconium oxide, known commercially as Rokide-Z, over the throat and
nozzle that further helped cool the motor. More significantly, neither the
Taifun nor the X-4 were constructed for long-term storage, in terms of propellant choices and construction. Propellant-loading inlets were welded over at

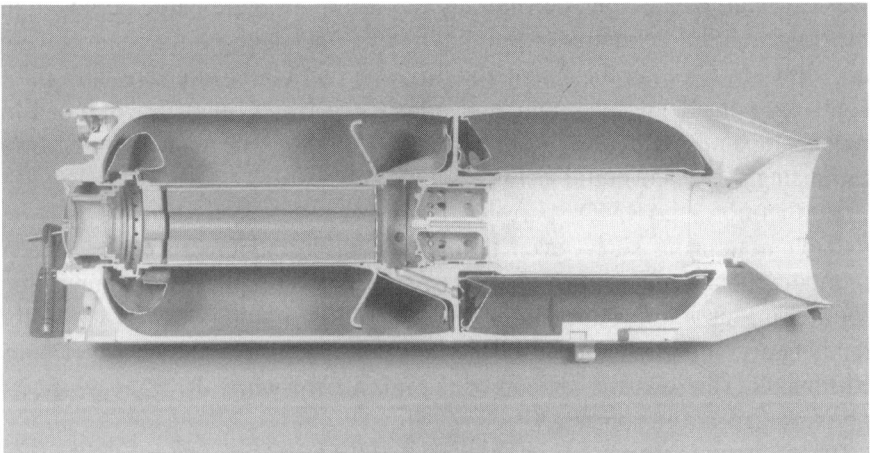

Fig. 6.7   Cutaway of the LR-58 packaged liquid-propellant rocket motor for the Bullpup
A missile. (Courtesy Smithsonian Institution, 81-5775.)

the RMD factory for the Bullpups; that guaranteed a storage life of five years over a temperature spread of −80°F to 165°F.

The single moving part in the basic Bullpup motor was a pistonlike initiator cartridge that moved only a fraction of an inch in order to shear off small seals, allowing the propellants to flow and activate the gas-generator charge. Full thrust was achieved within a tenth of a second after the electrical squib triggered the initiator.

Nevertheless, rocket engineer Sutton rightly remarks that the development of the Bullpup motor was not entirely trouble free. He cites fuel leaks early in the development of the motor that caused the plastic polymer material for sealing the fuel-fill plug to dissolve. This was solved when RMD engineers fortunately found a more resilient polymer. The sloshing of fuel had likewise complicated matters, but baffles within the tanks soon alleviated that problem as well. Apart from such growing pains, very rigorous quality checks were instituted throughout the Bullpup manufacturing processes, and hundreds of developmental static tests were run "without major hurdles," according to Davies.[27]

That said, there was one serious accident that very nearly ended the program—and potentially, RMI/RMD itself.

It seems that during a test launch of a Bullpup A from a Douglas A-4D Skyhawk in flight in the Point Mugu area, the missile broke up upon ignition while it was still on the launch rails attached to the aircraft. The Skyhawk was severely damaged, including most of the aircraft's flaps on one side, although the pilot was uninjured and with great coolness and skill was able to return safely to base. Debris from the rocket engine was recovered and revealed that the burst had been caused by a structural failure. Because static testings of the engine had gone well, the near-fatal accident was an alarming shock to both RMD and the Navy. A formal inquiry was made, and for a few weeks the LR-58 program was hanging in the balance, as was RMD. But the end result was that RMD recommended the substitution of a higher-strength aluminum alloy over a weaker one as part of the motor that did not increase the weight. This proposal was accepted by the Navy, resulting in a new variant, the LR58-RM-4, and flight tests after this were entirely successful.

As for the manufacturing of the Bullpup engines, the Navy provided funding for retooling the Bristol facility—a former World War II shell-loading factory—for producing the LR-58 motors. At its height, this 116,000-sq-ft plant was devoted almost exclusively to the mass production of the packaged liquid-rocket engines. In fact, a *Wall Street Journal* article for 27 August 1958 reported that the Bristol plant was specifically acquired by Thiokol that month—very shortly after the Thiokol–RMI merger—for the production of the first of mass-produced packaged liquids, the LR-44 Sparrow III motors.

This plant was adjacent to Thiokol's corporate headquarters, also in Bristol. Alan R. Maier became the program manager for the packaged liquids. Again, quality checks were highly rigorous; randomly selected production samples were subject to everything from 40-ft drop tests and 30-minute bonfires, to being fired upon by 45-caliber tracer bullets to simulate battle damage. The engines characteristically passed all these arduous trials. In fact, the Bullpup had attained the distinction of being the first mass-produced, air-surface, command-guided missile.

BULLPUP B

In 1960, BuWeps awarded RMD a $2-million contract for the engine for Bullpup B, also known as ASM-N-7b, which was a significantly enlarged version of the missile. Notably, it was armed with a 1000-lb warhead and featured the new, more powerful Thiokol-RMD LR-62 rocket engine, or Guardian III, of 30,000-lb-thrust for 2.3 s. The engine measured about 5 ft long, was 17.3 in. in diameter, and weighed 205 lb dry and 563 lb loaded. Because Bullpup B was far heavier than Bullpup A, it was drop-launched like a bomb from its carrier aircraft, although it had been connected to the plane by a lanyard to enable ignition safely at 15 or 20 ft below the aircraft. At the instant of ignition, the lanyard was disconnected, and the missile took off.

Whereas Bullpup A motors had been produced at Bristol, the Bullpup B LR-2 engines were fabricated in two rented buildings at RMD's additional

Fig. 6.8   Production line of the LR-62 packaged engines for the Bullpup B missile. (Courtesy the Frederick I. Ordway III collection, U.S. Space and Rocket Center, Huntsville, AL.)

Fig. 6.9 Mounting Bullpup missile on a North American FJ Fury fighter-bomber aircraft. (Courtesy Smithsonian Institution, 917050.)

new facilities at Rockaway, New Jersey. For safety's sake, "fill and load" operations were carried out at RMD's test area at Lake Denmark. Lake Denmark had also been the site of the extremely intensive earlier initial testing phases of both the LR-58 and LR-62, the latter commencing its testing in 1962.

As a truly remarkable measure of the extent and success of this testing, out of a total of some 2,695 LR-58 firings carried out between 1961 and 1964 (which included both ground static trials and flight-launch tests), there were only three malfunctions. Testing figures for the LR-62 are not available, but must have been just as dramatic. For certain, reliability for both LR-58 and LR-62 achieved a nearly perfect rating of 0.9972 percent.

The Bullpup B missile became operational in 1964, and its new power plant afforded the weapon a still larger dynamic range, although its effective range remained essentially the same. Bullpup B was eventually also adopted by the U.S. Marine Corps and the U.S. Air Force.

WHITE LANCE INPUTS INTO THE BULLPUP

Meanwhile, the Air Force had earlier contracted with the Martin Company to develop an advanced derivative of Bullpup, designated the GAM-79 White Lance. This missile was to use an RMD LR44-RM-2 engine and to have a nuclear warhead option. But as the development of White Lance would consume some time, the Air Force procured the unmodified ASM-N-7 Bullpup as an interim missile and designated it GAM-83. The GAM-79 White Lance design was gradually incorporated into the ongoing standard Bullpup development. Therefore, this Air Force version was essentially the same as the Navy's ASM-N-7a Bullpup A, differing only in its improved guidance system.

The GAM-79 White Lance designations were thus withdrawn, and this missile became the Air Force's GAM-83A Bullpup, to distinguish it from the Navy's version. To further complicate matters, the Air Force evolved its "Nuclear Bullpup" that received the designation GAM-83B.[28]

BULLPUPS PROCURED BY OTHER NATIONS

Apart from this expansion of the use of the Bullpup by the U.S. military services, other nations came to purchase the weapon as well. It was acquired by Britain's Royal Navy Fleet Air Arm and also the naval air forces of Australia, Denmark, Greece, Israel, Norway, Taiwan, and Turkey. No wonder the enormous total of 50,000 RMD Bullpup engines were manufactured—the largest number ever in the history of rocketry—although changing military needs saw the eventual slowing, then halt, of Bullpup-engine orders by the mid- to late 1960s. The overall Bullpup engine venture for RMD had thus been exceptionally lucrative for them and the orders came in, in different "buys." For instance, the *Wall Street Journal* of 6 July 1965 reported the very last large buy of Bullpup B rocket engines at $10,600,000.

LARGER PACKAGED-LIQUID EFFORTS

Back on 29 October 1958, RMD had presented a well-publicized demonstration firing of the 50,000-lb-thrust version of what they hailed as "the largest member of RMD's packaged-liquid family."[29] They also stated, perhaps presumptively, that this engine was "a significant contribution to the national defense effort," although this big engine never entered service.[30] This super-advanced liquid-packaged engine was hardly the largest mission on RMD's horizon, because the next month, engineers Hartmann J. Kircher and J. Rossetto co-authored a confidential theoretical study titled, *A 2,400 Mile Range Polaris Missile with a Prepackaged Liquid Propellant Rocket Powerplant.* But the study did not lead anywhere, and the submarine-launched, medium-range Polaris ballistic missile retained its solid-propellant system.[31]

THE BOMARC BOOSTER

Meanwhile, a few other, far less spectacular Navy, Air Force, and Army missile projects came and went at RMD. One was the Air Force's supersonic, long-range, ramjet-powered, area-defense anti-aircraft Bomarc. This important missile and missile-defense system had a long prehistory. Aerojet came to produce its powerful LR-59 liquid-propellant booster, even though when it became operational in 1959, there was great concern that a liquid-propellant booster arrangement was not an optimal solution. It took time to fuel up the boosters, and also the propellants were hypergolic (hydrazine and nitric acid);

these were very risky to handle and caused some accidents. Therefore, even at this juncture, plans were underway for a Bomarc-B, utilizing newly emerging large-scale solid-propellant technology.

Yet, RMD's predecessor, RMI, had already been deeply involved in the liquid-propellant development phase of the Bomarc in the early 1950s in securing Air Force contracts for a potential alternate booster of greater simplicity and storable propellants. This led to the development of the 50,000-lb-thrust, 25-s, XLR-77-RM-1 engine—one of their largest. Their exotic propellant choice for this engine was inhibited red-fuming nitric acid (IRFNA) and anhydrous ammonia converted into a hypergolic combination with the additive of lithium. Beyond this, RMI also made a rare deviation from their years of devotion to regenerative cooling. Instead, they opted for an uncooled, ceramic-lined, steel-shell thrust chamber and other approaches, all undertaken with close coordination between RMI and Bomarc's main contractor, the Boeing Airplane Company.

Then, when the XLR-77 had reached its full-scale engine phase, about 1954, the large 555-lb (dry weight), almost 4-ft-long, 22-in.-diam engine was test fired repeatedly at Lake Denmark more than 200 times. Ultimately, the contractor and the Air Force selected the Aerojet LR-59, which was also ceramic-lined and featured full-engine gimbaling, greatly facilitating the missile's control and probably helping Aerojet win its contract. Even so, the liquid-propellant rocket-boosted version of the Bomarc did not last long. It was soon supplanted by Bomarc B with its huge and uncomplicated—although highly effective—solid-propellant Thiokol M5 50,000-lb-thrust booster with a duration of 30 s. Here, it is important to note that prior to acquiring its Reaction Motors Division in 1958, Thiokol had also been a leading pioneer in the development and production of large-scale, high-performance solid-propellant rocket motors. Nonetheless, as Chapter 10 will show, the history of the XLR-77 did not end there. It briefly surfaced again a few years later as a potential powerful "building-block" engine for a supersonic Air Force rocket sled, although that project, too, never materialized.

Corvus

An all too short missile program involving RMI was the air-to-surface, long-range Corvus. In 1956, the Navy contracted the Temco Aircraft Corporation to develop the weapon, followed soon after by a separate contract to RMI for the missile's storable motor, later known as the XLR-48, or Patriot. Despite considerable progress and huge sums spent on the Patriot, however, the Navy cancelled the entire project on 18 July 1960. Their explanation was brief: "Corvus was more limited in its application than other systems now under development."[31] Today, an XLR-48 Patriot missile engine is on exhibit at NASM's Udvar-Hazy Center.

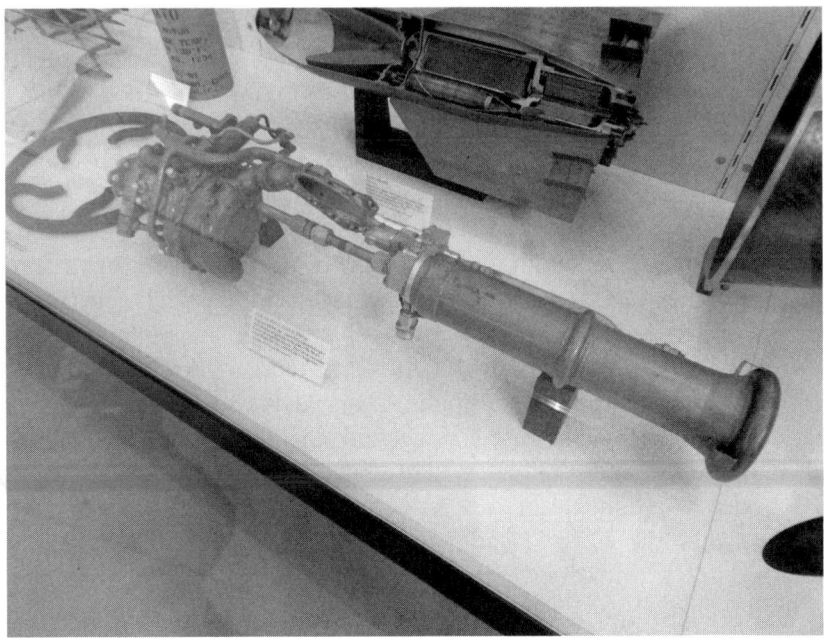

Fig. 6.10   The XLR-48, or Patriot rocket motor, for the Corvus missile, on display at
NASM's Udvar-Hazy Center. (Courtesy Frank H. Winter.)

LANCE

Another quickly passing missile-engine project undertaken by RMD was
the Army's short-range Lance. From the late 1950s, the Army Ballistic Missile
Agency (ABM), at Huntsville, Alabama, started conceptualizing this mobile
field-artillery, tactical surface-to-surface (tactical ballistic) missile, originally
designated Missile A. This evolved into the more advanced Missile B—later
called the Lance—that became a replacement for the old Honest John and
Sergeant solid-propellant missiles. But it was not until 1962 that the Ling-
Temco-Vought Aerospace Corporation (LTV) was awarded the primary contract for this complex weapon system. Facing stiff competition for a
packaged-liquid system for Lance, RMD was beaten out in the end by
Rocketdyne, with their own variable-thrust version of a prepackaged engine.
But the Lance project was somewhat salvaged, as RMD continued to receive
a series of lesser contracts for further refinements for the Lance propulsion-
expulsion and related systems, and this additional work on the Lance lasted
for several years.

CONDOR

Finally, there was the Condor program, which was among the more exotic
and ambitious of RMI and RMD's missile projects. Condor was originally

conceived as a high-altitude boost-and-cruise missile. Later, this program was switched to a low-altitude cruise development. In mid-October 1966, a very substantial, $15-million contract was awarded by North American Aviation, Inc. to RMD to develop "an advanced pre-packaged, liquid-propellant rocket engine," that was officially designated the XLR-68-RM-2.[32] RMD entertained high hopes that this engine would be superior to those in the Bullpups.

What made this project unique was its high-energy chlorine trifluoride and mixed-amine propellant that produced a new rocket flame color—green (although earlier experimental boron propellants tried by RMI also emitted beautiful green flames). RMD corporate managers were initially so excited about the promise of this innovative propellant that they invested several hundred thousand dollars in a new test stand that had unique scrubbing chambers to remove the highly acidic hydrogen fluoride (HF) from its exhaust gases.

But the oxidizer's excessive corrosive character was not the principal problem. Rather, as pointed out years later by RMD's Harold Davies, who had been assigned to the Condor program, it was "the amazing ability of chlorine trifluoride at raised temperatures to combine vigorously with everything in sight: fuel, chamber walls, injector, or what have you. In a packaged liquid it presented the worst of both worlds, not only a very difficult injector and thrust chamber design, but the need to completely isolate the oxidizer from the hot pressurizing gas."[33] Ed Govignon, another RMD engineer who worked on this project, recalled that the high-altitude boost-then-cruise thrust ratings for the Condor engine were 3,000–4,000 lb down to 200 lb, respectively. He added that the motor had a metallic reversible toothpaste-tubelike bladder for expelling the fuel, but "we were never able to get it it to squeeze properly."[34]

As may be expected, all these vexatious chemical and technological challenges were overwhelming, soon forcing the Navy to give up on the storable-liquid approach to this particular missile. In the fall of 1967, the Navy thus transferred the propulsion-system development to North American's Solid Propellant Rocket Division after that company's Rocketdyne (liquid-propellant) Division had experienced like problems with this "extremely energetic oxidizer."[35]

It turned out that Condor was RMD's last major missile effort. Former Thiokol administrator Harold W. Ritchey later sadly lamented that at this point in the late 1960s, since the highly lucrative Bullpup program had ended, the "big boys" in rocketry—Rocketdyne and Aerojet on the West Coast—were producing all the country's larger engines,while RMD was now reduced to conducting only minor projects. Condor, he asserted, was "the last gasp of the original innovators of rocket technology—the first in, and now to be the first out."[36]

He was not far wrong. RMD would end its history about five years later. Yet, it is also correct that they had also been among "the first in." RMI/RMD's veterans could very proudly look back at several outstanding firsts—not only in aviation and missile technology—but in space exploration. As covered in Chapter 7, back in the early 1950s, before the launch of Sputnik 1 opened the Space Age, RMI made tremendous strides in contributing to the development of America's first rockets to reach space.

ENDNOTES

1. Delmar S. Fahrney, "The Genesis of the Cruise Missile," *Astronautics & Aeronautics*, Vol. 20, Jan. 1982, p. 38.
2. Delmar S. Fahrney, "Guided Missile Development by [the] Navy's BuAer During WW 2," manuscript (also known as "The Fahrney Manuscript"), unpublished, Part 2, p. 468, copy in NASM Archives (copy also available in the National Archives).
3. Fahrney, "Guided Missile Development," p. 477.
4. "Appendix 'A' to Enclosure 'D'—Project Gorgon," 17 May 1945, National Archives, RG 218, Entry 92, box 47, folder unknown.
5. Letter from John A. Pethick to BuAer, 10 Aug. 1944, in National Archives, RG 72, BuAer, General Correspondence, box 4172, file 8586.
6. Ibid.
7. Ibid.
8. Ibid.
9. Lloyd Mallan, "Breaking the Star Barrier," typed article in author's collection, original source unknown.
10. [RMI] "Shop Weekly Report," Louis F. Arata to John Shesta, 21 Feb. 1946, "Reaction Motors, Inc." collection, NASM, Box 1, folder 2.
11. Ibid.
12. Ibid.
13. Ibid.
14. Shesta, "Reaction Motors, Inc.: A Memoir," in Frank H. Winter and Frederick I. Ordway III, *Pioneering American Rocketry: The Reaction Motors, Inc. (RMI) Story, 1941–1972*, AAS History Series, Vol. 44, Univelt, Inc., San Diego, 2015, p. 116; Sheets interview by Winter.
15. Robert C. Truax, "Development of the LR6-RM-2 Rocket Engine," paper presented at the ARS Semi-Annual Meeting, San Francisco, 10–13 June 1957, paper 440-57, p. 1, copy in "Reaction Motors, Inc.—LR-6" file, NASM.
16. Memo from P.H. Palen to M.W. Nesbitt, 1 Jan. 1947, for "Disclosure of Invention..." in Lovell Lawrence Jr. papers, NASM, box number and file number unknown.
17. Ibid.
18. Ibid.
19. Space does not permit further details on the MX-774 and its XLR-35 engine, but some of these may be found in Winter and Frederick I. Ordway III, *Pioneering American Rocketry*, pp. 117–120, et seq.
20. [RMI], "New Contracts...April 1, 1951 thru April 30, 1951," Lovell Lawrence Jr. Papers, NASM, box 5, folder 5.
21. Telephone interview with Arthur Sherman by Frank H. Winter, 6 May 1987.
22. Ibid.
23. Ibid.
24. Ibid.

25. George P. Sutton, *History of Liquid Propellant Rocket Engines*, AIAA, Reston, VA, 2006, p. 321.
26. Winter and Ordway III, *Pioneering American Rocketry*, p. 175.
27. Harold Davies, "Reaction Motors (Thiokol) Family of Packaged Liquid Rocket Engines," *Journal of Spacecraft and Rockets*, Vol. 44, Nov.–Dec. 2007, pp. 1278–1279. For further detailed descriptions of the Bullpup motors consult Sutton, *History of Liquid Propellant*, pp. 321–324, Winter and Ordway III, *Pioneering American Rocketry*, pp. 176–180; and especially the firsthand, very comprehensive article by Davies, "Reaction Motors (Thiokol) Family," pp. 1271–1284.
28. Unfortunately, space does not permit a discussion of the operational history of the Bullpups during the Vietnam War, but among the many works that may be consulted are: Jacob Staaveren, *Gradual Failure: The Air War Over North Vietnam 1965–1966*, Air Force History and Museum Program, U.S. Air Force, Washington, DC, 2002; Peter Mersky and Jim Laurier, *U.S. Navy and Marine Corps A-4 Skyhawk Units of the Vietnam War 1963–1973*, Osprey Publishing, Oxford, U.K., 2007; and Maj. A.J.C. Lavall, gen. ed., *The Tale of Two Bridges and The Battle for the Skies Over North Vietnam*, Office of Air Force History, United States Air Force, Washington, DC, 1985.
29. "RMD Fires Its Largest Packaged Liquid Rocket Engine," *The RMI Rocket*, Vol. 9, Nov. 1958, pp. 1, 6.
30. Ibid.
31. Specifically, this report is S. Lehrer and H.J. Kircher, *Study of a Novel Liquid Rocket Launch Vehicle Concept – TR 3767*, Reaction Motors Division, Thiokol Chemical Corporation, Dover, NJ, 21 Dec. 1960.
32. Ordway III, in Winter and Ordway III, *Pioneering American Rocketry*, p. 151.
33. Letters from Harold Davies to Frederick I. Ordway III, 9 July and 21 Aug. 1986.
34. Telephone interview with Edward C. Govignon by Frank H. Winter, 17 April 1987.
35. See also "Condor Engine," *Aviation Week*, Vol. 85, 17 Oct. 1966, p. 30; and "Problems Balk Condor Production Plans," *Aviation Week*, Vol. 90, 17 April 1969, p. 18.
36. Letters or telephone interview, Harold W. Ritchey by Frederick I. Ordway III, 1985–1986, cited in Winter and Ordway III, *Pioneering American Rocketry*, p. 170.

Chapter 7

RMI ENTERS SPACE: THE VIKING

"Takeoff! The rocket has broken loose! The rocket is flying!"

–NRL official report on unexpected launch of the tied-down
Viking No. 8.

V-2 ROCKET FIRST INTO SPACE

The first successful launch of the wartime German A-4 rocket (later called the V-2) on 3 October 1942 is often claimed as the first rocket launch into space. But it went up to only about 60 mi in altitude, although the officially accepted boundary between air and space is 100 km, or 62 mi. Thus, the distinction of being the first ever space rocket more likely may belong to a later and far less heralded test of an A-4 launched on 20 June 1944; that rocket made a vertical flight up to 174.6 km, or 108.4 mi. The A-4, of course, was designed to normally fly in a horizontal trajectory to achieve great range as a bombardment missile.[1]

After the war, a captured V-2 with scientific instruments in its nose in lieu of a warhead was launched from the White Sands Proving Ground in New Mexico on 28 June 1946. It traveled up to 66.8 mi—and therefore into space. This was number 6 of some 67 captured and rebuilt V-2s fired by the Americans during the early postwar period in order to gain knowledge and experience in launching large-scale liquid-propellant rockets.

These rockets also provided the U.S. scientific community with its most fortuitous opportunity to send up scientific instruments to explore the upper atmosphere and above into space itself to record atmospheric compositions and densities, pressures, the nature of cosmic rays, etc. This was the beginning of space science. Indeed, several other V-2s in this series also penetrated space.

The term "rebuilt" is used here because the Americans discovered that most of the V-2s they captured were far from complete and lacked parts that had to be rebuilt based on American (nonmetric) standards. Also, not all the captured V-2s were launched at White Sands; a few were fired at the newly opened Florida Missile Testing Range from Cape Canaveral on 24 and 29 July 1950 as part of Project Bumper. They were the first launched from this now very famous launch site. Additionally, practically all U.S. V-2s were fired vertically, toward the upper atmosphere, and were not tested for range.

In the meantime, the Russians were doing exactly the same thing with *their* captured wartime German V-2 rockets. Their flights were then top secret, and those in the West learned about them many years later. One of their slightly modified V-2s, known as the R-7 and fitted with two side containers carrying instruments, reached an altitude of 68 mi on 24 May 1949. The basic V-2, then, was the world's first single-stage rocket to reach space.

On 24 February 1949, a captured V-2 with a much smaller, American-built WAC-Corporal liquid-propellant rocket on top of it as its second stage set another major aerospace milestone by becoming the first two-stage liquid-propellant rocket to reach space. This Project Bumper flight attained the then-astounding record of a 250-mi (some say 242-mi) altitude, putting it into deep space; this peak-altitude record stood for many years. The separation of the two rockets took place 20 miles up.

First Viking Rocket into Space

On 11 May1950, the Viking sounding rocket No. 4—powered by a 20,000-lb-thrust RMI XLR-10 rocket engine—reached some 105 mi, technically becoming the first American single-stage, liquid-propellant rocket to reach space. This was truly an astonishing achievement, considering that RMI

Fig. 7.1   Flight of Viking rocket No. 4 into space, 11 May 1950, from USS *Norton Sound*. (Courtesy Frank H. Winter collection.)

was less than 10 years old and had grown out of the weekend hobby of the four young men who had founded it. Moreover, the Viking project was RMI's very first venture into developing the power plant for an upper-atmospheric sounding vehicle, to say nothing of one that could penetrate space.

Up until now, of course, all of RMI's products were essentially terrestrially based applications. The Viking rocket went on to reach far greater heights than its first space flight of May 1950. Its last two iterations served as test vehicles for America's first artificial satellite-launch vehicle, Vanguard—preparation for America's entry into the Space Age.

ORIGINS OF THE VIKING ROCKET ENGINE

Lawrence's existing *Daily Log* reveals that the X-10 Viking project began with an office visit to RMI at precisely 10:30 a.m. on 4 March 1946. The brief *Log* entry reads: "Two men from the Navy Dept.—Cmdr. [L.M.] Slack and Cmdr. R.E. Doll, and two civilians from the National Research Laboratory [that is, the Naval Research Laboratory, or NRL]—Mr. Milton W. Rosen and Mr. C.H. [Carl Harrison] Smith—flew in from Washington to discuss possibilities of a new high altitude rocket. Conference was attended by Mr. [John A.] Pethick, Mr. [John] Shesta and Mr. [James H.] Wyld."[2]

Shesta and Wyld must have been delighted, because in their earliest ARS days they had worked independently toward the designs of their own modest-sounding rocket projects; Wyld's project was directly linked to his evolution of the regeneratively cooled motor. In 1935 and before he had even joined the ARS, Wyld did not think a meteorological rocket would have been feasible at all without a motor that could maintain a useful duration. Cooling such a motor as efficiently as possible, he therefore felt, was *the* key. Without a proven method of cooling, the liquid-propellant rocket was not at all viable for any useful application. "Everything," Wyld wrote to a friend in a letter of 28 January 1936, "depends on getting a motor that will stay together for a full minute and burn the fuel with maximum efficiency. The motor design is not yet entirely worked out, but is to have 'regenerative cooling.'"[3]

ORIGINS OF THE SOUNDING ROCKET

Neither Carl Smith nor "Milt" Rosen—as he was called by his friends—nor Shesta, nor Wyld were the true originators of the sounding-rocket idea. Rather, the fundamental concept was far older. Goddard had conceived a plan before World War I for a preliminary vehicle to explore near space and beyond the usual 20-mi limit of sounding balloons, as preparatory to exploring space itself with a much larger and more sophisticated rocket. His later, more polished design that was based upon his theories and extensive experimentation became the main subject of his seminal treatise, *A Method of Reaching Extreme Altitudes.* As discussed in Chapter 1, this classic work was published in 1919 by the Smithsonian Institution and released in January 1920. The upper-atmospheric sounding rocket described in Goddard's treatise was a solid-fuel type.

Quite a few years earlier than Goddard's efforts, the French physician Dr. André Bing was granted Belgian patent No. 236,377 of 10 June 1911 for an "apparatus for exploring upper atmospheric regions however rarified."[4] Thus, the physician Bing may well have conceived the world's first sounding rocket. Bing's application went further. He considered "traveling beyond the Earth's atmosphere" with "successive [that is, staged] rockets" that used nuclear energy, although he did not discuss in detail how this would work.[5] There is no evidence that Bing or Goddard knew of each other.

Goddard's very cautious approach to exploring space, starting with a sounding rocket, seemed entirely logical. But Goddard was very secretive, and there is some controversy as to whether he was surreptitiously really using the Smithsonian to help him fund his foundation work, as his longer-range life-long goal had always been to find a method of achieving space flight.

O*RIGIN OF THE* V*IKING* V*EHICLE*

In any case, the Navy's—or rather, NRL's —"new high altitude rocket" proposal had emanated from their young American physicist Dr. Ernst H. Krause, who had headed the Laboratory's Communications Security Section; during World War II he became an expert on guided-missile countermeasures and then worked on electronic countermeasures to prevent German V-2 and other missiles from reaching Allied targets.

By late 1944, Krause had examined captured German air-to-ground missiles; after the war he made an official visit to Germany, where he helped gather more data on the V-2, especially on his main interests: guidance techniques and the accuracy of the missile. At that point, he hoped to convert his old Communications Security Section into a U.S. Navy guided-missile research-and-development center. From this background there evolved the formation on 17 December 1945 of a Rocket Sonde Research subdivision, soon upgraded to a section of the NRL, with Dr. Krause as its chief. "Sonde" is the French word for "probe"; therefore, "rocket sonde" is an alternate term for sounding rocket.

As pointed out by David DeVorkin, a senior curator in space history at NASM, Krause "remained deeply committed to guided missile research at NRL," but "his staff advocated upper atmospheric research only within the context of missile development."[6] By "context," DeVorkin meant that the NRL needed to seek more precise data on how upper-atmospheric properties might affect communications and guidance systems in missiles, and therefore sounding rockets could serve as ideal tools towards these ultimate ends. The remaining V-2s could have been workhorses for these goals, but the number available were quickly dwindling. Krause directed Rosen and Smith to begin planning for a large sounding rocket in early 1946 as a replacement for the V-2. Moreover, it was understood that the principal aim of the sounding-rocket program was to extend knowledge of "the Earth's atmosphere to as great an altitude as possible."[7]

At first, Rosen and Smith sought out Aerojet. The small and efficient solid-fuel rocket-boosted Aerobee with its 2,600-lb-thrust, Aerojet liquid-propellant motor would soon be available. But as useful as it promised to be, its payload capacity and peak-performance capabilities could nowhere near match those of a V-2 class-size vehicle. The Aerobee was thus retained by the NRL for smaller upper- atmospheric experiment packages and did become an extremely invaluable and very long-serving tool for a wide variety of upper atmospheric researches by the Laboratory and many other scientific and educational institutions. Aerobee subsequently evolved into nine increasingly more powerful model variations. The Aerobee family lasted up to 1985—marking almost 40 years of service since its introduction in 1947. Yet in 1946, in accordance with Krause's directive, Rosen and Smith had also pursued the development of a larger vehicle for their heavier scientific payloads, although they well knew this project would be considerably more expensive.

At an 18 February 1946 meeting of the Navy's Office of Research and Invention (ORI), Krause and others decided that because Aerojet was tied up with its development of the Aerobee, they would "try and develop some other source" for their prospective larger vehicle.[8] Krause had also been informed that the Glenn L. Martin Company of Baltimore, the well-established aircraft firm, had also expressed a wish to get "into the game."[9] That April, the chief of naval research therefore initially approved $2 million for a project involving BuAer to construct 10 large-scale high-altitude sounding rockets. The Martin Company won the prime contract to build the airframes in August 1946, and RMI was granted the contract for the engines the same month; six years later, in June 1952, 4 additional Vikings were added, for a combined total of 14 rockets. Rosen was chosen as the NRL's project manager for the Viking vehicle. Smith served as the codirector. Interestingly, Rosen was an electrical engineer by training and had earlier designed the guidance system for the Lark and other missiles among his earlier tasks at NRL.

Milt Rosen does not offer fine details on how he learned of RMI. He merely explained that at the time there were then only two established liquid-rocket engine manufacturers in the United States—Aerojet in California and RMI in New Jersey—and that the latter was chosen because of their closer proximity to the NRL in Washington, D.C. It is true that RMI was closer. However, there is no question that there was always a Navy connection in the Viking project; RMI received BuAer contract NOa(s) 8531 for this project. The Naval Research Lab therefore must have already been aware of the existence of RMI via BuAer. In any case, the first Viking contract was followed by contracts issued between 1949 and 1952 for an "Improved Design" and other related developments. Following their usual pattern, RMI originally gave the engine their company designation of 20000C1—meaning simply 20,000 lb of thrust from a single chamber. The engine was later officially designated the XLR-10-RM-2.

The vehicle itself had first been named after Neptune, the powerful ancient Roman god of the sea, in deference to the Navy as the project's sponsor. However, this name was discarded by 1948 to avoid confusion with the Lockheed P2V Neptune, developed for the Navy as a maritime patrol and antisubmarine warfare aircraft. The rocket was then rechristened the Viking, after the early Scandinavian seafaring adventurers. Rosen credited Thor Bergstrahl with the name choice. Of Swedish parentage, Bergstrahl was one of Rosen's physicists who had earlier been involved with atmospheric-reentry heating analyses of flights of captured V-2s. He was also one of those responsible for adapting aircraft cameras to V-2s. He would later serve on the Viking project, using a little analog computer to help determine the location of flown and crashed Vikings toward the recoveries of their experimental payload packages.

Originally, the Viking vehicle was officially designated the HASR-11, for High Altitude Sounding Rocket-11. Next, it was redesignated the RTV-N-12, meaning Navy Research Test Vehicle 12. The rocket was designed to carry payloads up to 500 lb to an altitude of 100 mi, "higher than the highest [captured] V-2 flight thus far," according to an *Aviation Week* article appearing in June 1947.[10] Most significantly, what is not mentioned in any of NRL's public releases on the rocket, nor apparently in any of the press reports at the time, is that in effect, this designed altitude was in space. That is, the Viking appears to have been America's—and perhaps the world's—first single-stage rocket designed specifically for flight into space, although most of its scientific data were gathered en route in the farthest reaches of the upper atmosphere. The simple reason why this was recognized at the time is that the Kármán line, or Karman line—named after the famed Hungary-born aerodynamicist Theodore von Kármán—had not yet been established; this occurred some years later, as late as after the launch of Sputnik 1 that opened the Space Age in 1957.

A few weeks after the visit of Rosen and Smith to RMI on 18 March 1946, James Preston "Pres" Layton—the key person from the Martin Company, as he was their chief of propulsion staff, Pilotless Aircraft Section—also visited Lawrence at RMI. But the notation in Lawrence's *Daily Log* is most curious. Pres had come, it reads, "to discuss [a] 3,000–4,000 lb. thrust controlled missile."[11] Could this have been for an entirely different project, because as far as is known, the Viking was not contemplated as a missile this early? Second, as this thrust level was far too low for a high-altitude sounding rocket, could the Neptune (that is, Viking project) really have been first explored as a more modest missile, with sounding missions as secondary priorities? It may never be known for certain.

INITIAL DEVELOPMENT OF THE ENGINE

Shesta claimed he made "all the necessary calculations and initial drawings" before engineer Ed Neu Jr. was assigned to the project.[12] This is corroborated in an article in *The Rocket*, the house organ of RMI, in its issue of July 1956 that says: "John Shesta...laid down the basic design features" of the Viking power plant.[13] Shesta further pointed out that his original specifications called for nickel for the nozzle and propellant lines for the Viking motor, "but because of some procurement foul-up, the material was not available in time for the first unit. It [the nozzle] was therefore made of stainless steel which is not as good a heat conductor as nickel. After that, nickel was used."[14] In his later book, *The Viking Rocket Story*, Rosen gave additional credit to Shesta, who "drew principally on previous RMI experience and so some extent from the newly learned information about the V-2...he made [that is, designed] the chamber overly large to insure complete combustion and he chose an injector, an array of conical spray-spray nozzles."[15]

Wyld, with his penchant for calculations, did his share of them—mainly thermodynamic—as well. Lawrence characteristically worked out various preliminary wiring schematics for the vehicle. But according to Rosen, RMI really did not begin to develop the engine until "the fall of 1946."[16] The article in *The Rocket* specifically says, however, that in September 1946 engineer Harry W. Burdett Jr. "was placed in charge of all the thrust chamber development [of RMI] which included the development of the...thrust chamber for the Viking."[17] Amazingly, assisting Burdett in making "thousands of calculations that transformed Shesta's ideas into an engineering design," the article continues, was Ann Dombras, a fellow engineer at RMI and therefore perhaps the first known American woman rocket engineer. Dombras was educated at Goucher College in Towson, Maryland, as a mathematician, although she was hired as an engineer and mainly worked with Shesta.[18]

In addition to Neu, who completed the design of the thrust chamber and injectors, Albert G. "Al" Thatcher designed the hydrogen peroxide-driven turbopump, while Maurice E. "Bud" Parker managed the valve and controls, which borrowed heavily from the MX-774 work (as related in Chapter 6). But the chamber was quite different from that for the MX-774. Peter H. Palen, senior project engineer, and Jimmy Wyld played major roles, both indirectly and directly, in the XLR-10 development. Most notably, the gimbaling system devised by them for the MX-774 was borrowed in principle and adapted to the Viking power plant; both systems used control pistons to affect the gimbaling. The big difference was that the gimbaling in the MX-774's XLR-35 was applied to that engine's four different chambers, permitting each chamber to swivel about in one axis. In the XLR-10, the swiveling was designed for one much-larger chamber only. The Viking's gimbaling system is explained in detail in a later section.

MARTIN COMPANY INPUTS TO THE VIKING ENGINE

The Martin Company furnished all the tankage, wiring, and piping for the Viking as associated with its engine. The use of hydrogen peroxide to drive the turbopump, as used in the V-2 and in the pump-fed version of RMI's 6000C-4 for the X-planes, was already becoming well-established technology, although the Viking system featured several innovations. For instance, the Viking's tank that held the 90% concentrated hydrogen peroxide was made into a pipe that coiled around the turbine housing to save space. Also, the exhaust steam from the turbine was put to work to correct the roll of the rocket through a valve arrangement. As in the MX-774, the fuel tank was integral with the fuselage of the Viking rocket. There were many other innovations that set it apart from the V-2, including a key one that the airframe was made of aluminum instead of steel, making the Viking the first large rocket built entirely of aluminum. This made it far lighter, and the substantial weight savings enabled the vehicle to reach much higher altitudes with a lower thrust compared to that for the V-2.

Behind the scenes at Martin was Robertson Youngquist, already mentioned as another early ARS experimenter, who was to join RMI in 1949. He had joined the Martin Company in 1946 and came to serve as the head of their Liquid Rocket Propulsion subgroup. In this capacity, his responsibilities included "the development of all Martin elements...of the Viking project."[19] This put Youngquist in close collaboration with the Naval Research Laboratory, BuAer, and RMI. Because rocket propulsion was always a main passion, it is understandable that he later came to work for RMI directly on improvements on the Viking engine.

From its propulsion to its aerodynamics, the overall technical history of the Viking as it developed was complex. On the aerodynamic side, the configuration on the eighth Viking and later models was markedly different because the diameter was greatly increased and the length shortened; the lengths of some of the earlier rockets were slightly lengthened from vehicle to vehicle. The purpose of the major dimensional changes was to afford a 50% increase in propellant capacity. Consequently, the former trapezoidal fins of the first seven rockets were replaced by smaller triangular fins for the later rounds.

These dimensional alterations also slightly affected the thrust of the engines. It is said that no Viking engine was like any other, including in thrust output, but it was generally around 20,000 lb. More specifically, the engine generated an average of about 20,450 lb for the first flights, although that increased to nearly 21,430 lb for the later ones. The propellants always remained LOX and ethyl alcohol. Originally, it had been planned to use an alcohol-water mixture, but tests showed that straight alcohol would increase the specific impulse and offer other advantages. The overall engine changes, including the distinctive configuration changes, were dictated by flight data as it became available and then incorporated into the construction of each vehicle, especially as each motor was to be an improvement over the previous one. By far, the rocket's stabilization, or gimbaling, was its most novel feature and this system, too, underwent its own evolutionary changes.

VIKING'S GIMBALING SYSTEM

The rationale for choosing gimbaling and its adaptation to the Viking vehicle from the MX-774 was a lot more complex than first appears. The choice was definitely not arbitrary. In his comprehensive 1955 article, "Development of a Stabilization System for the Viking Rocket," Norris E. Felt Jr., operations manager of the Martin Company, wrote that the anticipated "unconventional flight path of the Viking, viz., a high altitude nearly vertical trajectory," resulted in "some unusual problems in the design of an automatic stabilization system."[20] Felt then laid out how these problems were solved using three separate subsystems.

"First," he explained, "there is the motor control system which obtains corrective moments by altering the line of thrust of the liquid rocket engine [that is, the gimbaling]. The second system, the aerodynamic roll system, employs moveable tabs attached to two of the [four] fins. Finally there is a [small] thrust reaction system which obtains control moments by the use of small [hydrogen peroxide] jet motors."[21] Computers, then a novelty introduced into American aerospace technology, "and nonlinear methods were required for [working out] the third system. Results of flights to date [up to 1955] indicate that the basic approach was sound."[22]

Initially, it had been proposed that the vehicle's control moments for pitch and yaw could be obtained by the deflection of vanes placed in the exhaust path, similar to what was developed for the V-2. Much earlier, Goddard's own stabilization system, which he devised and used in his rocket flight models during the early 1930s, was quite different; exhaust gases passed by deflector vanes that were linked to gyroscopes that righted, or corrected, the stabilization fins. Due to the secrecy of the late Dr. Goddard, who had died in 1945, it is unlikely the Viking team then knew such details of Goddard's system, but they were definitely familiar with what the Germans did on the V-2, especially because they were concentrating upon creating a large vehicle that would be a great improvement over the V-2.

In fact, Shesta and Mason "Whit" Nesbitt journeyed to Fort Bliss, Texas, and on 29 May 1946 interrogated half a dozen former "Peenemünders" then stationed there—including Wernher von Braun himself. Either Shesta or

Nesbitt specifically asked von Braun about "motor steering by means of winging the motor in gimbal rings."[23] Evidently, the Germans had never tried this directly in their own prewar and wartime rocketry programs because von Braun responded that he "thought the idea quite practical for smaller units... [and] one had actually been designed [but not built] in Germany."[24] No details were offered on this design. But for "rockets the size of the V-2," the interrogation report goes on, von Braun reasoned that "the flexible mount would be rather heavy and the feed lines would give trouble."[25] For all intents and purposes, it thus appears that the Germans of this period had never adopted gimbaling, although they knew of it. However, in all three cases of rocket stabilization—Goddard's, the German V-2, and the American Viking—the stabilization was linked to onboard gyros in order to correct the respective rocket's flight path in relation to the center of gravity. The gyro always spins one way and the sounding or other rocket must fly in a straight line. The use of the gyro was a logical approach, and there had even been the development of a gyroscopic stabilizer apparatus by American aviation pioneer Lawrence Sperry as early as 1914 to improve the stability and control of aircraft, long before all these rocket developments.

At any rate, early in the history of the conceptualization of the Viking vehicle, the Martin Company conducted an investigation of alternate control methods, comparing the V-2 approach to a potential gimbaling mode for the Viking. Youngquist, who strongly advocated gimbaling, was the author of this study for the Martin Company. The report, according to Felt, showed that the former presented several disadvantages, including a weight penalty, a lengthy development program, the question of vane durability, "a reduction of specific impulse due to vane drag in the jet stream," and complexity in the jet-vane system. The gimbaling approach also "offered greater promise for future developments."[26] This led to the adoption of the gimbaled arrangement in which the motor was mounted to a gimbal structure and the control moments—two-axis pitch and yaw, if not roll—were obtained by "deflection of the entire motor."[27] The overall remarkable success of the Viking program, Felt concluded, had fully vindicated the correctness of such early planning of advanced design features, like gimbaling.

During the early testing phase of the rocket, Felt added: "It was discovered that the rocket body was a fairly efficient transmitter of the vibration of the motor's gimbal structure at the latter's resonant frequency...The result was a rather violent vibration of the entire rocket."[28] Considerable effort was thus spent "in the design of filters to minimize the chatter problem," including the use of computers to analyze and find solutions to the problem.[29] Fortunately, these just required minor changes in the hardware.

A few years later, in 1951, Youngquist, with his colleagues Howard R. Merrill and Irwin R. Barr, applied for a patent that eventually was granted on 17 January 1961. Patent No. 2,968,454 for a "Rocket Control System" was a further refinement on the Viking's gimbal control system, but by then the

Viking had long ceased to be operational. This patent mainly embodied the later refinement of the turbopump exhaust being channeled through a steam jet valve to enable roll control of the rocket in the rarified (space) environment. It also included the Viking's basic gimbal system, which was never patented by RMI, even though they had clearly originated it. This patent was assigned to the Martin Company.

EARLY TESTING OF THE VIKING ENGINE

It was providential that RMI had moved to Lake Denmark when it did, in 1946, because the large Viking engine could never have been tested in the populous area of Pompton Plains. Due to its size, the XLR-10 was then America's largest and most advanced rocket power plant; testing it, and cooling the motor while testing, presented challenges to RMI. In a letter of 29 July 1947 to BuAer, Lawrence wrote:

> Cooling water for use [for the 20000C1, as it was then still called]…is estimated at 1,000 gallons per minute as a pressure of approximately 100 psi. This water will be taken from the reservoir [at Lake Denmark] adjacent to the test stand, passed through the water box and returned in a closed system to the reservoir. There will be an estimated loss of approximately 1% of the total water used which will be used for film cooling of the water box…The exact requirements for this purpose will depend on the difficulties which are experienced during the development of the 20000C1 engine and pump. Based on experience, however, firefighting equipment [additional, high-pressure water pumps] will be required very seldom but must be available in the event of emergency requirements.[30]

Lawrence went on: "If further information is desired, please advise the writer immediately since the delay in moving the water box and the reaction balance [scale] to the test area, which is an inter-related problem with the forgoing, will result in a serious [schedule] delay." Lawrence enclosed "calculations on heat transfer on the…water box…These calculations have been prepared by Mr. J. Wyld, Chief Research Engineer."[31]

In his book *The Viking Rocket Story*, Milton Rosen described the complete Viking test facility at length. The main stand was built by the Navy at Lake Denmark and operated by RMI, although its construction "was started before Viking, for the Navy its need for large rocket power…This test stand, of which the rotatable platform was the most unusual feature, was built so that a large rocket motor, mounted on the platform, could be fired at any angle."[32] The planned test program, he continued, consisted of three phases: the motor alone was to be tested, but using a pressure-fed system and mounted on an A-frame; the motor, pump-fed, was also tested on an A-frame; and the final tests using the rotating platform. On the platform was mounted "almost an entire rocket, the actual tanks, the valves and regulators, the turbo pump and the motor, all in line…This was to be as close to the actual Viking rocket as we could get at Reaction Motors."[33]

Thanks to Lawrence's *Daily Log*, it is possible to pinpoint the date and approximate time of the initial test, which was about noon on 17 October 1947. The entry for 9:55 a.m. that day read: "Mr. [Laurence P.] Heath to see Mr. [Lovell] Lawrence re personnel to witness testing of 20,000 lb motor."[34] This is followed by the entry for 11:20 a.m. that recorded: "Mr. Lawrence left for the Test Area."[35] The next entry, which had no time given but was probably close to noon, read: "Mr. Lawrence at the Test Area."[36] Subsequent entries show that Lawrence "left for Test Area #2" for other firings of the engine successively on 29, 30, and 31 October.[37]

One RMI data record, dated 1953, reported that the "reliability and durability" of XLR-10-RM-2 "have been checked by more than 200 test and flight findings" although it is difficult to interpret this.[38] The same document stated that one engine was "fired 873 seconds [14.5 minutes total] before it propelled a Viking…for 59.6 flight seconds."[39] The latter may be referring to the fact that every Viking was static-fired prior to launch. Payloads were not included in the vehicle during the static tests.

"Motor A," as Rosen termed it, likely the one fired on 17 October 1947, "fired on the first attempt. Although it did not develop full power and was cut off after fourteen seconds, the results of this first test were heartening."[40] "After the second run," he added, "a large area inside the nozzle was found to be burned through."[41] Similar burn-throughs were experienced in other early runs because the liner had been made of steel, "which is not a good heat conductor," Rosen continued.[42] Therefore, he concluded, "The RMI engineers abandoned steel for nickel as a liner material [as Shesta had commented above], and the cooling problem was solved."[43]

As expected in the rocket business, other problems were bound to crop up. One concerned the cooling water box again. In Lawrence's memo of 20 February 1948 to Dr. Paul F. Winternitz—the brilliant Austria-born chemist who then headed RMI's Chemical Laboratory—he noted: "I have observed a number of tests run on the 20,000 pound test stand…in which the water box was subjected to overheating."[44] The burn-through was not serious. On the other hand, Lawrence recognized that "the phenomena of burnout taking place where it could be viewed was of interest to me…where ease of heat transfer measurements and visual observation can be made without difficulty."[45] Lawrence the perennial pragmatist and opportunist thus astutely suggested to his chief chemist (Winternitz) that the insertion of "measuring elements"—in this spot and "without an actual engine"—offered a promising "idea for future research on heat transfer and materials of [engine] construction."[46] This instance also represents an ideal example of other, sometimes unexpected, duties undertaken by the rocket chemist beyond propellant research.

Finally, on 21 September 1948, the official acceptance test firing for the big engine was made, delivering 21,000 lb of thrust for 66 seconds. As Rosen put it, "There was much rejoicing in the test area."[47] Meanwhile, the dynamic tests of the steering (gimbaling) system had been fully underway at the Martin plant by the spring of that year. At the same time, the first airframe of the rocket was taking shape, and the field crew, under the leadership of Pres Layton, were immersed in an intensive six-month training course before they would be sent to White Sands to be stationed there for the launch preparations.[48] Pres is given credit for "meticulous planning for static and flight test and attention to safety."[49]

Layton is likewise credited for possibly formulating the modern "X-time" schedule—"X minus 15 minutes," for example—in which every single rocket-test-procedure sequence to be performed is called out in seconds or minutes and verified before the next operation in sequence, up to the X moment for the actual time of firing. If a procedure was not—or could not be—performed, the countdown was held up at that X-reading until it was able to be resumed after the problem was fixed. Such postponements were characteristic of Viking's launch history.[50] Similar X-timing callout sequences have since become extremely familiar throughout the U.S. space program.

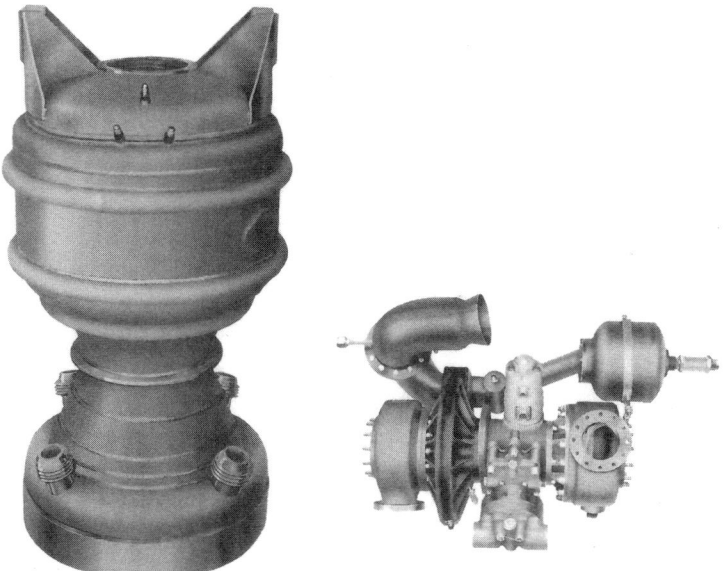

Fig. 7.2 RMI's XLR-10 rocket engine and pump for the Viking sounding rocket. (Courtesy Frank H. Winter collection.)

VIKING MODEL 1 FLIGHTS

The flight of Viking No. 1 took place on 3 May 1949. Although not recognized at the time, this was also the first flight of the so-called Model 1, or the longer and skinnier configuration of the vehicle. Standing 45.25 ft long and with a 32-in. diam, the pencil-like, sleek rocket with a conical nose was constructed to rest on its four clipped-delta fins. A 65-s burn had been planned, but the engine shut down prematurely at 54.5 s due to turbine leaks. The vehicle thus ascended to only 51.5 mi, or barely half the planned distance.

The flight of the slightly longer Viking No. 2 on 6 September suffered a similar fate, cutting off at 49.5 s and just reaching 32.3 mi. Viking No. 3, now 47.4 ft long, experienced a control system malfunction, causing the range-safety officer to shut off the engine by a radio command after a 59.6-s burn; like No. 1, this round reached 50 mi.

By stark contrast, the historic Viking No. 4 was the first U.S. liquid-propellant rocket launched from a seagoing vessel, the experimental guided-missile ship USS *Norton Sound*, in the vicinity of Christmas Island in the Pacific Ocean on 11 May 1950. On top of that, the 48.6-ft-long rocket flew flawlessly. It carried up its 959-lb scientific payload—the heaviest in the Viking series—into space, up to 106.4 mi, and attained a maximum velocity of 3,520 mph. Its motor burned for 74 s. Unfortunately, the radiation-particle-count experiment of its payload was not satisfactory, although useful temperatures and pressures were gathered.

The shipboard launch of Viking No. 4 was also assigned the special military code name of Project Reach. It thus fulfilled another, lesser-known objective of the NRL's Viking program: "to advance the art of rocketry in the United States in the interest of National Defense."[51] That is, the launch provided the Navy the ideal opportunity for this branch of the military to gain experience in launching a large-scale liquid-propellant rocket from aboard a ship at sea, in the event the Navy would later pursue the development of a vehicle as a missile.

About the same time—and undoubtedly directly linked to little known Project Reach—an NRL report marked "secret" proposed that the Viking might be furnished with the addition of a guidance system to convert the rocket into a 150-mi-range missile. Later, this Super Viking concept was upgraded to a potential 500-mi-range missile, although neither plan was carried out.

After the solitary seagoing launch, the remaining Viking launches were resumed at White Sands, New Mexico. Viking No. 5 was launched six months later on 21 November, carrying a 675-lb payload; it traveled slightly higher at 107.5 mi. Its instruments successfully transmitted very useful counts on electron density in the upper atmosphere, making it the first Viking in a series to return good scientific data—including among the first photos of Earth from space.

Viking No. 6 was not as lucky. It carried the lightest payload yet at 373 lb, mostly pressure gauges to more thoroughly gather pressures in the upper atmosphere. But its lighter weight, which was further reduced with new aluminum fins instead of steel ones, created a faster boost liftoff than expected, but with greater aerodynamic heating and structural stress. Consequently, at 25 mi, the fins buckled, causing the vehicle to loop, then straighten out until it finally reached just 40-mi peak altitude before falling back to Earth.

But Viking No. 7, which may be considered the last of the Model 1 series and had a long-and-skinny configuration, was launched on 7 August 1951 and reached a new world's record of 135.6 mi, beating the vertical record of the V-2. After the disaster of the previous flight, the fins for No. 7 had been strengthened, although the X-ray and cosmic ray plates on this flight were damaged upon recovery.

VIKING MODEL 2 FLIGHTS

From that point on, the Viking Model 2—shorter, at 45 in. long, although fatter, with a 41.6-in. diam for Viking No. 8—came into service. These dimensional changes had been dictated midstream in the Viking program due, in part, to Viking 4 exceeding the design altitude of the rocket and the promise of further increase. In addition to the alterations in the dimensions, as mentioned, Model No. 2 also featured four smaller triangular fins. The 50% gain in propellant capacity allowed greater access to the engine components. Internally, the hydrogen peroxide coil was gone, replaced by a separate tank, and the controls were also modified, primarily with the addition of small hydrogen peroxide jets. They controlled the rocket's attitude outside of Earth's atmosphere—that is, within the space environment. Again the Viking had been designed for space flight, even if for a limited duration; its anticipated longer duration in space saw this design aspect slightly more marked with Model 2.

The Model 2 incorporated other changes. For preflight static tests, the vehicle was now held down by only two bolts. In the case of Viking No. 8, though, this turned out to be disastrous, when an unexpected thrust surge occurred a few seconds into the usual static firing on 6 June 1952, in preparation for its upcoming launch. At first the rocket wobbled. Then, at X+13 s, the vehicle tore loose and flew off until X+60 s, when a cutoff signal was transmitted to the wayward rocket that crashed a minute later. It had broken apart at a peak altitude of about 4 mi.

The Viking No. 9 test on 15 December went very well, almost exactly comparable to Viking No. 7; it reached 136 mi, although theoretically it could have gone higher. Moreover, the vehicle obtained excellent cosmic-ray-emulsion recordings and took more photographs from space.

On 30 June 1953, the firing switch was pushed for the launch of Viking No. 10, but the motor exploded and the rocket caught fire; emergency water nozzles were immediately turned. When the fire reached the alcohol tank, one of the courageous firefighters, Navy program officer Lt. Joe Pitts, rapidly moved in with a hose and doused the fire. But the tank began to buckle, threatening to topple the rocket and create a far bigger disaster. Leaks were already dripping through a small pipe. Fortunately, the quick-thinking and acting Lt. Pitts borrowed a carbine from a guard and shot a bullet into the alcohol tank. This instantly vented it—relieving its pressure—and the rocket was spared from the worst damage.

Astonishingly, there were enough salvageable parts left to rebuild it. After the terrible mishap, the remnants were shipped back to Martin's plant in Maryland where the new vehicle—now designated the Viking 10R, probably meaning Viking 10, Rebuilt or Refurbished—was indeed prepared. The rebuilding delay accounts for the fact that there were no Viking launches in 1953.

Number 10R was finally shipped back to White Sands, along with Viking No. 11, as it was to be fired within days of No. 10. On 7 May 1954, Viking No. 10,

Fig. 7.3   Launch of Viking rocket No. 11 to 158 mi. (Courtesy U.S. Navy.)

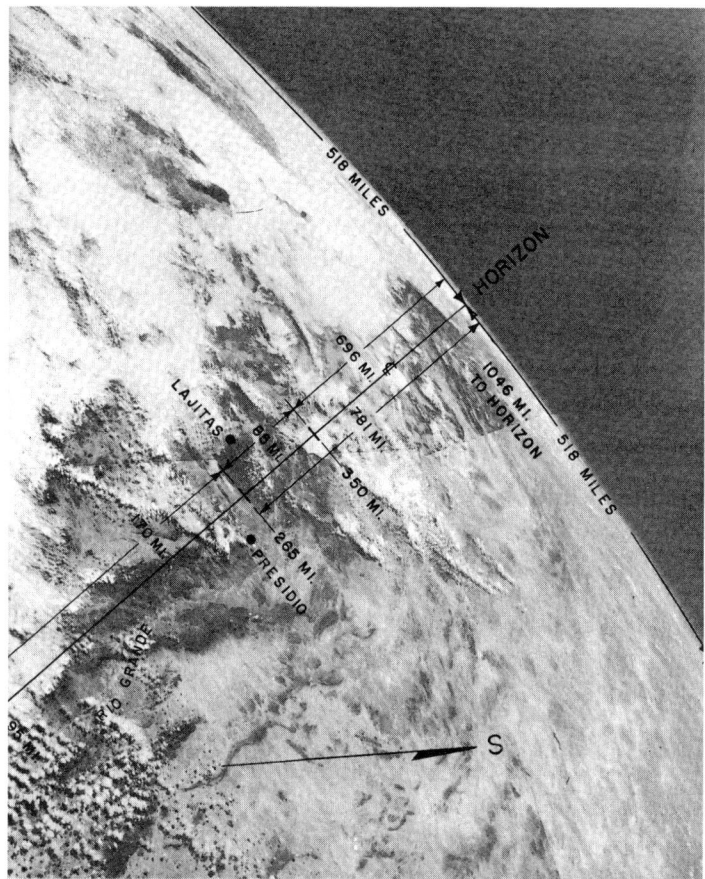

Fig. 7.4   Composite of two photographs of Earth taken from Viking No. 11. (Courtesy U.S. Navy.)

as it was still called, was successfully launched and almost exactly matched the altitude reached by No. 7: 136 mi. The vehicle's ion and other experiments also performed flawlessly. Then, just 17 days later on 24 May, sister vehicle Viking No. 11 beat it out and established a new altitude record of 158.4 mi for a single-stage rocket—the highest of all the Viking rockets.

Viking No. 11 was spectacular in other ways. It attained the highest velocity of the series, at 4,300 mph, and conducted perhaps the first U.S. "live" reentry test from space. Upon reaching its apogee, the nosecone was pointed down-ward and separated, using small hydrogen peroxide thrusters. This first-ever maneuver of its type was photographed throughout the procedure; the nosecone then reentered the atmosphere. Photographs later retrieved from this flight revealed for the first time Earth as a sphere. These are hailed as the first photographs from space itself. With such a singular accomplishment, this flight alone was a dramatic harbinger of the Space Age.

Viking No. 12, launched on 4 February 1955, also performed very well, with a flight up to 144 mi. It was to repeat the separate-nosecone reentry experiment, although the small gas-jet control system malfunctioned, causing the nosecone to reenter facing point up, not point down.

This was the last of the Viking sounding-rocket series; at several hundred thousand dollars each plus the cost of $400,000 for each launch, they were far too expensive. Each engine cost about $70,000 in late-1940s/early-1950s dollars; today (in 2017), this would be about $716,000. The far smaller and cheaper Aerobees, along with many other, simpler, and even less costly, all-solid-propellant types of rockets, would form the bulk of the country's future sounding vehicles.

THE VIKING ROCKET IN THE VANGUARD PROGRAM

Viking 13 and 14 had been planned, but after the Martin Company won the contract to develop and build the Project Vanguard satellite launch vehicle—greatly helped by its Viking experience and gained expertise—it was decided to cancel these missions and adapt the remaining pair of Vikings for Vanguard test vehicles. America's artificial-satellite program had been announced by the White House in July 1955 as part of the coming International Geographical Year (IGY) of 1957–1958, which astronomers had calculated as a period of maximum solar activity. In addition to a satellite, numerous sounding rockets would also be launched to capture as much upper-atmospheric and near-space data as possible.

Meanwhile, there had been other plans formulated for the final two Vikings. In July 1954, the Air Force entertained the possibility of converting these last Vikings into reentry test vehicles, designated as the M-10 and M-15 Vikings, respectively, to signify their expected Mach numbers, benchmarks toward the development of ICBMs. Rosen recognized the M-10 Viking as a potential satellite launcher by itself. These ambitious plans continued for some time and on 8 July 1957, a meeting of the Guided Missiles Committee of the Department of Defense Research and Development Division was even held at the Martin plant on the M-10 concept. In other words, the Viking with added stages was in the running for a while as a contender among early U.S. satellite launchers. But these plans never came to fruition.[52]

In the end, the last two Vikings were converted as test vehicles for Project Vanguard. On 8 December 1956, No. 13, redesignated TV-0 (Test Vehicle 0), was launched as a dramatic night shot from the Air Force Missile Test Center at Patrick Air Force Base in Florida—the first such vehicle for Project Vanguard. The vehicle carried a telemetry system for that program. Climbing up to a peak altitude of 126.5 mi, TV-0 fulfilled its mission to evaluate this system, ejecting a minitrack radio transmitter at a 50-mi altitude that was successfully tracked until it fell into the ocean. The launch additionally provided greater experience to the site's launch, range-safety, and tracking crews. This flight, observed the popular space writer Willy Ley, was "the first shot in the [U.S.] satellite program."[53]

The final Viking, No. 14—now redesignated TV-1—made another spectacular night launch on 1 May 1957. TV-1, or Test Vehicle 1 for Project Vanguard, carried a second stage, a prototype of the solid-propellant Grand Central Rocket Company third stage for the Vanguard vehicle. In addition, the rocket was topped by an instrumented nosecone.TV-1 reached a slightly lower height than its previous sister vehicle at 121.2 mi, although it otherwise met all test objectives. These included the separation of the Vanguard stage prototype and its firing.

Fig. 7.5 **An artist's rendering of the TV-0 and TV-1 vehicles. (Courtesy NASA.)**

THE VIKING AS A TECHNOLOGICAL CELEBRITY IN POPULAR CULTURE

Viking's story, replete with utter failures and outstanding successes, hardly ended there. A far-advanced, 50,000-lb-thrust Super Viking engine emerged from the early 1950s and played a major role in laying the foundations for the creation of the power plant for the X-15, as related in the next chapter. Beyond this, Viking became an early cherished American aerospace "celebrity," and entered into our popular culture in several interesting ways.

Back in 1954, wreckage from a recovered Viking, perhaps No. 11, was incorporated into a full-scale cutaway model of a Viking vehicle that was bestowed by the Martin Company on the Hayden Planetarium in New York City; it was complete with an XLR-10 motor. At the Hayden, it was placed horizontally on display and became the centerpiece of their Viking Rocket Hall. This highly publicized exhibit officially opened on 30 June 1955 for a two-year stay and opened to the public on 1 July. Apart from the full-length rocket, the exhibit featured a "dynamic" smaller display with "a push button actuating the launching of a miniature Viking, which with a *whoosh* and a cloud of smoke shoots into space."[54] The Viking—or at least parts of it— seems to have been the first real rocket to go into space that had ever been placed on exhibit in a museum.[55]

Included in the Hayden exhibit was a "spectacular photograph, taken by a Viking camera at an altitude of 155 miles," and a presentation on the "findings of Viking No. 11," as well as an extensive treatment of the nature of artificial satellites.[56]

Milt Rosen himself delivered the talk at the official opening ceremonies of the exhibit on the 30 June. "The question of 'Who owns the Viking rocket?' was never a matter of serious consideration," he began.[55] "The rocket has been referred to as the Navy Viking, as the Martin Viking, less frequently as the RMI Viking...All of these are...partially correct," he went on.[56] The Navy paid for it, but the Navy obtained its funds from Congress, "who in turn obtains them from the taxes of the American people."[57] Therefore, he concluded, "the Viking rocket belongs to the American people."[58]

Rosen was accorded much public distinction for his own top role in the Viking program. In 1954 he was awarded the first James H. Wyld Memorial Award, later called the Wyld Propulsion Award, by the American Rocket Society with RMI sponsorship. Rosen, who later played a key role in promoting the launch of an artificial satellite based upon the Viking that helped lead to Project Vanguard, died on 30 December 2014 at the age of 99. Viking's input into Project Vanguard is a whole other story.

Rosen's book was also promoted at the exhibit; it did very well and greatly helped spread the Viking rocket story—and the promise of space flight. Autographed copies of the book were placed on the tables for the dignitaries at the official opening of the exhibit; signed first editions are now sold for up to $600 each. The Viking rocket thus played its own very important role in generating and spreading the cause of space flight. Viking achievements even reached TV audiences; for example, a Viking documentary special aired on the "Armstrong Circle Theatre" on Channel 4 in New York City during the spring of 1956.

With its silver gray-painted RMI engine visible inside through a Plexiglas covering, the Viking rocket finally reached one of the world's most visited museums for all to see. In 1974, after years in storage at the Hayden Planetarium following the Viking Hall closing in 1957, the same cutaway Viking model was offered to the National Air and Space Museum. After refurbishment by the Martin Company, the vehicle was finally placed back on exhibit again, in time for the nation's Bicentennial opening of NASM in July 1976. It literally has been admired by millions.[59]

Fig. 7.6   Close-up of Viking No. 12, rebuilt, on exhibit at NASM with Plexiglas covering, showing the top of the engine pump. (Courtesy Frank H. Winter.)

ENDNOTES

1. The true height of the most historic A-4 flight made on 3 October 1944 is not known with certainty, and the quote of "nearly 60 miles" comes from Walter R. Dornberger, the military commander of the wartime Peenemünde rocket research center; others say the launch of 3 October went up about 85–90 km (53–56 mi). Dornberger also wrote: "For the first time in the history of the rocket we had sent an automatically controlled rocket missile to the border of the atmosphere at *Brenschluss* [burnout] and put it practically airless space." Walter Dornberger, *V-2*, Bantam Books, New York, 1970, pp. 12, 14, 17. The maximum horizontal range, or point of impact, reached by the V-2 was far easier to determine and was about 125 mi. Again, the 100-km definition of the beginning of space is known as the Kármán line, after Dr. Theodore von Kármán, and was later recognized by the Fédération Aéronautique Internationale (FAI), or International Aeronautical Federation.

2. Lawrence, *Daily Log*, Lovell Lawrence Jr. Papers, NASM, box 4, folder 16.

3. Letter from James H. Wyld to unknown recipient, 28 Jan. 1936, in "James H. Wyld" file, NASM.

4. Lise Blosset, "Robert Esnault-Pelterie: Space Pioneer" in Frederick C. Dunant III and George S. James, eds., *First Steps Toward Space*, AAS History Series, Vol. 6, Univelt, San Diego, 1995, pp. 8–9, 19.

5. Ibid, pp. 8–9.

6. David DeVorkin, *Science with a Vengeance*: *How the Military Created the U.S. Space Sciences After World War II*, Springer-Verlag, New York, 1992, p. 79; Kenneth W. Gatland, *Development of the Guided Missile*, Iliffe & Sons, Ltd., London, 1954, p. 184.

7. DeVorkin, *Science with a Vengeance*, p. 79.

8. Ibid, p. 75.

9. Ibid.

10. C.H. Smith, Jr., M.W. Rosen, and J.M. Bridger, "Super Altitude-Research Rocket Revealed by Navy," *Aviation*, Vol. 47, June 1947, p. 40.

11. Lovell Lawrence Jr., *Daily Log*, Lovell Lawrence Jr. Papers, NASM, box 4, folder 16; Yvonne C. Brill, "J. Preston Layton 1919–1992: A Guiding Light in Nuclear Space Power and Propulsion," IAF-93-R.1.420, paper presented at the 44th Congress of the International Astronautical Federation (IAF), 16–22 October 1993, Graz, Austria, preprint, p. 2.

    Layton, who had a bachelor's degree in aeronautical engineering, worked with both liquid- and solid-fuel JATOs with Goddard at Annapolis early in the war while he was serving as a U.S. Navy Reserve officer.

12. John Shesta, "Reaction Motors, Inc.: A Memoir," in Frank H. Winter and Frederick I. Ordway III, *Pioneering American Rocketry: The Reaction Motors, Inc. (RMI) Story, 1941–1972*, AAS History Series, Vol. 44, Univelt, Inc., San Diego, 2015, p. 75.

13. "Good Old Viking Days," *The RMI Rocket* (Reaction Motors, Inc., Denville, N.J.), Vol. 7, July 1956, p. 10.

14. Shesta, "Reaction Motors, Inc.," p. 75.

15. Milton W. Rosen, *The Viking Rocket Story*, Harper & Brothers, New York, 1955, pp. 56–57. Consult also Milton W. Rosen, "The Viking Rocket: A Memoir," in R. Cargill Hall, ed., *History of Rocketry and Astronautics*, Univelt, Inc., San Diego, 1986, AAS History Series, Vol. 7, Part 2, pp. 429–441.

16. Ibid.

17. "Good Old Viking Days," p. 10.

18. Ibid; Telephone interview with Ann Dombras by Frank H. Winter, 26 Feb. 2017.

19. Robertson Youngquist résumé, in "Robertson Youngquist" file, NASM.

20. N.E. Felt, Jr., "Development of a Stabilization System for the Viking Rocket," *Journal of Jet Propulsion* (American Rocket Society, New York), Vol. 25, Aug. 1955, p. 392.
21. Ibid.
22. Ibid.
23. "Report of Interrogation of German Scientists at Fort Bliss—May 29, 1946," Lawrence Papers, NASM, box 9, folder 15, p. 8.
24. Ibid.
25. Felt, "Development of a Stabilization System," p. 393.
26. Ibid.
27. Ibid.
28. Ibid, p. 394.
29. U.S. Patent No. 2,968,454.
30. Letter from Lovell Lawrence Jr. to BuAer, Naval Ammunition Depot, Lake Denmark, New Jersey, 29 July 1947, copy in Lawrence Jr. Papers, NASM, box 10, folder 6.
31. Ibid.
   For Rosen's account of a serious water-shortage crisis that threatened to curtail the Viking test firings, see Rosen, *The Viking Rocket Story*, pp. 61–62.
32. Rosen, *The Viking Rocket Story*, pp. 58–59.
33. Ibid.
34. Lawrence Jr., *Daily Log*, NASM, box 4, folder 19.
35. Ibid.
36. Ibid.
37. Ibid.
38. "The XLR-RM -20,750 Pound Thrust Liquid Propellant Rocket Engine," entry in RMI, *Liquid Propellant Rocket Engines Developed for the U.S. Military Services by Reaction Motors, Incorporated*, 1953, n.p., copy in author's collection.
39. Ibid.
40. Rosen, *The Viking Rocket Story*, p. 59.
   Consult Rosen, *The Viking Rocket Story*, pp. 59–62, for details on other Viking tests and how they affected the design of the engine.
   Based on an illustrated Norton Refractories Company advertisement in *Jet Propulsion*, Vol. 25, for Oct. 1955, p. 509, it is likely that Rokide "A" aluminum oxide refractory coating was also applied within the Viking motor as further protection against excessive heat, although we do not know when this improvement was adopted.
41. Rosen, *The Viking Rocket Story,* p. 59.
42. Ibid.
43. Ibid.
44. Memo from Lovell Lawrence Jr. to Dr. P.F. Winternitz, 20 Feb. 1948, in Lawrence Papers, NASM, box 10, folder 6.
45. Ibid.
46. Ibid.
47. Rosen, *The Viking Rocket Story*, p. 61.
48. Ibid.
49. Brill, "J. Preston Layton," p. 3.
50. Consult Willy Ley, *Rockets, Missiles & Space Travel*, Viking Press, New York, 1958 (and other editions), pp. 264–276, 278, 283 for instances of X-delays with the Viking.
51. Charles DeVore, "Project Reach," *Navigation: Journal of the Institute of Navigation*, Vol. 2, July 1950, p. 277.
52. Consult Constance McLaughlin Green and Milton Lomask, *Vanguard: A History,* Scientific and Technological Information Division, NASA Historical Series, NASA, Washington,

DC, 1970, pp. 43–45, 58–59, for details on the Viking as a contender among early U.S. satellite-launch vehicles. See also, John P. Hagen, "The Viking and the Vanguard," *Technology and Culture*, Vol. 4, Autumn 1963, pp. 435–451.

53. Ley, *Rockets, Missiles & Space Travel*, p. 336.
54. "Aero News Digest," *Aero Digest*, Vol. 71, July 1955, p. 5.

    The simulated Viking launching display at the Hayden is described and depicted in *The RMI Rocket*, Vol. 6, July 1955, in a two-page spread, on pp. 4–5, 7. This is not to ignore that for many years a Viking engine had been part of RMI's own company museum, although that one does not appear to have been a flown engine, and the museum was likely not open to the general public. Apart from the vehicle, or its engine, being on exhibit in museums, there are also instances of the appearance of Viking hardware in short-term displays or shows. In 1957, for instance, an exhibited Viking engine was viewed by more than 200,000 visitors at the Berlin International Trade Fair, as hosted by the United States Information Agency.

55. Milton Rosen, "Remarks at the opening ceremonies of The Viking Rocket Hall," copy in Milton Rosen Papers, NASM, box 2, folder 53.
56. Ibid.
57. Ibid.
58. Ibid.
59. The Viking rocket in NASM is catalogued as Cat. #1976-0843, but the XLR-10 motor within it is not catalogued separately. For the scientific results of the Viking launches, consult DeVorkin, *Science with a Vengeance*, cited above in note 6. There are also several NRL Viking "Rocket Research" reports in the "Viking" file, NASM.

Chapter 8

Last Major Triumph: The X-15

"The goal was to come up with a dream engine."

–Scott Crossfield. *Always Another Dawn*,
The World Publishing Company,
Cleveland and New York, 1960, p. 228.

Earliest Proposals for Black Betsy in the X-15

RMI's development of the power plant for the hypersonic X-15-rocket research vehicle was most ironic in two ways. First, it utilized the old 1940s RMI Black Betsy X-1-type engine for the aircraft's initial flights. Second, the X-15 was finally powered with the XRL-99 Pioneer engine—the most sophisticated and revolutionary engine ever produced by this truly pioneering company.

Back in Chapter 5, it was seen that the last of the Bell-X1E aircraft flights, on 6 November 1958, appeared to mark the end of the operational life of the Black Betsy engine that had evolved from its use in Chuck Yeager's historic Bell X-1. In the mid-1950s, however, when the design of the far more powerful and complex power plant for the X-15 was not fully defined and presented immense technological challenges, Walter C. "Walt" Williams, chief of NACA's Flight Research Center at Edwards Air Force Base, recommended the use of the proven and reliable X-1-type engine as a substitute "until the final [X-15] powerplant could be developed and tested," according to one aviation historian.[1] The veteran engine—now designated the XLR-11—seemed to show promise for yet another chapter in its already highly illustrious career, at least as a backup until the final selection and maturation of a dedicated engine for the aircraft.

The X-15, it must also be stated, was a highly advanced joint project for NACA (later NASA), the U.S. Air Force, and North American Aviation, Inc. for a hypersonic rocket research aircraft. North American was the main contractor. As for the word "hypersonic," it means aerodynamic flight above Mach 5. That is, it covers the flight regime above supersonic, with speeds ranging from 3,840 to 7,680 mph.

But Williams's recommendation was not accepted. The project went ahead and led to a highly competitive X-15-engine competition, discussed later. Here, it is enough to say that RMI was declared the winner, because on

21 February 1956, the Air Force officially assigned the new designation for the planned engine, the XLR-99-RM-1. On 7 September, the final contract was assigned to RMI. Right from the start, however, RMI was plagued with recurring developmental delays for the "99"—as the engine was sometimes called by the company's employees. According to the later recollections of test pilot Scott Crossfield, who was primarily an engineer and had joined North American to help develop the aircraft's power plant, by February 1958 "the XLR-99 engine was exactly one year behind schedule and considerably heavier than originally planned."[2] In addition to the weight issue, engine costs had sharply risen; the delays thus complicated matters all around. Indeed, the slippages threatened the entire multimillion dollar X-15 program. That same month of February, mention of the XLR-11 surfaced again for a quite different and far more active role.

The Power Plant Laboratory of the Air Force's Wright Air Development Center (WADC) at Wright-Patterson Air Force Base near Dayton, Ohio, recommended the use of the XLR-11 for the initial flights of the X-15. The recommendation was was speedily adopted. There was thus a marked distinction between Williams's earlier suggestion about the XLR-11 to allow trials of the aircraft before the XLR-99 was even conceived, and its new application as a temporary substitute for the 99 until the bigger engine could enter active duty.

There are several possible answers to the fundamental question of "Why the XLR-11?" But it is difficult to determine which is correct. One is related by Crossfield, who said that after countless meetings about this early engine crisis, L. Robert Carman—then of North American and later with NACA—one day announced suddenly: "I've been doing a little figuring here. Suppose that instead of waiting for the XLR-99 engine we substitute, pending its arrival, two X-1 type engines. They could be [re-]built in a few months at most."[3] Crossfield was at first very reluctant to pursue this solution on the grounds that the X-1 engine (the XLR-11) was inferior, and would damage the integrity of the advanced technology of the X-15. On the other hand, Charles H. Feltz of North American's design team strongly supported the idea, and with a couple of fellow engineers they worked out its feasibility.

Feltz reasoned that the same fuel tanks could be used, and the combination of eight chambers from two XLR-11s could afford greater throttleability. One of Feltz's colleagues also pointed out another big advantage. "Those engines have a lot of time on them," he declared. "They ought to be reliable."[4] The men then calculated that the XLR-11 produced a theoretical maximum (that is, vacuum) thrust of 8,000 lb—actually, 8,200 lb, or 16,400 lb for a pair. Theoretically, this would enable the X-15 to reach Mach 3.5 at 150,000 ft. This output was sufficient enough for preliminary demonstration flights "as well as re-entry, ballistic controls...and so forth," added the engineer.[5]

As North American had "already considered exactly the same idea," Crossfield noted, "they approved it at once, and Edwards [Air Force Base] got busy building up a dozen 'proven' X-1 engines from old parts. We planned to put two each in the first two X-15s, holding the third X-15 [there were three of the aircraft] in the factory for the first XLR-99 engine and other improvements which flight tests would generate."[6] These decisions also, he said, were accepted with no problem.

Crossfield's account agrees in most respects with the official (NASA) history of the X-15 program, *X-15: Extending the Frontiers of Flight* by Dennis R. Jenkins, although it contains more details and qualifications. For one, in addition to North American, the Air Force's Propulsion Laboratory had also considered the XLR-11 as a temporary substitute possibility. Crossfield had not mentioned that the Propulsion Laboratory had arrived at its own recommendation, sent on 17 February to Brig. Gen.Victor R. Haugen of the Air Research and Development Command (ARDC) and other Air Force officials. Moreover, the Laboratory's findings had been based upon a through evaluation of RMI's program on the XLR-99.

Another detail expressed in the official history by Jenkins is a further reason why Crossfield originally did not favor the XLR-11 in the X-15. This was his initial grave concern that if the Air Force approved the change, "we'd be making a big mistake."[7] The Air Force, he felt, would give up on the troublesome XLR-99, and North American would be left with the XLR-11, thereby turning the X-15 into a permanently Mach 3+ aircraft, instead of one capable of twice the speed.

One other important detail involves Harrison "Stormy" Storms, the highly influential and sometimes volatile aeronautical engineer who had successfully led North American's bid for the contract to design and build the X-15. Storms had also been present during the animated talk by Feltz proposing the XLR-11 substitute idea. Later better known for directing North American's development and construction of the Project Apollo Command/Service Module, Stormy had agreed with Feltz on the merit of the XLR-11 concept. In any case, after mulling over the advantages, Crossfield was finally persuaded after a few weeks to go along with the plan. He now remarked philosophically, "We should learn to crawl before we enter the Olympic hundred-yard dash."[8]

ORIGINS OF THE *XLR-11S* IN THE *X-15* PROGRAM

There is some confusion as to whether the dozen or so XLR-11s assembled at Edwards's own rocket shop for the X-15 program's interim engines were from prior Air Force or Navy programs, or both, as both surplus XLR-11and LR-8 models are mentioned in the available literature. All things considered, it does seem to have been both.

ADAPTATION OF THE *XLR-11s* TO THE *X-15*

Examples of some of the X-15 interim engines are in the collection of NASM, including one originally joined pair, coupled by a steel framework arrangement called an "XLR-11 engine mount," that was used in the first powered flight X-15 #2 made on 17 September 1959. This pair consists of an XLR-RM-9 Mod 5, Serial No. 6 (Cat. #1963-0363); the formerly adjoining engine is the same model, but Serial No. 5 (Cat. #1963-0362). Two XLR-11s were bolted to the engine mount and the completed mount then bolted inside the aft end of the aircraft.

After the decision to temporarily adopt the XLR-11s, a joint Air Force, BuAer, and NACA Technical Advisory task group was set up. Their first meeting was held at RMI on 24 February 1958, although this team seems to have been more focused upon the vastly more complicated further development of the main XL-99 power plant for the X-15.

The actual adaptation of the XLR-11s—and possible LR-8s—to the X-15 was relatively simple with only minor structural modifications required, in addition to the uprating of the thrust to about 8,000 lb per engine. This uprating, incidentally, does not seem to have been related at all to the earlier 1954 proposal, mentioned in Chapter 5, of the potential adaptation of an 8,000-lb-thrust version to the Douglas D-558-2 Skyrocket; that proposal never came about. Also, while the XLR-11 and LR-8s used standard LOX/alcohol propellants, the X-15 was built only to accommodate the more exotic combination of LOX/anhydrous ammonia for the XLR-99. But "the two liquids," Jenkins explains, "had a similar consistency and temperature" that also facilitated the XLR-11/LR-8 matings with the X-15.[9]

Fig. 8.1 Dual XLR-11 motors for the X-15 as its interim engine, in the collection of NASM. (Courtesy Smithsonian Institution, 89-1868.)

Fig. 8.2   Rollout of X-15 with its interim engine, 15 October 1958. (Courtesy, Smithsonian Institution, 88-17874.)

ROLLOUT OF *X-15* WITH *INTERIM ENGINE AND XLR-11 FLIGHTS*

The X-15 #1 featured dual interim engines, one stacked above the other and recessed within the rear of the supersleek black aircraft. Its rollout took place in front of very excited crowds at North American's Inglewood, California, plant on 15 October 1958. The visitors also were treated to a look at a nearby mockup of the forthcoming and much publicized XLR-99. The following day the aircraft itself was transported to Edwards.

Following captive flights—comparable to those undertaken years before by the Bell X-1, although now it was the X-15 that was suspended beneath a B-52—a single-glide flight of the X-15 was made. The first of the X-15-powered flights with its interim engine was flown on 17 September 1959. This was X-15 #2, with Crossfield as the pilot.

On this debut flight for the XLR-11 interim version of the plane, the engine's turbopump suffered a minor failure, although the resulting small fire was easily extinguished. Otherwise, this powered aircraft attained a maximum speed of Mach 2.11 (1,393 mph) and altitude of 52,341 ft. There were a few other near-harrowing—and occasionally, truly harrowing—experiences with the interim engine in this phase of the X-15's history. Notably, on the aircraft's fourth powered flight on 5 November 1959, there was an inflight explosion and fire in the lower engine in which the pilot—again, Crossfield—was forced to shut down the engine and make an emergency landing on Rosamond Dry

Lake. But on the whole, the engines performed well enough up to the last of the dual XLR-11 flights made on 7 February 1961. Robert M. White was the pilot for this mission. This was also the fastest of the interim engine flights, at Mach 3.

Throughout this series of 29 interim engine flights, there were undisguised and understandable frustrations among those in the X-15 program over the XLR-99 delays. They were extremely impatient that these X-1-type engines from another era were, as Crossfield had feared, inadequate to demonstrate the full potential of perhaps the most ambitious aircraft ever created. However, aviation historian Jay Miller has placed the value of the XLR-11 flights of the X-15 into the proper perspective. He asserts that the earlier flights were each invaluable and *necessary* data-gathering and learning experiences. Each flight, he writes, "had expanded the X-15s speed and altitude envelopes to the point where new records were being set during virtually every mission."[10] Moreover, the XLR-11 flights "served to familiarize...pilots with the X-15s unique flight characteristics," while the "the biomedical aspects of the [especially the early] X-15 program proved extraordinarily productive."[11] In addition, the "last XLR-11 flights also served to allow installation and low-speed testing of the new Northrop-developed 'hot nose' (sometimes referred to as the 'Q-Ball' nose)."[12]

TECHNOLOGICAL ROOTS OF THE XLR-99

The early history of the XLR-99 itself was troubled, but the troubles did not prevent the program's many record-breaking triumphs, some of which still stand to this day. This remarkable engine also fulfilled the extraordinarily high expectations placed upon it that make it stand out in the annals of aerospace history.

The XLR-99 Pioneer benefitted from several of its very important technological roots. One of these was the "spaghetti" configuration of its thrust chamber. Back in about 1947, the highly gifted and innovative young New Jersey-born engineer Ed Neu, Jr. (introduced in Chapter 7), who had recently joined RMI, arrived at the ideal way to not only greatly improve rocket-motor cooling, but to considerably lighten the motor at the same time. The solution was to construct the walls of the chamber with the cooling tubes themselves. The bundle of tubes were ingeniously shaped in the form of the contour of the chamber, then joined tightly together. Consequently, the weight savings for the entire chamber was now also substantially reduced by as much as 50%, compared with a standard chamber. This approach appeared to be a great breakthrough and promised to be a more efficient and practical way to build larger-scale rocket motors.

To some of Neu's RMI colleagues, a motor constructed in this fashion resembled a stack of spaghetti. Hence, the name stuck—the "spaghetti-type rocket chamber."

"Spaghetti" still did not seem quite appropriate, though, and as early RMI pioneer Harry Burdett Jr. recalled, there was actually a friendly tongue-in-cheek debate as to whether it was more semantically correct to apply the term "spaghetti" or "macaroni" to this new technology.[13] Macaroni noodles are hollow and would have technically been more accurate, he concluded, but for some inexplicable reason—perhaps merely because the word sounded better—the term "spaghetti" won out and it became an accepted standard term from that point on.

RMI went on to try out this idea and built such a chamber, or at least a nozzle section or sections, based upon a standard 400-lb-thrust motor. This would have been a Lark-type sustainer power plant. The system worked, although its construction posed formidable technical challenges, like finding the right material for the tubes, insuring a uniformity in overall shape, and precisely welding the tubes together. Among the materials tried were copper, nickel, stainless steel (type 347), Inconel, aluminum, and carbon steel. In each case, spaghetti motors formed of these materials were test fired to see how they held up. The various joining techniques included various types of welding and furnace brazing. The effort became so intense that RMI dubbed this activity "Project Spaghetti"—or more officially, Project 251—an inhouse development, although unfortunately the original documentation on this highly significant development no longer appears to exist.

Eventually, the Solar Aircraft Company of San Diego, California, one of the best-known firms for producing manifolds, heat exchangers for aircraft engines, and other aircraft parts, was contracted by RMI to braze the tubes for the first such rocket motors developed for commercial sale. Solar was also singled out because they also possessed an advanced furnace brazer and the expertise to operate it. RMI did not have such equipment, nor this specialized level of expertise. Just how RMI learned of Solar—especially because this firm was situated all the way across the country—is unknown, although they did their share of searching for such a company throughout the aircraft industry of the time.

RMI's pioneering first spaghetti motors started with their XLR-22 Bleed Turbine Rocket Engine of 5,000-lb thrust, developed under a BuAer contract awarded in 1949. It included other advanced features, notably a turbopump powered by gases bleeding off from the combustion chamber to which was attached a then-unique "bell" nozzle that further offered far greater exhaust efficiency. In addition to these features, their XLR-22 ran on the exotic new choice of liquid ammonia and LOX as propellants. This unusual propellant combination for the day delivered a higher specific impulse—an important measure of the efficiency of rockets or rocket propellants—than many other propellants, and was a pioneering achievement by itself. Ammonia as a rocket fuel was used experimentally earlier by others, although it seems to have been introduced to RMI by their new chief chemist, the Austria-born Dr. Paul F. Winternitz, first mentioned in Chapter 7, who came aboard in 1946. Winternitz worked out calculations for the propellant as well as other more highly energetic fuels.

The bell-shaped nozzle, as found in the small XLR-22, was perhaps another significant first for RMI. The bell, or contour shape of the nozzle, was designed to impart a larger angle expansion of the rocket's exhaust gases, therefore increasing the overall efficiency and performance of the motor. However, the prominent early Rocketdyne rocket engineer and historian George P. Sutton credited Indian-American Gandicherla E.V. Rao with working out the mathematics of this configuration as adapted to large-scale liquid-propellant engines by the 1950s. In September 1953, the rocket section of North American Aviation, which became their Rocketdyne Division in 1955, successfully test-fired their huge, 120,000-lb-thrust XLR-43-NA-3 engine featuring the spaghetti-type chamber construction. However, North American called it a "tubular" or "tube-wall" configuration, although it lacked the bell-shaped nozzle. At any rate, the XLR-43 was a building block and the start of a lineage that evolved through all of Rocketdyne's large-scale engines thereafter, all the way up to the mighty engines on the Saturn V launch vehicle that took the first men to the moon. (Other engines in the lineage are discussed later.)

In the meantime, during the mid-1950s, North American developed their experimental large-scale 135,000-lb thrust MB-1 engine featuring both a bell contour nozzle and the tube-wall chamber construction in a single engine. Both of these early rocket advances—the bell-shaped chamber and the spaghetti, or tubular engine chamber configuration—saw their continued, respective, side-by-side evolution. The bell-shaped engine type configuration served in subsequent North American and later Rocketdyne engines up to the Space Shuttle Main Engine (SSME). At the same time, all these engines included spaghetti cooling tubes—as pioneered years earlier by RMI's Ed Neu.

Back in the early 1950s, RMI went on to promote its XLR-22 as a potential power plant for short-range missiles or high-altitude sounding rockets, as JATOs, or as an engine for piloted aircraft, although it never became operational for any of those applications. At the same time, to further strengthen his claim over perhaps the most significant feature of the XLR-22, its spaghetti chamber, Neu filed a patent for the concept on 5 April 1950. Unfortunately, however, it was not until some 15 years later, on 22 June 1965, that the patent was finally granted as Patent No. 3,190,070 for a "Reaction Motor Construction." It was assigned to the Thiokol Chemical Corporation, then the parent company for the Reaction Motors Division. Sadly, Ed Neu never lived to see the issue of his patent, as he succumbed to health problems and died in 1963 at the young age of 43.

Another early RMI pioneering rocket motor that incorporated the spaghetti configuration was their 5,000-lb-thrust XLR-26-RM-2 Topping Turbine Turbo Rocket. It burned white-fuming nitric acid and kerosene. The XLR-26 was designed as a superformance (extra boost) aircraft power plant and also dated to the early 50s, but it too failed to be picked up by the military. Finally, there was the XLR-30 spaghetti motor, then RMI's largest and most powerful of this type, that produced 50,000 lb of thrust. Meanwhile, Ed Neu's spaghetti concept one way or another became widely circulated throughout the rapidly growing postwar American rocket industry; for proprietary reasons, however, it was not always called by that name. In 1950, North American Aviation's Aerophysics Laboratory, the name of their rocket section before it became Rocketdyne, dubbed their first large spaghetti-type motor design the "Light-Weight Tubular Rocket Motor." Rated at 3,000 lb of thrust, it was scaled up soon after to 75,000 lb, The highly important milestone in U.S. rocketry led up to the XLR-43, which was converted into the power plant for the Redstone missile. It then dynamically evolved over the years in association with other Rocketdyne tubular, or spaghetti, engines, including those for the Jupiter, Thor, and Atlas missiles, moving on to the F-1, H-1, and J-2 engines for the Saturn family of launch vehicles, and after that, the Space Shuttle Main Engine. Still other rocket-engine developers, like General Electric, preferred the term "tube bundle" for this configuration of engines. Aerojet favored "tubular."[14]

As for how North American may have picked up the idea from RMI in the first place, all the details may never be known. According to William W. Mower, who had joined North American in 1947 and worked on their earliest rocket project (the liquid-propellant booster for their projected very-long-ranged Navaho missile), in those years the government let rocket contracts to both North American and RMI, in addition to other companies engaged in rocket work. In doing so they simply "passed on information to the various competitors to speed things up and not pay more to learn the same thing."[15]

THE XLR-30 ENGINE: DIRECT ANCESTOR OF THE X-15 POWER PLANT

The XLR-30 had previous names: initially, the Viking II, then later, the Super Viking. It was a greatly uprated though spaghetti-configuration version of RMI's XLR-10, a more conventional LOX/alcohol 20,000+-lb-thrust engine for the Viking liquid-propellant sounding rocket discussed in Chapter 7. But like its much tinier XLR-22 predecessor, the XLR-30 burned the more exotic combination of LOX/liquid ammonia as propellants. Additionally, its turbopump was powered by gases bled from the main combustion chamber.

A June 1953 RMI data sheet on the XLR-30-RM-2 model therefore rightly claimed it was "essentially an enlarged version of the 5,000 pound thrust XLR-22-RM-2."[16] The XLR-22 features were now enlarged "to incorporate, on a larger scale, the high performance, low weight, high thrust to weight ratio advantages of the smaller...XLR-22-RM-2."[17] The larger scale XLR-30 was experimental by this date, the sheet continues, although it was intended to be used in the high-altitude Viking sounding rocket.

Apart from regarding the XLR-30 as a more ambitious follow-on to the Viking sounding-rocket program, RMI envisioned it as powering "long range bombardment vehicles" (that is, missiles). What is not mentioned on the data sheet, though, is that RMI foresaw a missile up to the 500- or 600-mile-range class for the engine. Earlier, as the engine research was supported by BuAer grants, the Navy had even considered the XLR-30 for propelling a submarine-launched version of the rocket. In retrospect, this made it an interesting pre-Polaris, liquid-propellant, long-range-missile concept, appearing some years before the advent of the large-scale solid-propellant rocket technology that made the solid-fuel Polaris possible.

Fig. 8.3   RMI's XLR-30 "spaghetti" rocket engine and its pump. (Courtesy National Archives.)

The XLR-30 concept had originated as a paper study in June 1949 under BuAer contract NOa(s) 10613, but its hardware development was initiated in July 1951 and continued until at least the mid-1950s. Interestingly, Ed Neu played a major role in its development. According to his obituary in *RR Topics* of the Reaction Research Institute, Inc., of Glendale, California, for December 1963, Ed "was really in his glory when the Navy initiated the XLR-30 engine."[18] He appears to have been the lead engineer on the engine and authored one of RMI's major reports on it, dated December 1952, although Harold Davies was later bumped up to chief design engineer on the project.

Neu was especially anxious to see his spaghetti approach applied to large-scale engines. For this reason, he left RMI about 1954 and went to California, working for Aerojet. Hence, he may well have facilitated Aerojet's adoption of this major technical advance for their Titan ICBM then under development.

In the efforts by both RMI and the Martin Company to scale up the Viking vehicle as a potential missile power plant and perhaps also as a much-higher-climbing sounding rocket, XLR-30 thrust chambers were fabricated and test-fired at Lake Denmark, although technical problems arose. This was one of RMI's largest undertakings to date after all, and they simply could not afford to spend huge sums to complete the project. Consequently, while a Super Viking seemed an exceptionally promising goal, RMI planners had no choice but to drop the program, and this was the recommendation of RMI's *Final Report* on their XLR-30, dated 30 June 1956. Ironically, the report's 15-page appendix dwelled upon the theoretical possibility of applying the XLR-30 to a piloted aircraft. Whether this appendix, or an earlier version of it, had played any role in the X-15 engine competition that took place a year earlier is not known.

THE X-15 AIRCRAFT ENGINE COMPETITION

The earliest history of the X-15 and its crucial engine competition is well documented, notably in the official history of the X-15 program by Dennis R. Jenkins.[19] Therefore, details do not need amplification here. But it is important to know that the contending engines in the competition for the aircraft were Bell Aircraft's XLR-81, Aerojet's XLR-73-AJ-1, North American's NA 5400, and either RMI's XLR-10—the standard Viking power plant—or their newer, experimental XLR-30.

Bell's XLR-81, rated at 15,000 lb of thrust, was under development. It was later cancelled, but evolved into what became the Agena upper-stage power plant. The XLR-73, of 10,000-lb thrust, was also under development and employed white-fuming nitric acid and jet fuel as its propellants. The nature of the NA 5400 is still obscure, but it was apparently not considered sufficient enough for the task and was withdrawn. The 20,000-lb-thrust X-10 as used in the Viking has already been covered in Chapter 7, although it was quickly abandoned for the X-15 project. RMI's XLR-30 has already been introduced.

The more powerful XLR-30 was neither throttleable nor man-rated—two absolutely essential requirements for the X-15 power-plant selection. Nonetheless, RMI submitted its proposal for this engine on 9 May 1955, as did the other contending companies with their bids; they all emphasized that they would follow the provisos for the selection process. This meant that their respective engines were to be modified to emphasize safety, to be throttleable at all altitudes, and capable of at least five successive restarts. Also implicit in these provisos was that the engine had to be man-rated. On 26 October 1955, RMI was notified by the Air Force that it had won the competition.

SCOTT CROSSFIELD'S ROLE IN THE X-15 ENGINE DEVELOPMENT

In addition to the thoroughly documented treatment of the competition and post-competition phases in Jenkins's official history of XLR-99, the existing papers of Scott Crossfield in the NASM Archives now provide a few rare insights about the post-competition XLR-30 evaluation stage, as he was then employed as a design specialist by North American and was heavily involved in developing the aircraft, including its engine. In short, Scott was far more than a pilot, as mentioned earlier in this chapter; he had earned a B.S. degree in aeronautical engineering in 1949 from the University of Washington and a master's in aeronautical science in 1950.

Crossfield's papers consequently reveal that he had direct and important interactions with RMI, including meetings with Robert W. Seaman Jr., RMI's head project engineer for the XLR-30—and soon to head the new XLR-99 development. Crossfield likewise met with Harry A. Koch, who became the XRL-99 program manager. In addition, Crossfield more closely examined and evaluated the XLR-30 shortly after it had been selected in late 1955. He was concerned with such issues as the engine's minimum thrust of 17,500 lb that would "not admit less," and surmised that if thrusts were reduced to 10,000 to 12,000 lb, the engine would operate "with very low efficiency."[20] He was likewise not wholly satisfied with the XLR-30's level of cooling. "Cooling," he observed, "is a tough one with RMI and is tied up with fuel flow [and] hence at low flow rates it appears that sweat cooling on wet walls will be necessary."[21] He thus strongly felt that the XLR-30s regenerative cooling system was not fully adequate for the far higher standards required for the X-15; it would need supplemental modes of cooling.

A few of Crossfield's terse though essential recommendations are likewise discovered in his notes, such as: "Frangible discs should be added to all relief valve installations"; "Recommend emergency shut off valves for main pump in fuel lines"; and "Titanium impact strength very low at low temperatures— delete from lox tank."[22] There are also revealing self-reminders: "Investigate insulation of the LOX tank"; and "High freq[uency] vibration effects upon transmitters and/or instrumentation [of the aircraft] Has RMI got any dope [on this possible problem[?].]"[23]

Also included in the notes is a marvelous large North American schematic drawing, dated 30 December 1955, of the X-15 fitted with an XLR-30, complete with a layout of the propellant pumps, other connecting lines, and main components. This was a start—on the way to designing the far more sophisticated replacement for this engine, the XLR-99, though of similar thrust.

EARLY DEVELOPMENT OF THE *XLR-99*

The development of the XLR-99 Pioneer was planned to take two years. As part of the specifications, it was to be a single chamber of 50,000-lb thrust at sea level and 57,850 lb at 100,000 ft, with a duration limited by propellant supply and throttling—usually 90 seconds at full power. The dry weight was set at 915 lb, and the engine was to restart up to six times during a single flight. At higher altitudes with less air, the nozzle could expand gases more completely. Yet despite RMI's excellent reputation, the XLR-99 presented tremendous technical challenges, particularly in man-rating it and achieving full throttleability with restart capabilities. These requirements were infinitely more stringent than any they had ever faced. According to Robert W. Seaman, "The XLR-99....was the first rocket engine with this [man-rating] requirement...that no single malfunction of the of the engine shall result in hazard to the pilot's life."[24] Consequently, the fully developed XLR-99 bore almost no resemblance to its XLR-30 forebear.

According to Davies, there were two main elements in the first phase of the XLR-99's development that "demanded special attention."[25] These were the

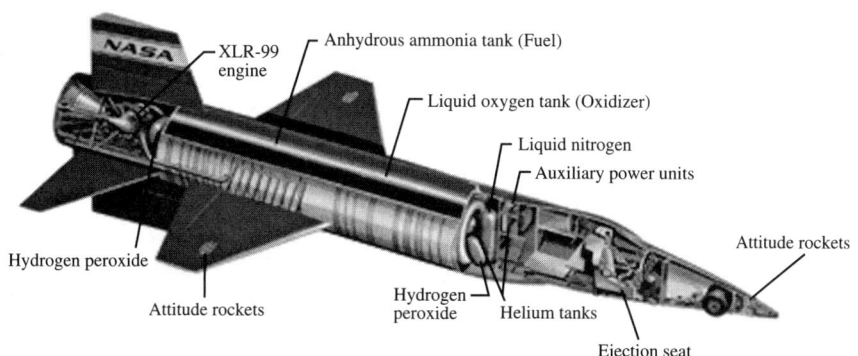

Fig. 8.4   Schematic drawing of the X-15 showing the placement of the XLR-99 engine. (Courtesy NASA.)

LOX/ammonia propellant combination and the means of achieving absolute combustion safety during starting and shutdown.

Despite the high specific impulse of liquid ammonia as a fuel and RMI's long experience with it going back to the XLR-22, there were some initial aversions to adopting it for the X-15 due to its perceived toxicity, corrosion to certain metals, and handling problems. During and after the engine-selection process, though, it was determined that while liquid ammonia was toxic, the exhaust gases were not, and the corrosion and handling problems were annoying but tolerable. Apart from these factors, conversion from ammonia to a hydrocarbon fuel would have been excessively time-consuming for the program. Ammonia was thus retained as the rocket fuel, but a host of other problems cropped up in handling the stuff.

"The great stability of the ammonia molecule," according to rocket propellant historian John D. Clark, "made it a tough customer to burn and from the beginning they [the RMI engineers in the XLR-99 program] were plagued with rough running and combustion instability. All sorts of additives to the fuel were tried in the hope of alleviating the condition...The combustion problems were eventually cured by improving injector design, but it was a long and noisy process."[26] To this, key XLR-99 RMI engineer Edward C. "Ed" Govignon, added: "We had a lot of trouble developing the injector...The injectors had many small parts and each had to be machined to extremely close tolerances. This took time."[27] Engineer Harold Davies recalled that in all, some fifteen injectors were tried, and during this very difficult phase a major explosion occurred—on 23 January 1958—part of the "noisy process" alluded to by Clark.[28] Eventually, a multispud (spuds were injector elements), showerhead type of high-strength Inconel X alloy was chosen, and this solved the problem.

A second major accident in the XLR-99 development program took place exactly a year later, in 1959, and was due to exploding fuel that had escaped into a heated environmental chamber. This was another instance in which much was learned from the disaster and in the area of achieving combustion safety. As a result of the accident, it was determined that preventing the accumulation of the propellant in an unburned state could be resolved "by the use of of a large, continuously operating igniter which would initiate the combustion at a low level, run continuously during operation and over-run main combustion after shut-down," according to Davies.[29] A "generous source of hot gas" was thus made available to vaporize any liquid propellants that might be present in the combustion chamber, also making sure that the flammability of LOX/ammonia were not exceeded and "safe ignition would be obtained without elaborate controls."[30]

For extra safety's sake, ignition occurred in two stages. In the first, three spark plugs ignited the incoming LOX and ammonia for combustion at a low level, while the heat of the igniter also vaporized and expelled any fuel that may have accumulated. The propellants were then routed to the second-stage igniter. Pressure switches in both stage igniters were sequenced so that the propellants did not enter the main thrust chamber unless the first and second stage igniters had built up the necessary pressure. These safeguards also ensured there was no explosion in the main combustion chamber, nor throughout the propellant flow system. Along these same lines, helium was introduced at engine shutdown into the drain cavities to clear any possible leakage.

Still another newly created major safeguard was RMI's incorporation of a system of several electrical circuits that automatically shut the engine down in the event of malfunctions such as over-speed of the turbopumps or excessive vibrations. This system, explained Davies, "required special development and this work was carried out in our Component Development Laboratory."[31]

The thrust chamber itself passed through many modifications. This, said Davies, was because "The weak spot of the [XLR-30] design was the chamber"; regenerative cooling and the spaghetti configuration pioneered by RMI were retained but were entirely redesigned.[32] For the X-15, the "two pass type" of cooling pattern was chosen in which the coolant, liquid ammonia, "passes down one tube and returns in the adjacent tubes."[33] The preformed tubes were made of AISI (American Iron and Steel Institute-certified) stainless steel and individually and carefully flow-checked with very high pressure water and selected prior to the assembly of the tube bundle that made up the chamber. The finished bundle contained 196 tubes furnace-brazed together. Nickel, aluminum, and stainless steel were all considered in the early developmental stage as tube materials. Some 30 different tube variations were tried and closely analyzed for their heat transfer characteristics, added Davies, until the ideal one was chosen. Stainless steel won out due to its exceptional strength, high melting point, and excellent thermal conductivity.

All the while, there was great anxiousness to speed up the very long and drawn-out developmental process. Therefore, the Air Force funded Rocketdyne to assist RMI in the developments of alternate injectors and a thrust chamber, based upon Rocketdyne's XLR-105-NA-1 Atlas ICBM sustainer engine. Matters became further complicated when RMI management and employees became temporarily distracted by the merger of RMI with Thiokol that became approved by mid-April 1958.

In addition to the great deal of attention paid to redesigning the chamber, there was an extra effort to enhance the engine's overall durability. For this purpose, a high-heat resistant ceramic coating was sought to add to the interior of the chamber. The selection was in the same family of coating used in the Bullpup missiles—Rokide. According to Harry Koch, "We went through Rokides A to Z and finally decided on the right ceramic."[34] This was literally the truth as they ended up choosing Rokide-Z, so-called for the zirconium oxide within its chemical makeup. Kenneth "Ken" Gaddis, who had earlier worked on the old 6000C-4 engine back in the X-1 days, had helped design and fabricate a modified type of oxyacetylene gun to more evenly apply the sprayed-on Rokide inside the XLR-99 chambers.

"The XLR-30 fuel pump," Davies continued, "was taken as is, for the XLR-99, with a new oxidizer [LOX] pump to permit an axial flow inducer. The bearings and seals also followed the earlier [XLR-30] design, although the XLR-30 fuel pump had been "quite radical" in its day.[35] The LOX/ammonia propellants were fed into the engine by the two respective centrifugal pumps at flow rates of more than 10,000 lbs per minute; these pumps were driven by the high-pressure decomposition products of 90% hydrogen peroxide. Haakon O. Pederson was RMD's principal designer of the engine's turbopumps. The development of the pump system was another troublesome and very time-consuming aspect of the history of the XLR-99; it experienced constant structural and burst failures, due in part to the much higher pressures encountered in the engine than RMI and RMD engineers had ever faced. These difficulties also took time to analyze and resolve because of newly instituted, stringent quality control measures.

The XL-99 turbopump system, according to a *Missiles and Rockets* write-up at the time, was virtually the "key component in the [XLR-99] rocket engine" and was "the most significant flight feature of the...engine, its controllability."[36] The throttling was "controlled by altering the turbo pump's speed, thereby affecting the propellant flow to the thrust chamber."[37] The speed was altered by an ingenious throttling valve that permitted variable-thrust ratings; originally, this was to be from 13,500 lb up to the full 50,000 lb, or about 25% to 100%, but was finally settled on at from 30% to 100%.

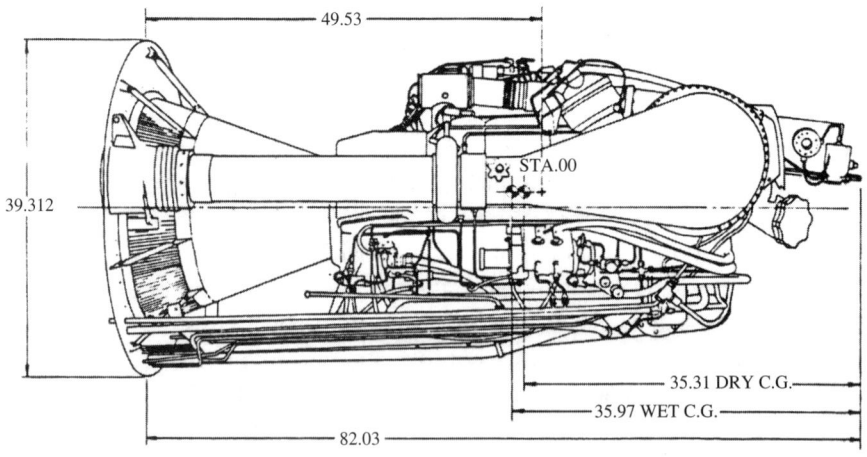

Fig. 8.5   Drawing of the XLR-99 engine. (Courtesy Frank H. Winter collection.)

FIRING TESTS OF *XLR-99*

Naturally, the first XLR-99 static-firing tests were carried out at Lake Denmark. Once again, however, there arose the problem of noise complaints—an endemic headache for RMI and RMD alike. Although their Lake Denmark location initially seemed to make the company finally immune to such complaints—because the testing facilities were on remote areas of a military arsenal that was used to detonations of all kinds—urban sprawl now reached the environs of Picatinny Arsenal. Moreover, the constant test firings of 50,000-lb rocket engines for three-minute runs were something quite different. They were exceedingly more intense and longer than the local populace were accustomed to, recalled former RMD manager Ed Seymour. The long blasts from the test firings certainly "must have been annoying to nearby residents," he said in an understatement of the situation.[38] "We tried to cooperate, but with development behind schedule, we had to run whenever the engine was ready—which sometimes meant at night."[39] Lawsuits inevitably unfolded, which are discussed in Chapter 11.

Fig. 8.6   The X-15 with the XLR-99 engine installed in it. (Courtesy Smithsonian Institution, 83-16760.)

Test firings were not confined to Lake Denmark. They were just the beginning of an extremely extensive testing program for probably the most sophisticated American rocket engine development up to that time. Ten runs were made, for instance, in a simulated 70,000-ft-high-altitude environment in a large wind tunnel at the Air Force's Arnold Engineering Development Center at Tullahoma, Tennessee. Then, on 7 June 1959, the first ground-test version of the XLR-99 (Serial #101) arrived at Edwards Air Force Base and underwent its first hot test on 26 August. Notwithstanding, there remained many more months of testing by both Air Force and North American crews on the Propulsion Test Stand (PSTS) at Edwards Air Force Base before the Air Force could at last grant their approval for flight status for the XLR-99.

However, it was not until 2 June 1960, that X-15 #3—the only one of these aircraft never to have been fitted with the dual XLR-11 interim engines—made its beginning ground run with the "Big Engine," as they now called the XLR-99. The purpose of the ground tests was to demonstrate and test throttling and restart capabilities of the new engine. But on 8 June, during the third ground run of X-15 #3, with Crossfield at the controls, a catastrophic explosion took place. Most fortunately, he was uninjured.

MAIDEN FLIGHT OF THE X-15 WITH THE "BIG ENGINE"

The accident investigation later revealed that the engine was not at fault. The failure was traced to a minor pressure regulator that had been custom built for ground tests. Consequently, X-15 #3 was rebuilt and the regulator reconfigured to prevent a repeat of the mishap. Then, following a few more glitches that were easily repairable, Crossfield achieved the distinction of performing the first XLR-99 flight on 15 November 1960 in X-15 #2. This flight demonstrated the engine at 50% thrust and the plane reached Mach 2.97 (1,960 mph). There were still some growing pains with the XLR-99, such as incompatible seal materials, that were fortunately easily fixed.

THE OPERATIONAL XLR-99 TEST FLIGHTS

The last of the XLR-11 flights was made on 7 February 1961. Exactly a month later, on 7 March, the X-15's first XLR-99 NASA mission was carried out with Major Robert M. "Bob" White at the controls. His X-15 #2 became the first aircraft ever to fly past Mach 4 (Mach 4.43, or 3,399 mph) and reached 77,450 ft. Eight months later, on 9 November, White was the first to hit Mach 6, when he flew to 4,603 mph. He was likewise the first to go hypersonic—a speed of Mach 5 and above.

From there on, the X-15 continued its incredible flight program until its last flight, No. 199 on 24 October 1968—after nine years of test flying. The

Fig. 8.7   The renowned X-15 #2 in flight, rocket engine ablaze. (Courtesy NASA.)

XLR-99 Pioneer thereby fully lived up to its name and enabled the hypersonic aircraft to write many more glorious pages in aerospace history. This included flights above 50 miles—literally into the fringes of space—and earning some of its pilots astronaut wings. Although as stated in earlier chapters of this book that space is technically defined as 100 km (62 mi) and above, the X-15's flights were so exceptional that the U.S. Department of Defense started awarding astronaut badges to military and civilian pilots who flew aircraft higher than 50 miles (80 km).

On 17 July 1962, Bob White became the first to top the 50-mile mark, when he took X-15 #2 up to 314,750 ft, or 59.6 mi. Joseph A. "Joe" Walker went higher. On the X-15's famous Flight 90 of 19 July 1963 Walker climbed to 65.87 mi, making his X-15 the first aircraft to technically fly into space when he passed the "Karman line," explained in Chapter 7. The X-15 also became the world's first reusable spacecraft. The next month, Walker reentered space. On 22 August he soared to 67 mi, the highest flight ever achieved by the X-15.[40] On this flight, the pilot ran his XLR-99 engine at 100% power for 85.8 s, with burnout occurring around 176,000 ft on the way uphill. He then coasted to apogee, where he reported that the Earth was truly round. Joe Walker was thus the only X-15 pilot who entered space twice. To William J. "Pete" Knight goes the added distinction of achieving the X-15's fastest flight. That occurred on 3 October 1967, attaining Mach 6.72, or 5,156 mph at 102,100 ft.[41] To this day, more than half a century later, none of these outstanding aircraft records have

Fig. 8.8 The X-15, rear view, on exhibit in NASM, showing the XLR-99 engine. (Courtesy Frank H. Winter.)

been broken. The next chapter chronicles still further triumphs of Reaction Motors—when their rocket motors literally reached the moon.

ENDNOTES

1. Jay Miller, *The X-Planes*, Aerofax, Inc., Arlington, TX, 1988, p. 117.
2. A. Scott Crossfield, *Always Another Dawn*, The World Publishing Company, Cleveland and New York, 1960, p. 293.
3. Ibid, p. 294.
4. Ibid.
5. Ibid.
6. Ibid.
7. Dennis R. Jenkins, *X-15: Extending the Frontiers of Flight,* U.S. Government Printing Office: Washington, DC, 2007, NASA SP-2007-562, p. 203.
8. Ibid.
9. Ibid, p. 229.
10. Jay Miller, *The X-Planes*, p. 119.
11. Ibid.
12. Ibid.
13. Telephone interview with Harry W. Burdett Jr. by Frank H. Winter, 25 June 1982.

14. This is just a very brief overview of the later development of the spaghetti-type motors. For a more complete account, consult Frank H. Winter, "On the Spaghetti Trail: The Story of a Revolution in Modern Rocket Technology," in Otfried F. Liepack, ed., *History of Rocketry and Astronautics,* AAS Series, Vol. 34, Univelt, Inc., San Diego, 2011, pp. 293–340.

15. Telephone interview with William W. Mower by Frank H. Winter, 30 Jan. 2001.

16. RMI, data sheet, XLR-30-RM-2 (originally from untitled RMI looseleaf notebook), circa June 1953, n.p.; copy of this sheet in "Reaction Motors, Inc.—XLR-30" file, NASM.

17. Ibid.

18. "Ed Neu, RRI Trustees, Passes Away," *RRI Topics* (Newsletter for Research Associates of the RRI, Glendale, CA), No. 9 (Dec. 1963), pp. 2–3, courtesy George S. James.

19. Consult Jenkins, *X-15*, pp. 189–23, for his thorough treatment of the X-15 engine competition.

20. Scott Crossfield, handwritten notes, "RMI Comments," and other notes, n.d., but circa Nov. 1955 to Jan. 1956, in A. Scott Crossfield Papers, NASM Archives, box 16, folder 7.

21. Ibid.

22. Ibid.

23. Ibid.

24. Letter from Robert W. Seaman Jr. to Frank H. Winter, 30 May 2009.

25. Harold Davies, "The Design and Development of the Thiokol XLR-99 Rocket Engine for the X-15 Aircraft," *Journal of the Royal Aeronautical Society* (London), Vol. 67, Feb. 1963, p. 79.

26. John D. Clark, *Ignition! An Informal History of Liquid Rocket Propellants,* Rutgers University Press, New Brunswick, N.J., 1972, p. 104.

27. Telephone interview with Edward C. Govignon by Frank H. Winter.

28. Clark, *Ignition!*, p. 104.

29. Davies, "The Design," p. 80.

30. Ibid.

31. Ibid, p. 87.

32. Frank H. Winter and Frederick I. Ordway III, *Pioneering American Rocketry: The Reaction Motors, Inc. (RMI) Story, 1941–1972,* AAS History Series, Vol. 44, Univelt, Inc., San Diego, 2015, p. 281; Letter from Harold Davies to Frank H. Winter, 26 June 2007.

33. Davies, "The Design," p. 81.

34. Telephone interview with Harry Koch by Frank H. Winter, 15 April 1987, notes in "Reaction Motors, Inc." file, NASM.

35. Davies letter.

36. William Beller, "Turbopump Key to New X-15 Engine," *Missiles and Rockets*, Vol. 7, Oct. 1960, pp. 33–34.

37. Ibid.

38. Interview with Dr. Edward H. Seymour by Frederick I. Ordway III, 25 Jan. 1986, and other communications; quoted in Winter and Ordway III, *Pioneering American Rocketry*, pp. 146–147.

39. Ibid.

40. Ironically, it appears that Joe Walker never officially received astronaut wings for this highest flight.

41. Plans were entertained for an even faster and higher-flying orbital-mission X-15 model, designated the X-15A-2, for flights up to Mach 8, using an uprated engine of approximately 64,000-lb thrust and the conversion to a fuel blend of hydrazine and ammonia, although this was never carried out. See "Engine Changes Proposed for X-15A-2," *Aviation Week*, Vol. 80, 23 March 1964, p. 24.

Chapter 9

SURVEYOR ON THE MOON

"We were involved in something wondrous, something monumental, and when man first landed on the Moon, there was a little bit of each of us with him."

–Harry W. Burdett Jr.

HIGH HOPES FOR SMALL ENGINE MARKET

In the heyday of America's space program during the mid-1960s, Thiokol's Reaction Motors Division (RMD) sought to capture the vernier and other small auxiliary-rocket market because the big boys—Rocketdyne and Aerojet, on the West Coast—now dominated the large-scale rocket-engine business. Those two companies not only possessed superior resources in terms of finances and sprawling facilities, but were also blessed with enormous and remote expanses of land upon which to test fire their giant rocket engines. In addition to these concrete advantages over Reaction Motors, they enjoyed year-round ideal weather to carry out their tests. Rocketdyne's Santa Susana test site, affectionately known as "Suzy," had been in operation since the late 1940s. Situated some 40 miles north of Los Angeles, it was naturally and acoustically protected within southern California's Santa Susana Mountains by a bowl of massive boulder formations.

With the exception of their Bullpup packaged missile-motor manufacturing plant in Bristol, Pennsylvania, Reaction Motors's operations and testing had always been confined to rural northern New Jersey. Hence, contrasted with the big California rocket companies, they were always far too close to urban population centers or to ever-threatening encroaching urban creep, even on the outskirts of Lake Denmark. As seen in earlier chapters, this made them especially prone to noise and damage complaints from their neighbors. These geographical factors had a compound effect, because location also hindered their growth potential.

Despite the disadvantages of their placement, Thiokol headquarters still envisioned a highly lucrative market for RMD with smaller, extremely precise, and reliable vernier and other rocket motors that were critically required for attitude and velocity adjustments for the really big vehicles up to the manned Apollo Saturn launch vehicles and beyond. In those years, too, America's space program seemed to be optimistically limitless in its future goals and prospects. Post-Apollo projects of planetary probes and manned Mars expeditions were regularly proposed, and all these types of ambitious projects would need tiny verniers to help finely adjust their courses through deep space.

VERNIER ORIGINS

The vernier rocket motor was not named for the visionary novelist Jules Verne, as might be expected. Rather, the name references the French mathematician Pierre Vernier (1580–1637), who invented an extremely precise caliper, or measuring tool. It is ironic that Verne must be accorded credit for being the first to conceive the application—even just a clever literary one—of firing off small firework-type rockets to adjust the course of a spacecraft. This highly fanciful idea appears in his 1870 book *Around the Moon*, the sequel to his 1865 novel *From the Earth to the Moon*. In the former, Verne has his fictional astronauts use onboard skyrockets to help alter the course of their space capsule back toward Earth to prevent a crash as they closely approached the lunar surface or flying off forever into space. Verne also certainly deserves recognition for accepting another fanciful idea at the time—that a rocket can work in the vacuum of space. As Chapter 1 describes, for centuries most people believed that a rocket moves because its exhaust gases push against the atmosphere.

START OF THE SURVEYOR PROJECT

A century after Verne's novels appeared, the fantasy of man's first flight to the moon became a reality. But it was realized in a completely different way than the French novelist could ever have imagined. The Space Age began on 4 October 1957 with the Soviet launch of the first unmanned artificial satellite, Sputnik 1. At the same time, the beeping Sputnik initiated the Space Race between the United States and the U.S.S.R. Fewer than four years later, President John F. Kennedy delivered his famous speech of 25 May 1961 to a joint session of Congress in which he set as a national goal the achievement "before this decade is out, of landing a man on the Moon and returning him safely to Earth."[1] This was the start of Project Apollo.

Project Surveyor was a necessary precursor to the manned Apollo landings. Its primary mission was to demonstrate the feasibility of making soft landings on the surface of the moon, while its secondary objective was to survey the best potential manned lunar mission landing sites. In July 1964, RMD succeeded in securing a subcontract from NASA's Jet Propulsion Laboratory (JPL), managers of the Surveyor program for NASA, to develop and produce the verniers for the unmanned Surveyor lunar probes, destined to become America's first spacecraft to soft-land on the Moon.

Hughes Aircraft Company was the main contractor to develop and build the spacecraft itself. The almost $500 million program called for seven spacecraft to be built and flown. Each craft was to travel directly to the moon on an impact trajectory on a journey lasting 63 to 65 hours, then conclude with a deceleration of just over three minutes to its soft landing. In addition to the primary goal of accomplishing soft landings, the mission of each craft was to demonstrate its ability to make midcourse corrections en route between the Earth and the moon. This meant that the verniers were required to make multiple inflight restarts on instant commands from either an onboard autopilot or from Earth stations. During the final landing maneuver, the function of each vernier was to very finely control the landing velocity—to trim this velocity after the descent retro had fired.

On the moon, instruments on the landers were designed to help evaluate the suitability of landing sites for the manned Apollo landings to follow. The verniers were thus to play several crucial roles in every mission to fully enable each landing.

A Closer Look at the Surveyor Verniers

Physically, the verniers were quite small yet compact: Each weighed only 5.9 lb dry weight (minus tankage) and was 9.3 in. long, with a throat diameter (internal) of 5 in. Despite their small size, they were immensely important and could not fail. Each had to function perfectly in every mission.

Within RMD they were simply called "Surveyor verniers," although the final version was technically designated the TD-339 motor, or TD-339 TCA (Thrust Chamber Assembly). Earlier developmental models are known, however, such as the TD-280. Harry Burdett Jr., then chief of RMD's Advanced Design, was responsible for the overall design of the vernier. According to Harold Davies, who headed RMD's Project Engineering Department when the Surveyor vernier project began, he (Davies) supervised the engineering effort and "saw it through the development phase."[2] Donald Zimmet was the project engineer, but Hughes came to insist upon an independent program manager, and Zimmet was thus assigned to that position. Oscar E. Holt was in charge of the Surveyor vernier components. The main development of the Surveyor vernier took place at RMD's Denville facility, while the motor testing was carried out at Lake Denmark.

Surveyor, the Spacecraft

The Surveyor stood about 10 ft high and 14 ft across when its three legs were extended. The hinged legs were folded within the top stage of the Atlas-Centaur vehicle as it was launched. The basic spacecraft structure consisted of a tripod of thin-walled aluminum tubing and interconnecting braces providing

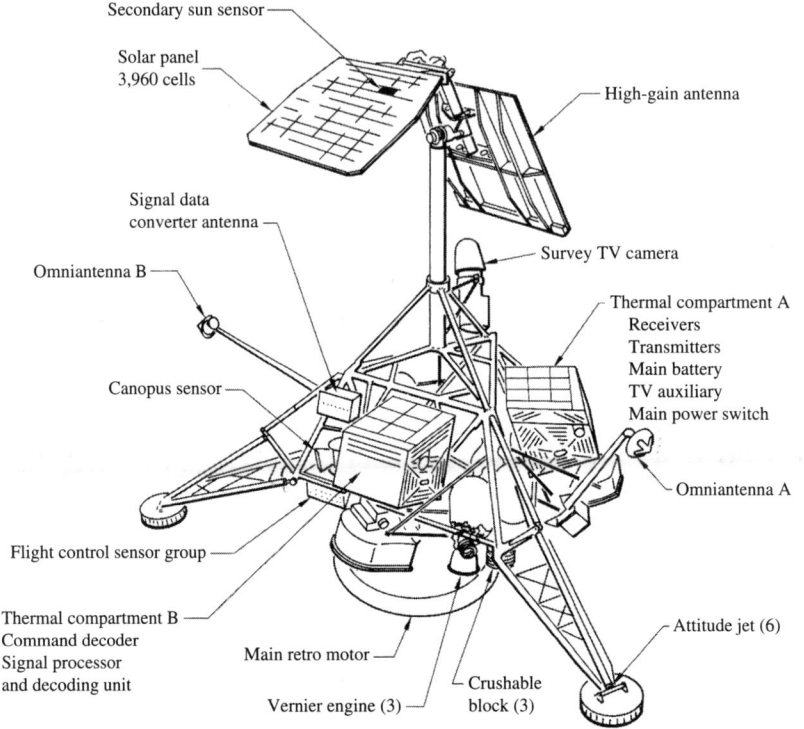

Fig. 9.1   Drawing, Surveyor spacecraft, showing the locations of the vernier motors. (Courtesy NASA.)

mounting surfaces and attachments for the power, communications, the Thiokol solid-propellant retrorocket propulsion in the center (made by Thiokol's Elkton, Maryland division), flight control, and payload systems. The spacecraft's legs held shock absorbers—crushable, honeycomb aluminum blocks—and terminated in footpads with crushable bottoms.

THE VERNIERS ON THE SPACECRAFT

Every Surveyor was also fitted with a set of three RMD verniers; placed equidistantly, each vernier attached to the "knee" of each of the spacecraft's legs. The basic vernier was comprised of three main components: a bell-shaped thrust chamber and injector assembly, a dual-throttle valve, and a dual-propellant on-off valve. The propellants of the vernier consisted of a space-storable hypergolic (self-igniting) combination and, therefore, an igniter was unnecessary. The oxidizer was nitrogen tetroxide with 10% nitric oxide—overall, called MON-10. The fuel was monomethylhydrazine hydrate that contained 28% water, which helped cool the combustion. This exotic fuel was selected for its excellent thermal stability and capability of operating over all

Surveyor's throttling modes in the super coldness of the space environment. The vernier was regeneratively cooled, had a ceramic throat insert, and was also gold-plated on the exterior of its "plumbing" situated above the nozzle.

Every Surveyor spacecraft carried three individual pairs of oxidizer and fuel tanks for the verniers, although a common tank filled with pressurized helium forced the propellants into each of the vernier chambers. The tank was fabricated of high-strength 7078 aluminum alloy, which has high resistance to stress and corrosion. Each propellant was contained in a bladder. Throttle and off-on solenoid valves for each vernier were controlled by electrical signals beamed to the spacecraft by JPL from Earth, as was the helium-gas pressure valve. Each vernier was also independently controlled and throttled with a thrust range from 30 lb to 104 lb. The total duration for each vernier was 4.8 minutes, although the operational firings amounted to a few seconds, or even split seconds at a time. The total thrust level was controlled by an accelerometer at a constant acceleration equal to 0.1 Earth gravity. Trajectory errors, down to minute ones, were sensed by gyros that enabled individual verniers to alter their thrust levels to correct for pitch and yaw errors. In addition, one of the verniers on each spacecraft was built to swivel to correct for roll errors.

Operationally, after about 16 hours into the launch, the verniers were to execute their first function: to help effect a midcourse correction maneuver, as directed by an autopilot with linking gyros that caused the individual engines to change thrust levels to correct for pitch and yaw. Another maneuver corrected for roll, before the spacecraft returned to its cruise mode. Flight controllers at NASA's Goldstone tracking station located in the Mojave Desert near Barstow, California, also monitored the trajectory and could send their own

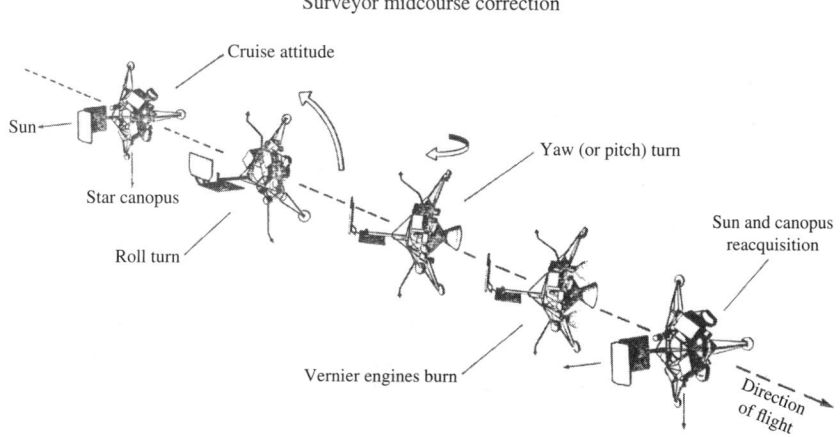

Fig. 9.2 Schematic showing the lunar touchdown sequences of the Surveyor. (Courtesy NASA.)

correction commands. Normally, two maneuvers were required, a roll-pitch and a roll-yaw.

As a Surveyor approached the moon—at about 60 miles from the lunar surface—an altitude-marking radar initiated a flight-control programmer, which in turn activated a clock that counted down and ignited the larger solid-fuel retrorocket mounted underneath the lander. This braked the spacecraft to near-zero velocity. The verniers had also been ignited. Once the work of the retro was done and burned out after 40 seconds, its spent rocket case was jettisoned and allowed to crash on the moon.

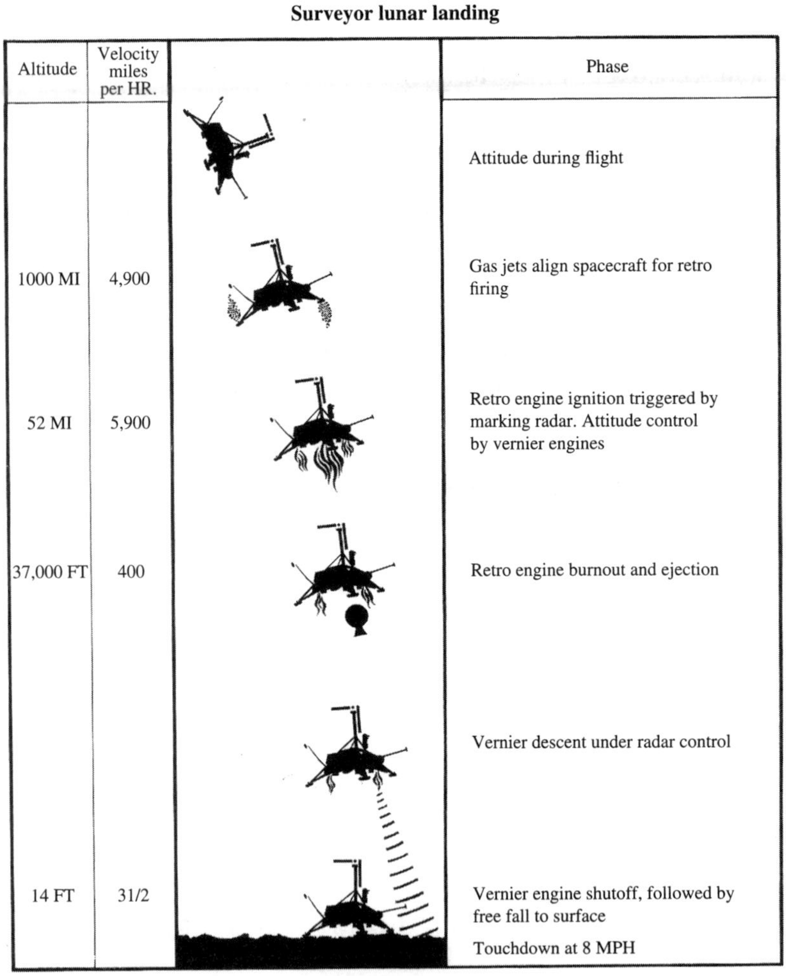

Surveyor lunar landing

Altitude	Velocity miles per HR.		Phase
			Attitude during flight
1000 MI	4,900		Gas jets align spacecraft for retro firing
52 MI	5,900		Retro engine ignition triggered by marking radar. Attitude control by vernier engines
37,000 FT	400		Retro engine burnout and ejection
			Vernier descent under radar control
14 FT	31/2		Vernier engine shutoff, followed by free fall to surface
			Touchdown at 8 MPH

Fig. 9.3   Schematic showing the lunar touchdown sequences of the Surveyor spacecraft. (Courtesy NASA.)

Meanwhile, the three small verniers continued to burn evenly as the spacecraft's Radar Altimeter and Doppler Velocity Sensor (RADVS) beamed signals to flight-control electronics to correctly adjust the throttling of these verniers. At about a 13-ft altitude from the moon, when the Surveyor now traveled at less than 1.2 to 3 mph, all the verniers were cut off. From here on, the spacecraft gently descended in free fall and settled upon the lunar surface.

DEVELOPMENTAL CHALLENGES

Because the verniers were to provide completely reliable, instantaneous starts and very fine throttling in the below-zero coldness of the space environment, these small motors presented a number of highly complex heat-transfer technical challenges to RMD's engineers. There was, for instance, the potential internal danger of boiling and possible decomposition of the propellant. One of RMD's innovations to help solve such problems was a special fuel regulator to maintain constant high pressure that inhibited fuel boiling within the chamber cooling jacket. The motor also featured a "Voramic chamber," described in a later section of this chapter. Cooling was further enhanced by a heat-resistant, thin molybdenum-alloy nozzle extension.

The purpose of the polished gold plating on the motor? The plating formed a layer only one ten-thousandth of an inch (0.00254 mm) thick and had been designed to provide passive thermal control, or thermal balance, on each of the motors to balance heat absorbed from the sun and the amount lost by radiation to space and to keep the lowest levels. That is, the gold plating prevented propellant freezing or boiling during the spacecraft's lengthy Earth-moon trajectory, when it was fully exposed to deep space-temperature extremes. In essence, the gold plating acted partly like a sun screen.

DEVELOPMENTAL HEADACHES

A project with such demanding requirements and exactness was bound to suffer tough developmental difficulties. A small sampling of examples among the lesser problems RMD's engineers encountered in the Surveyor vernier project includes the detrimental effects of the original greases on seals and the oxidizer, paint blisterings, fuel-injector distortions, fuel-regulator clogging, leakages through seals, faulty solenoids, and nozzle cracking.

In fact, the final TD-339 model of the vernier had evolved from a number of earlier configurations, including the TD-280 that had faced far more severe problems, notably loss of thrust control, erratic restarts, throat-insert corrosions, limited component endurance, throttle-valve instability, and throttle-valve leakage. Matters were particularly bad during the programs's earliest days. In its issue of 3 June 1963, the respected aerospace magazine *Aviation Week* singled out the latter set of problems when it reported: "The up-stream throttling [of the Surveyor] vernier engine has experienced some difficulties in its throttling system during development by [the] Reaction Motors Div.[ision]."[3] Proper cooling of the motor became another major concern. In this case, the motor's regenerative cooling presented its own share of complications because this mode produced complex heat-transfer problems with the throttling, with consequent low coolant flow and lower thrusts.

According to historian Clayton R. Koppes, who authored a prizewinning history of NASA's Jet Propulsion Lab, by the fall of 1963 the vernier development "had become the 'pacing element' in the [Surveyor] project...RMD had encountered a host of problems, particularly in the motor-cooling system... Technical disagreements were virtually continuous...between Hughes and RMD."[4] JPL was so concerned that already in April 1963 it had awarded a contract to Space Technology Laboratories (STL) for research on a backup engine.

Finally, things came to a head. As vividly remembered by former RMD manager Edward H. Seymour, one night he received a telephone call from Dr. Robert K. Roney, chief scientist of Hughes Aircraft. "Ed," the call began, "you're going to get this telegram tomorrow terminating the contract! I just want to let you know ahead of time."[5] Seymour thought to himself that there was really "something to this Friday the 13th [of March 1964]."[6] "And on Saturday, the 14th," he went on, "we officially got the word of the cancellation."[7]

Koppes's version of this occurrence, however, differs markedly from that of Seymour, in which he says that it was JPL who, on Saturday, 14 March, "shocked everyone involved with a unilateral order to terminate the RMD subcontract and upgrade the STL effort to a hardware development program."[8] But it may be that Dr. Roney of Hughes had received prior word of JPL's intention earlier on that fateful Friday the 13th.

Another small item in *Aviation Week*, in its issue for 27 April 1964, verifies the basic facts. It reported that the Space Technologies Laboratories Division of Thompson Ramo Wooldridge (TRW) had now been "selected by [the] Jet Propulsion Laboratory to provide throttleable vernier engines for the Surveyor spacecraft."[9] The selection, they went on, was made "after evaluation alongside the Thiokol Reaction Motors Div.[ision] engine originally planned for the spacecraft. Better throttling capability of the STL engine was an important consideration in the selection."[10] In more ways than one, of course, this was a huge setback for RMD, as the decision was made after RMD had reportedly spent about $8 million in developing the engine "before it was discarded," concluded *Aviation Week*.[11]

A few days after the startling phone call that he had received, Seymour journeyed to southern California to visit both JPL and Hughes Aircraft to better judge what had really gone wrong—and, most important of all, to see if anything could still be done about it. Among other things, he learned that out that apart from all the technical troubles RMD itself had encountered in the Surveyor vernier development, Hughes was experiencing a whole group of problems with most of their other Surveyor subsystems contractors.[12] His visit to JPL was especially fruitful.

According to Koppes, after RMD was so "shaken by the termination" that they "adopted many of JPL's suggestions" and also mounted a serious effort at their expense "to right the project. JPL and RMD worked intensively on the technical bottlenecks. By the fall of 1964 RMD convinced JPL and NASA that its motor would be the best after all."[13]

The upshot was that five weeks after perhaps the most distressing phone call Seymour had ever received in his career, he answered another one from JPL. This time, the message was altogether the opposite. "Hey, Edward," the caller began. "Would you consider taking the [vernier] job back?"[14] Once more, Seymour was stunned, though for a different reason, and could respond only with the words: "Just give me time to think."[15] This just took him half a minute, after which which he answered in the affirmative. RMD, it seemed, was miraculously granted a new chance.

A new contract was issued, although it entailed some important managerial and other changes. Mainly, says Koppes, JPL "assumed management of the contract directly without going through Hughes."[16] In addition to following JPL's guidelines and advice, RMD also realized another way out of their main technical hurdles. They learned a lot from their problems and decided upon a complete switch to a new model vernier designated the TD-339. This regenerated vernier development project was picked up in October 1964 and continued—and all worked out smoothly from then on. NASA headquarters later reported that "The performance of RMD since development has resumed under JPL direction has been exemplary."[17]

The development was completed by March 1965; however, the design had been "frozen" in December 1964. One technical solution to one of the earlier problems, by the way, was RMD's innovative Voramic chamber, although it is not known when it was introduced. Voramic referred to a combination of a vortex injector and ceramic throat insert and thrust chamber. As mentioned, a heat-resistant thin molybdenum-alloy nozzle extension was also included, as well as a fuel regulator that maintained constant high pressure to inhibit fuel boiling within the chamber cooling jacket. At any rate, the vernier project

Fig. 9.4   Installing a vernier on the Surveyor spacecraft. (Courtesy Smithsonian Institution, 84-1007.)

passed without major difficulties through its approval test phase, lasting from June to August 1965.

THE NASA/CONTRACTOR PREOPERATIONAL QUALIFICATION TESTS

Both NASA and Hughes held stringent trials for the verniers, like the simulation tests of the overall Surveyor spacecraft. These included important simulations of Surveyor lunar descents, made with scaled-down T-2N test vehicles and conducted at the Air Force Missile Development Center at Holloman Air Force Base, New Mexico. In these runs, the models were raised to altitudes of 1,450 ft or less by tethered balloons. The model was released, simulating a lunar descent maneuver while three onboard TD-339 verniers were activated and throttled from low to high thrust levels and shut down at specified limits before the model touched down upon the desert soil.

Overall, the rigorous tests and qualifications programs involved tens of thousands of seconds of firings and many hours in high-altitude vacuum simulations in addition to the tethered-balloon simulated ascents. Still other trials in both the NASA and main-contractor qualifications phase involved the VPS (Vernier Propulsion System) being subjected at the Hughes Space Simulation Laboratory to vibrational tests up to 3 g's at a certain level, followed by vibrations up to 6.75 g's at another level, after which it was shipped to JPL's Edwards Test Station. At Edwards, the system underwent exposure to a vertical-axis environment on a "shaker table" when the main retro was fired.[18] During this test, the verniers were also fired. Endurance and other analyses were likewise performed. All the tests confirmed the flight worthiness of the verniers and of the spacecraft.

THE SURVEYOR MISSIONS

Finally, the day came on 30 May 1966, when Surveyor 1 was launched from the Air Force Test Station at Cape Canaveral, Florida, by an Atlas-Centaur-D booster toward the moon. En route, as planned, the spacecraft successfully completed its midcourse corrections utilizing the RMD verniers on electronic commands relayed from an antenna near JPL. Radio signals indicating that one of two antennas on the spacecraft may have failed to deploy were the only indication of a problem. Yet, JPL Surveyor project manager Robert J. Parks optimistically commented to the press that this problem would not preclude the possibility of achieving full success from the mission because one antenna was sufficient to receive radioed commands. Surveyor's second antenna had been properly deployed and was working perfectly.

On 2 June, Surveyor 1 successfully soft-landed in the Oceanus Procellarum (Ocean of Storms), a vast lunar mare, or plain, on the near side of the moon— becoming the first such landing by an American space probe onto any extraterrestrial, or astronomical, body. RMD had very much been a part of this history-making event. This period was the height of the Space Race with the Soviet Union, however, and this event occurred just four months after the first U.S.S.R moon landing by their Lunar 9 probe that touched down on 3 February 1966.

The Soviets had therefore won this part of the race because it was Luna 9 that was thus the first ever spacecraft to achieve a soft landing on the moon, or any planetary body other than Earth, and transmit photographic data back to Earth, although the Soviet spacecraft was far less complex than the American one. It was crash proof, with shock absorbers and a hermetically sealed instrument package jettisoned by its carrier rocket immediately before impact. Then it free fell under the pull of lunar gravity. Also, it did not employ a precise vernier system for touchdown, but used four outrigger engines to slow the craft toward its approach. Upon landing, the package transmitted just a dozen photographs.

By contrast, upon its own landing, the American Surveyor craft began beaming the first of its astounding 11,237 still photographs of the lunar surface to JPL's Deep Space Facility at Goldstone, using a television camera in conjunction with a sophisticated radiotelemetry system.

All seemed to bode very well for the next spacecraft in the series, Surveyor 2, and its very smooth launch about three months later on 20 September. But the next day, a midcourse-correction failure resulted when one of the three verniers did not ignite. Asymmetrical thrust caused the spacecraft to roll violently at about 0.85 revolutions per second. This spinning rate was soon multiplied many times over. Frantically, 30 additional attempts were made to start all the verniers; in all cases the same pair of engines ignited while the third did not. It was now abundantly evident that no soft landing could be made. All three verniers were required for the task to stabilize the Surveyor during its decelerating descent to the lunar surface. Surveyor 2 did arrive at the moon and the retro fired. After this, all communication was lost and the entire mission was deemed a failure.

Harold Davies explains that Hughes "was quick to blame RMI for it [the loss of Surveyor 2]."[19] But later events, he adds, showed that the mishap was traced to "an electrical problem unrelated to the [vernier] engines. Hughes was pleased to let the matter drop. The engines on the other [Surveyor] shots seemed to have worked flawlessly."[20]

Surveyor 3 was luckier, although it ran into its own share of unexpected problems. Launched on 17 April 1967, the spacecraft landed on 20 April within the Mare Cognitum (Known Sea) area of the Oceanus Procellarum (Ocean of Storms). However, the site turned out to be located in a crater of highly reflective rocks that confused the spacecraft's lunar-descent radar. This caused the verniers to fail to cut off at 14-ft altitude to slow down the craft to about 3 mph as called for in the mission plans. In turn, the delay caused the lander to bounce twice on the lunar surface. On its third bounce, Surveyor 3 finally settled down to a soft landing.

This Surveyor was able take photographs and was the first of the craft to carry a surface-soil sampling scoop that was used to dig four trenches in the lunar soil. Samples were placed in front of the Surveyor's TV cameras and the pictures radioed back to Earth. When the first lunar nightfall came on 3 May, Surveyor 3 was shut down because its solar panels were no longer producing electricity. Due to the extremely cold temperatures it had experienced, the hapless spacecraft could not be reactivated. Therefore, Surveyor 3 transmitted about half as many images back to Earth as Surveyor 1—just 5,487.

Surveyor 4 carried out a flawless flight to the moon, although signals from the spacecraft ceased during the terminal-descent phase on 17 July 1967, about two and a half minutes before touchdown. Contact with the spacecraft was never reestablished, and the mission was another one lost. Afterward, it was theorized that the retrorocket may have exploded near the end of its scheduled burn.

On 8 September, Surveyor 5 was sent to the moon and was the first to carry an instrument to analyze the chemical characteristics of the lunar soil. Also, for the first time, it headed toward a landing site in the eastern portion of the planned Project Apollo touchdown zone. But soon after midcourse correction was accomplished, a helium leak in the vernier system was detected. Evidently, a regulating valve was not properly closed. The leak was the helium that pressure-fed the propellants to the verniers.

The spacecraft continued on its journey to the moon, affording time for the mission controllers to arrive at plan that could still salvage the mission. They devised an improvised landing sequence that started the retrorocket burn at a 26-mi altitude above the moon—about half the expected height. This allowed the verniers to still bring the spacecraft down, although in 106 s from a height of only 4,396 ft, or 10% of the usual. In other words, they barely conserved the remaining helium supply. Otherwise, the lack of helium pressure would have shut the engines down and totally aborted the mission.

Through this wise measure, the mission was saved. The landing was successful, Surveyor 5 touching down on Mare Tranquillitatis (Sea of Tranquility)—later becoming world famous as the landing site for the first manned touchdown on the moon carried out almost exactly two years later by Apollo 11 astronauts Neil Armstrong and Buzz Aldrin. The number of images gathered surpassed that made by Surveyor 1, totaling 19,049 transmitted to Earth, including stunning panoramas of the mare surface. In addition to these outstanding accomplishments, the miniature chemical-analysis lab obtained invaluable new data on the nature of the lunar surface soil.

Surveyor 6 was the fourth of the Surveyor series to successfully achieve a soft landing on the moon, on 10 November 1967 in a small mare in the Sinus Medii, (Central Bay) region. It obtained the most television pictures up until then: 30,027 images total. This region was the last of four potential Apollo landing areas. Surveyor 6 also determined the abundance of chemical elements in the lunar soil, acquired touchdown-dynamics data, as well as thermal and radar reflectivity data, and carried out a successful vernier engine erosion experiment.

In this highly unusual experiment conducted on 17 November, the spacecraft's vernier engines were fired for 2.5 s, lifting Surveyor 6 to a 10-ft altitude. The spacecraft then touched down 8.5 s later at about 8 ft from its original resting spot. This maneuver was the first ever liftoff of a spacecraft *from* the moon, even if the distance it carried was absurdly short. The whole purpose was to test the engine erosion effects upon the lunar surface; that is, to study soil-erosion effects and to determine lunar soil properties. The results were observed with the television camera. Determining lunar-surface bearing strength was one of the scientific objectives. Post-mission studies of resulting stereo photographs revealed cratering and scouring effects on areas underneath the engines blasted. Overall, Surveyor 6 accomplished all the planned objectives and fully satisfied the Surveyor program's obligation to Project Apollo.

The final spacecraft in the series—Surveyor 7—was another remarkable success. Launched 7 January 1968, the spacecraft touched down upon the moon on 20 January 1968, at the outer rim of the crater Tycho.

All systems operated perfectly. The first of two scheduled midcourse maneuvers by the RMD verniers was so precise that the second was declared unnecessary and the spacecraft was directed to a point only 1.6 miles from the center of its landing target. A total of 21,274 pictures of terrain were transmitted to Earth. They were dramatically different from those of previous Surveyor missions, and included Tycho's rugged landscape and three-story-high boulders. In addition, the spacecraft carried many more scientific experiments and gathered new details on the character and makeup of chemical elements of the lunar surface. Photographs of Earth and Jupiter were also taken, and laser beams were projected to test the feasibility of future uses of laser communications in space. And, of course, more invaluable scientific and engineering data was obtained in support of Project Apollo.

All told, Project Surveyor witnessed 113 successful space firings from 20 RMD verniers—including 40 successful firings by a single Surveyor vernier, as well as the reliable performance of 20 out of 21 verniers; only one did not function for unknown reasons. Surveyor 5's verniers were the first to restart on the lunar surface and the first spacecraft flight from the moon was made by the verniers on Surveyor 6.

Arguably, RMD came very close to entirely missing their chance to make several key contributions to the highly important Surveyor project, although this was saved by the good fortune of the resurrection of the RMD Surveyor vernier after its initial cancellation. RMD thus went on to significantly help further America's space program. Years later, Davies expressed his personal view that "neither Hughes nor RMI [that is, RMD]...received sufficient credit [for their Surveyor accomplishments and were], swept off the map by the much more dramatic Apollo shots."[21]

As a postscript to the Surveyor story, the Surveyor 7 site was planned to be visited by manned Apollo 20 mission, although subsequent budget cuts saw this mission cancelled. RMD's managers entertained hopes that their TD-339-type verniers might be picked up by NASA for the Voyager, Mariner, MOL (Manned Orbiting Lab), and other candidate space programs, but these prospects likewise never materialized. For all intents and purposes, the vernier contribution was RMD's final major rocket project.

Fig. 9.5 Astronaut Charles "Pete" Conrad Jr next to Surveyor 3 during Apollo 12 mission in 1969. (Courtesy NASA.)

Interestingly, on 19 November 1969, the Apollo 12 spacecraft Intrepid landed on an area of the Ocean of Storms, about 600 ft from the site where Surveyor 3 had landed two and a half years earlier in mid-April 1967. On one of their two moonwalks, Apollo 12 astronauts Charles "Pete" Conrad Jr. and Alan L. Bean visited this spacecraft and removed some parts for return to Earth for later analysis. These parts did not include any of the verniers. This was the first—and, to date, only—occasion on which humans purposefully encountered a probe previously sent to land on another world.

Should any future manned lunar landing take place and visit any of the Surveyor sites, any hardware they retrieve (say, for example, any of the RMD verniers) would become a potential artifact of NASM. Upon the termination of a NASA project like Project Surveyor, such a retrieved object would technically be eligible to become accessioned, if so requested, by the museum under the NASA/NASM Artifact Agreement of 1967.

THE C-1

As an outgrowth of the Surveyor vernier, RMD's C-1, or Common Engine, emerged during this period with its hardware development initiated in early August 1964. It, too, was part of Thiokol and RMD's push to capture the small auxiliary-rocket market. It was likewise another species of Radiamic Engine in that it featured the almost identical cooling means—regenerative-radiative cooling. Another designation for the C-1 was the TD-345. RMD engineer Raymond "Ray" J. Novotny was largely responsible for its design. Yet compared to the Surveyor vernier, there were multiple purposes envisioned for the C-1.

The C-1 was conceived as small liquid-propellant auxiliary rocket capable of adaptation to a number of launch vehicles and spacecraft up to the Saturn S-IV-B's stage—the second stage on the Saturn 1B and third stage on the Saturn V—and the Apollo Command and Service Modules. Plans were also projected for the Project Gemini spacecraft. RMD managers thus expended a sizable amount of money toward developing the motor. Among other things, this funding paid for a custom-made computerized high-altitude-simulation test stand that was, by itself, one of the most advanced test rigs in the rocket industry at the time.

C-1 gained its versatility via different types of detachable phenolic-nozzle extensions, or exit cones. The thrust of the basic motor was set at 91 lb, although other thrust ratings were available, such as the 85–100-lb-thrust models, to potentially serve as Radiamic Engines for the Gemini OAMS (Orbital Attitude Maneuvering System).

Fig. 9.6   C-1 motor cutaway in the collection of NASM. (Courtesy Frank H. Winter.)

The standard 91-lb-thrust C-1model, if applied, say, as auxiliaries to the Project Apollo Command Module, could serve as an aft-pitch engine, a forward-pitch engine, a yaw engine, a short-roll engine, and a long-roll engine. For all its applications, the C-1 usually burned nitrogen tetroxide as the oxidizer and a hydrazine fuel, although the latter could vary to an alternate fuel. To gain a picture of just how much effort RMD put into its C-1 prospects, a single prototype C-1 with its "Long Roll Extension" was subjected to to some 5,000 seconds of firing. Like so many other rocket programs, the C-1's development encountered its own technical problems. *Aviation Week* for 23 March 1964 in its regular "Industry Observer" column noted, for instance, that unspecified "difficulties" had been met at this point in working out the ablative thrust chambers.[22] But these were rectified and RMD went ahead to seriously promote the project to NASA.

RMD was initially successful, and they received a $1.5 million NASA Phase 1 (a so-called definition, or exploratory phase) contract from NASA's Marshall Space Flight Center, as announced in the *Wall Street Journal* for 14 October 1965. The objective was to further develop the engine to better adapt it to the needs of Project Apollo. The projected contract "target" value, if NASA would have gone beyond this and adopted the C-1, would have been quite lucrative for RMD, potentially reaching more than $16 million. The C-1 did become fully flight-qualified, but it never became operational. One reason is offered by Maurice E. "Bud" Parker, who worked on the project. The C-1 "development," he says, "was [simply] started too late by [the] Marshall [Space Flight Center] and failed to overtake the primary motors that [the] Johnson [Space Center] had [already] selected and developed earlier at Marquardt and Rocketdyne [for the Apollo program]."[23] In sum, the C-1 did not exceed the performances or advantages of comparable units already selected for Apollo.

NASM, incidentally, has several examples of C-1s in its collection, including cutaways, as well as several Surveyor verniers. A cutaway vernier is presently on exhibit at the Udvar-Hazy Center.

STARMITE

Last, there is one other small family of auxiliary rocket motors of the 1960s period that RMD also developed and promoted—the "Starmite." The Starmite series were teeny and capable of producing thrusts of 0.5 to 10 lb. They were designed as very precise, satellite-attitude controls.

RMD had actually announced the Starmite rockets earlier, in 1961. The industry magazine *Missiles and Rockets* for 10 April of that year presented an impressive illustrated feature on them titled "Compact Power Package." Also dubbed "micro-pulse motors," the compact and very diminutive liquid-propellant rocket motors were undoubtedly the tiniest either RMI or

RMD had ever developed and produced, and they actually delivered pulses as frequent as 22.5 per second. Such rapid and short bursts would have been entirely suitable to shift the attitude of a satellite. Continuous firings for durations as long as ten minutes were also recorded. According to RMD, in addition to their potential applications in satellites, the Starmites could also be deployed as "high performance attitude controls for missiles and space vehicles."[24]

The Starmite propellant was usually nitrogen tetroxide and monomethyhydrazine (MMH), a hypergolic that made sense because no igniters were required and the motors had to be extremely light. Any other storable hypergolics could substitute.

But like the C-1, no Starmite motors are known to have been chosen by NASA for any satellites or other spacecraft or vehicles. Later in 1961, Hughes had announced that the Starmites would be used as the attitude-control system for their Surveyor spacecraft, three per spacecraft to correctly position the spacecraft for landing on the moon. However, Hughes and NASA went with a much simpler and effective cold gas technique instead.[26]

As to why the Starmites, like the C-1, did not succeed with other projects, the answer is probably very similar. Tiny attitude station-keeping motors appear to have been already available and the market for this newer RMD product was not welcoming. The examples of both the C-1 and Starmite also underscore the highly competitive nature of the American rocket industry then and now, and that very many new product entries all too often fall by the wayside. The C-1 and Starmite examples are also representative of the waning years of RMD, which was simply running thin on viable projects once the amazingly successful and lucrative X-15 and Bullpup programs were over.

Again, the geographical factor also manifests itself as a huge and unavoidable problem facing RMD's fortunes during the later 1960s. "Beyond a shadow of a doubt," RMD old-timer Edward C. Govignon remembers, "we were severely restricted to testing large rockets and this [ultimately] contributed toward the demise of RMD."[25] And the huge financial requirements for a move of RMD to another location such as the West Coast for far more expansive testing and development prospects, were undoubtedly far too prohibitive. Psychologically, RMI and RMD had always been tied to their northern New Jersey roots. Here, too, the diminishing cadre of old-timers, who, until the autumn of 2016 still held their biennial reunions, look back with enormous pride and nostalgia to their pioneering days. For indeed, RMI-RMD was America's very first family in modern rocketry.[26]

They could also be proud of the myriad extraordinary and often exotic miscellaneous applications of rockets pioneered by their company, as covered in the next chapter. Some of these applications go back to RMI's earliest days, and also reflect the ingenuity and determination of their founders.

ENDNOTES

1. John M. Logsdon, ed., *Exploring the Unknown: Selected Documents in the History of the U.S. Civil Space Program*, NASA, NASA History Office, Washington, DC, 1995, p. 453.
2. E-mail from Harold Davies to Frank H. Winter, 30 Sept. 2002.
3. "Surveyor Verniers," *Aviation Week*, Vol. 78, 3 June 1963, p. 31.
4. Clayton R. Koppes, *JPL and the American Space Program*, Yale University Press, New Haven, 1982, p. 178.
5. Interview with Dr. Edward H. Seymour by Frederick I. Ordway III, 25 Jan. 1986, and other communications; quoted in Frank H. Winter and Frederick I. Ordway III, *Pioneering American Rocketry: The Reaction Motors, Inc. (RMI) Story, 1941–1972*, AAS History Series, Vol. 44, Univelt, Inc., San Diego, 2015, p. 156.
6. Ibid.
7. Ibid.
8. Koppes, *JPL*, p. 178.
9. "Surveyor Engines," *Aviation Week*, Vol. 80, 27 April 1964, p. 30.
10. Ibid.
11. Ibid.
12. What Seymour may not have learned on this trip was that according to Koppes, the JPL Surveyor project manager, Walker E. "Gene" Giberson, had been immediately summoned to NASA headquarters in Washington, D.C., for a Monday meeting on 16 March 1964 where he was severely criticized for his (Giberson's) "precipitous" decision that had been made "without analysis of the latest test results and without consideration of the effects of the decision on spacecraft interactions." For one, the sudden RMD contract termination threatened to lead to greatly increased costs "in the multi-million dollar category." Koppes, *JPL*, pp. 178–179.
13. Koppes, *JPL*, p. 179
14. Seymour interview.
15. Ibid.
16. Koppes, *JPL*, p. 179.
17. Ibid.
18. A. Broglio, Jr., "Development of the Surveyor Vernier Propulsion System," *Journal of Spacecraft and Rockets*, Vol. 4, March 1967, p. 342. Consult this entire article, pp. 339–346, for a complete account of the development of this system, although from the Hughes perspective.
19. Davies e-mail.
20. Ibid.
21. Ibid.
Like so many other RMI and RMD projects, the Surveyor vernier undertaking was vastly more complex than first appears, and the account here is merely a brief summary. As a hint of its enormity, Robert J. Briggs, who served as one of the testing technicians who worked extensively on this project, was the author of the work titled *Margin Limits Testing of the Surveyor Spacecraft*, as he recalls it (circa 1966), that details all phases of testing of the motor, including firings at below-zero temperature regimes. However, this work of more than 270 pages cannot be located. Telephone interview with Robert J. Briggs by Frank H. Winter, 29 Sept. 2016.
22. "Industry Observer," *Aviation Week*, Vol. 80, 23 March 1964, p. 13.
23. Interview with Maurice E. Parker by Frederick I. Ordway III, 26 March 1982. See also, "Thiokol Chemical Chosen to Develop for NASA 100-Pound Thrust Engine," *Wall Street Journal*, 14 Oct. 1965, p. 10.

24. "Compact Power Package," *Missiles and Rockets*, Vol. 8, 10 April 1961, p. 10; Thiokol Chemical Corporation, Reaction Motors Division, data leaflet, *Starmite Micro-Pulse Rocket Engines,* Denville, NJ, June 1961, passim.
25. Telephone interview with Edward C. Govignon by Frank H. Winter, 17 April 1987.
26. The 20th Reaction Motors Biennial Reunion was held on 11 September 2016 at the Zeris Inn in Mountain Lakes, New Jersey; the author attended this event, which has turned out to be their final reunion.

Chapter 10

MISCELLANEOUS PROJECTS: ROCKET BOATS TO JUMP BELTS

"to attain high running speeds, perform long leaps down, across, or up large obstacles, and to skim water at high speeds."

–RMD booklet on their Small Rocket Lift Device, February 1960.

ROCKET ASSAULT BOAT

From the start, RMI was always highly innovative. Over the years they generated a number of most interesting and often colorful miscellaneous projects involving unorthodox applications of rocket power. Not all were practical or successful.

One notable example showed up as early as January 1942, just a the month after the formation of the tiny company, when Jimmy Wyld drew a sketch of a "Rocket Landing Boat." Undoubtedly it must have been by inspired by the bombing of Pearl Harbor a few weeks earlier. But it was not a well-thought-out idea. In fact, the sketch has the look of a mere doodle, and it may have been casually executed during a brainstorming meeting or phone conversation, perhaps with the person who actually may have originated the scheme. Or it could have been the result of one of Jimmy's own brainstorming sessions. Either way, the actual circumstances may never be known.

The boat was to be rocket propelled and to carry as many as 15 men, in addition to a pilot and two gunners seated behind a light armor screen. The men were to be heavily armed with bazookas, machine guns, and other weapons. Potential missions of the craft were not spelled out, but it obviously was intended for use against enemy troops in a major beach or shore assault, either on Japanese-held islands or even against the Japanese mainland, should the progress of the Pacific campaign dictate that opportunity. The inclusion of bazookas (antitank weapons) presumes the inclusion of enemy tanks in this kind of combat. This was understandable, as the idea was dreamed up very early in the war when the memory of Pearl Harbor was still fresh.

Whether Wyld was the originator or someone else thought up the concept, the whole character of this scheme was overly optimistic and somewhat naive. To begin with, the proposed boat was to be powered by not one, but two 1,500-lb-thrust liquid-propellant rockets. One would be affixed on each side of the stern, and a standard radial auxiliary engine would be mounted at the vessel's stern. This arrangement appears excessively complicated and vastly over-powered, especially as this idea was conceived during the infancy of RMI. They had scarcely begun to work on their first task for the Navy—to develop a modified LOX/gasoline version of Wyld's regen rocket motor capable of producing only 100 lb of thrust.

There were additional major flaws to the plan. For instance, a single craft of this design—or many of them, for deployment in beach-assault operations—would have been prohibitively costly in development for both the craft and the rockets. It was also highly presumptive, from an early-1942 perspective, that a 1,500-lb-thrust liquid-propellant rocket could be fully developed and perfected enough to be considered for very early use in the war. From a current perspective, the choice of a liquid-propellant rocket motor itself was also wholly impractical. Even a pressure-fed system is relatively complicated compared with a far simpler and cheaper, short-duration/powerful-solid-propellant unit, or booster. But solid-propellant rocketry also had very far to go in its technological development at that time, and RMI specialized in liquid-propellant systems. Aerojet, formed two months later in March 1942, developed both liquid and solid-propellant rockets.

Other details of the plan for the landing craft were also overly simplistic: "(1), Start up with both rockets to attain speed; (2), Operate with one rocket to conserve fuel; and (3), Open up both rockets for extra speed when close-in, to avoid [Japanese] machine gun fire."[1]

Nevertheless, those were different times, and due to the pressures of the war, the "rocket landing craft" concept took on a life of it own, appearing to gradually evolve throughout the conflict to become something quite different. There were a few later iterations, including a craft described simply in Wyld's handwritten notes dated 29 January 1944 as a "Rocket Boat." By this juncture, he had fully taken up the idea and had gone to great lengths to work out calculations for two different versions. But the basic idea was still grandiose—or rather, it was more so.

One model was to have a power plant of some 545 HP that was theoretically capable of attaining 68.2 mph. The other was to be fitted with motor of 960 HP for a speed up to an incredible 120 mph. Wyld also calculated weight breakdowns of the basic boat, now based on holding 20 men, each man weighing about 200 lb—maximum, it is presumed, yielding a total of 4,000 lb. The liquid-propellant fuel added some 1,900 lb. The "motor outfit" was calculated at 1,200 lb, and the "auxiliary engine" at 1,000 lb. The hull, the boat proper, came to 3,000 lb, for a grand total of 11,100 lb. Wyld also determined that it would take precisely 20.2 seconds for the boat "to reach top speed (not allowing for drag)."[2] Based upon the level of technology of the day, Jimmy must have still naively believed that a liquid-propellant system was the most promising way to achieve such very high anticipated velocities and overall performances with a boat of this capacity.

At about the same time, an actual rocket boat was built and tested. But its dimensions and power were now considerably toned down. However, it appears that this model of the vessel was just the "test version" of the larger projected rocket landing boat as envisioned by Jimmy Wyld, designed to test the basic feasibility of the concept. This was an entirely logical approach. Even though the available documentation on this project is sparse, it contains invaluable, if brief, remarks on the smaller test boat, as found in G. Edward Pendray's book, *The Coming Age of Rocket Power* (1945). Pendray stated that "a number of experimental runs were made in February and March of 1944 on the Severn River, near Annapolis, Maryland" with a vessel that was "powered by a standard Reaction Motors 250-pound regenerative [sic.] motor, fueled with liquid oxygen and gasoline."[3] RMI's Pierce served as the pilot.[4]

Fig. 10.1   Rocket boat test run on the Severn River, near Annapolis, Maryland, 1944. (Courtesy Frederick I. Ordway III collection, U.S. Space and Rocket Center, Huntsville, AL.)

From these few clues, there is no doubt that the tests (Pendray does say "February and March," so there was probably more than one run) were conducted in the Navy's interest, as they were made on the Severn River, near Annapolis and their Experimental Research Station, where RMI's live JATO tests were conducted the previous month. There are also several photographs of these runs within NASM's "Reaction Motors, Inc." collection that appear to have been taken by—or acquired by—Wyld, as revealed by the captions in his unmistakable handwriting on the backs of several of the pictures. These images also confirm that the boat runs were made near the Navy's Experimental Station.

Pendray did not specify the purpose of the RMI boat, but he had envisioned that: "Entirely new types of craft might well be developed to skim the surface of the water—almost flying—by rocket power. In time of war, they could deliver a torpedo into an enemy craft...Or they could be launched from shore or special carriers or special carriers."[5] But he did not include the possibility of a superfast "rocket landing craft," much less one meant for part of a massive naval assault upon the Japanese homeland, either because he had simply not considered this wartime application, or because this part of the project was top secret at the time. (From his dated acknowledgments, it is known that Pendray had completed his book by early December 1944, although it was published in 1945.)

Pendray, who was much taken with the project, also went into potential peacetime applications: "In peacetime, fast rocket-driven boats of this sort could be used for carrying mail from ship to shore, for rescue work, [or] for sports events."[6] The Navy, of course, would hardly have been interested in those particular modes of use, only the wartime applications, especially during that critical time in the conflict before them. Indeed, by early 1944, long-range planning for a potential invasion of the Japanese mainland was already underway.

The other interesting clue in Pendray's description—that the rocket motor employed in the test was a standard Reaction Motors 250-lb-thrust LOX/gasoline motor—points to a modified Lark missile sustainer, although the Lark normally employed nitric acid and aniline. Louis Arata did recall that a Lark sustainer was involved. As Shesta remembered only that the boat's motor "burned a very long time, about five minutes," this is very likely because the sustainer ran for about 260 s or 4.3 min. Moreover, Shesta informed the author some years ago that a static test run of 4 min was made with the boat's motor and that: "The boat [itself] was provided by some one in Annapolis. I forget who. It was the worst possible craft. Flat bottom without keel; it had no longitudinal stabilizer. For [a] rudder it had two: one at each side of the stern that could be deflected away from the hull on one side or the other."[7]

Shesta continued with his remembrance of the trial run:

> The first time we tried it at Annapolis, Pierce driving, we [were] following another boat, [and it] was a ludicrous sight. He would start up the rocket[,] build up speed and disappear in a cloud of of steam. A few seconds later he would emerge from the cloud and he would be going at right angles to his former direction. A moment later he would be headed in still a different direction. The motion was so erratic that he could not build up any appreciable speed. No wonder the Navy was not interested [anymore].[8]

Pendray's own description of the run confirms the above and adds a detail or two: "Mr. Pierce found it no problem to reach speeds up to 40 miles per hour, but control of the boat became progressively more difficult as the velocity increased. No simple method of steering, as by rudder, seemed practical when full motor power was on, and he concluded that steering at high speeds would have to be accomplished by mechanically changing the direction of the jet [gimbaling the motor].[9] "It appears that the latter was never done.

Among the extant photographs of the ill-fated RMI rocket landing craft, there are a few that interestingly reveal that a preliminary test boat—or rather, a short, cutoff portion of a stern that approximated the one on the main rocket boat—had evidently been hastily fashioned just to test the motor on the riverbank prior to the first run. Or perhaps this setup was no more than a rowboat cut in half with the rocket motor and fuel tanks shoved into the corner of it for the static pretest. In addition to Pierce and Shesta, Lawrence was on the scene for the tests, along with another early RMI member, perhaps Frank T. Muth, the company's chief draftsman who had witnessed the live JATO trials. At any rate, this did not mark the end of RMI's connections with rocket boats.

GUY LOMBARDO'S INTEREST IN A ROCKET SPEED BOAT

After the war, on the afternoon of 16 May 1946, according to the *Daily Log* Lovell Lawrence Jr. kept on behalf of RMI, Lawrence received a surprise phone call from Guy Lombardo, the famous Canada-born bandleader of The Royal Canadians. Lombardo was inquiring about the possibility of RMI "installing [a] rocket motor in one of his racing boats."[10] "Mr. Lombardo," the *Log* continues, "can be located at [the] Capitol Theatre in NYC."[11] Lombardo very avidly raced hydroplane speedboats as a pastime, becoming a U.S. national champion.

The call led to a conference on 31 May between Lawrence and a "Mr. Meyers," who was either Lombardo's representative or Walter H. Myers, a test engineer with RMI. It is not known what transpired at the meeting, but as Lombardo called again on 29 July "about his boat and testing of the [rocket] motor" for it, it is assumed RMI went ahead with the project.[12] This was followed on 19 August, when Lawrence's wife called and left a message that Guy Lombardo had hit 65 mph in a championship race on the Shrewsbury River in central New Jersey, according to an article she had just read in the *New York Herald Tribune*. However, a perusal of the *Tribune* for that particular race says nothing at all about his boat, *Tempo VI*, being fitted with a rocket.[13] Nor does the *Log* or any other source make any further reference to Guy Lombardo.

A few years later, RMI's "Business Forecast" and "Proposal Priority Sheet," for 1 June and 27 June 1951, respectively, listed an "ERDL Rocket Propelled Boat," a project valued at the then-large sum of $50,000. ERDL may stand for the Army Corps of Engineers Engineer Research and Development Laboratories, based at Fort Belvoir, Virginia, or some other entity. But unfortunately, no details on this project are known; nor did anything come of this.

ROCKET-BOOSTED AUTOGIROS

RMI, and its successor, the Reaction Motors Division (RMD) of the Thiokol Chemical Corporation after RMI's merger with Thiokol in 1958, were more involved in a variety of rocket-boosted aircraft propulsion projects other than JATOs.

The 1 May 1944 entry of the RMI *Log Book* records the test of a "moment of inertia test" of an autogiro blade.[14] The test was part of an exploration for the Pitcairn Aircraft Company of the feasibility of a rocket-boosted autogiro. An autogiro is not a helicopter, as propulsion is by a conventional mounted engine, although lift is obtained by rotating horizontal vanes.

The 1 May test appears to have led to full propulsion tests of two 50-lb-thrust LOX/gasoline motors on the blade. But serious LOX leakage and other problems arose, aborting these kinds of efforts. The idea of a rocket-boosted autogiro resurfaced in 1946, when Paul H. Stanley of the Autogiro Corporation of America contacted Lawrence about the possibility of incorporating a 6,000-lb-thrust 6000C-4 engine—the same type of engine that was to play its famous role in powering the Bell X-1—in Stanley's own autogiro. A meeting was held on Stanley's proposal, although this project never materialized. It was just as well, because it would have been far too foolhardy for that much thrust to be considered for so light and fragile an aircraft.

PROJECT PINWHEEL

Likewise in 1946, the G & A Aircraft Company of Willow Grove, Pennsylvania, conferred with RMI on the potential of 50-lb motors to boost helicopters. It gave up the notion upon realizing that a LOX/gasoline rocket system would have been far too dangerous for this application.

A seemingly far safer and viable rocket-boosted helicopter concept surfaced in 1951 and was quaintly called Project Pinwheel. RMI initially produced its tiny 16F1 noncombusting 90% hydrogen peroxide (gaseous) 16-lb-thrust motor for this project. The operating gas was produced by catalytic action. Two of the motors—delivering a total thrust of 32 lb—were to be strapped to the blade tips of the one-man strap-on Rotor-Craft Corp. helicopter. This extremely lightweight (fewer than 100 lb), primarily tubular apparatus was strapped onto the pilot's waist and back. The man sat on an bicycle seat fixed to the aircraft. The two rotor blades were attached above the apparatus. The aircraft was the ingenious invention of Rotor-Craft's founder and president, Gilbert W. Magill of Glendale, California, who had initially approached Lawrence about the project. Magill envisioned an appealing assortment of military and civilian applications for his machine that included leapfrogging over entire armies, climbing inaccessible terrain, and so forth. He was so confident in its possibilities that he took out several patents on it and developed the more advanced RH-1 model.

While the project indeed appeared promising, it was relatively short lived. Pinwheel had gotten off to a good start, and the RH-1 model made its first free flight in 1954. RMI then improved the motors for throttleability, with thrusts reportedly from ranging from 2 lb to 17 lb. An early RMI 16F1 data sheet specified that it ran from 20% to 100% thrust. The RMI rocket units gave the Pinwheel apparatus a high rate of climb, and the Office of Naval Research

Fig. 10.2   Flight of an early model of a Project Pinwheel one-man helicopter with RMI motors on blade tips, circa 1951. (Courtesy Smithsonian Institution, 83-2858.)

(ONR) was very much taken with its possible deployment by the Marine Corps. Flights lasting up to nine minutes at 75 mph were also claimed. Despite this respectable performance for so small a system, after about 300 tests up to 1960, however, the whole program seems to have been abandoned; it was clearly underfunded from the start. The demise of Project Pinwheel also is attributed to the notion that the Army found it too complex for military operations. Perhaps Pinwheel ended due to a combination of both reasons.

ROCKET ON ROTOR (ROR)

A far more ambitious project of this kind was the ROR (Rocket on Rotor) program, which began in the early 1950s, perhaps as a spinoff from Pinwheel. ROR's beginnings are plausibly attributed to the difficult experiences faced by Marine helicopter squadrons during the Korean War, when they realized that they desperately needed more power, particularly for higher-altitude operations. Consequently, BuAer awarded RMI a contract in 1952 for the development of "Rocket Engines for Helicopter Blade Tip Applications" for their adaptation to full-sized helicopters.[15] Early in this venture, the Kellett Aviation Corporation's KH-15 one-man helicopter employed a TU-111 rocket motor for 131 initial test flights. Soon after, the ROR program switched to the standard Sikorsky HRS-2 helicopter used by the Marine Corps. In this case, RMI's XLR-32 hydrogen peroxide rocket motor of 32-lb thrust had been developed with tests undertaken at Picatinny Arsenal.

The ROR project generated an enormous amount of interest, both in aviation circles and in the popular media. RMI went as far as to produce an excellent 20-minute color documentary film on the system that was simply titled "ROR (Rocket on Rotor)." The 1955 film may still be viewed on YouTube. Interestingly, the film credits Jimmy Wyld for originating such an idea, referring to passing remarks in the conclusion of his article "The Problem of Rocket Fuel Feed" in the June 1936 issue of the American Rocket Society's *Astronautics*. An examination of the article, however, shows that Wyld really credits other inventors: "Finally, an ingenious scheme has been suggested by various inventors for what may be termed a 'rotor-motor.'"[16]

Wyld then followed with a brief description of a small propeller blade to which were mounted two small, simple, liquid-propellant rocket motors, one at each tip. A basic drawing of this arrangement was also provided. In truth, the idea was hardly new. The 19th-century British engineer William Henry Phillips had successfully flown a small 2-lb model of a reaction-propelled early species of helicopter-type aircraft as early as 1842 that was powered by the combustion gases from a potassium nitrate-based (gunpowder type) mixture.

Fig. 10.3   RMI's ROR (Rocket on Rotor), using Sikorsky HRS-2 helicopters, 1955. (Courtesy Frederick I. Ordway III collection, U.S. Space and Rocket Center, Huntsville, AL.)

The *New York Times* notably also reported that a demonstration of the ROR was held on 9 June 1956 at RMI's plant at Denville, near the Arsenal, which "nearly 400 rocket and helicopter enthusiasts" and their families attended.[17] Moreover, during development, a Sikorsky HRS-2 fitted with the XLR-32 units on its three blade tips underwent an incredible number of 4,078 flight tests total, with only 34 malfunctions, for a rocket motor reliability of 99%. Many other newspapers and publications such as the venerated *Aviation Week* likewise publicized the ROR.

Why then, despite RMI's huge expenditure of time, money, and publicity efforts in developing and promoting ROR, did it never become operational?

This is a difficult question to answer that is not found in the literature. One reason given is that ROR required a large amount of propellant to maintain a full system, including special RMI-developed storage and servicing trucks, as seen in the ROR film. But Reaction Motors Division pioneer Arthur "Art" Sherman presented a more compelling explanation. The Sikorsky helicopter, he said, was powered by a reciprocating engine. But when the newer generation of helicopters came into use with turbojet engines "with much superior power/weight ratios," this gave them far greater lift capabilities that obviated the need for extra rocket boosts.[18] "I am not sure when the ROR project ended," he concludes, "but this is why it did."[19]

ROCKET-PROPELLED ICE SLED

An earlier and more bizarre miscellaneous RMI venture was their rocket-propelled ice sled. This was not an official company project, it was an after-duty one—for fun. Back in late 1946, RMI's chief test supervisor, Ernest John "Buck" Pellington, a former World War II B-24 crew leader with 58 missions over Germany to his credit, headed the design-and-construction effort toward the stainless-steel tubing, rocket-propelled ice sled. At the aft end, with Lawrence's approval and encouragement, Buck attached an RMI 400-lb-thrust LOX/alcohol motor—most likely a larger chamber from the Lark missile power plant. Jimmy Wyld also became an enthusiastic contributor to the undertaking. He drafted detailed dimensional drawings of it and, characteristically for him, he meticulously calculated the sled's potential speed, which he determined would range from 100 to 150 mph.

Thanks to Jimmy's diligent note taking, exact dimensions and other data about the sled are available. The sled body measured 18 ft 2 in. in length. Its overall width, with runners, was 12 ft, and the nitrogen gas-pressurized motor actually produced 220 to 620 lb of thrust. The dry gross weight of the sled was 885 lb "without fuel or pilot."[20] The fuel added another 340 lb, and about 180 lb was figured in for the pilot—who was to be Pellington himself. A wonderfully illustrated article in both color and black and white, with the misleading title "Jet Iceboat," appeared in *Mechanix Illustrated* for April 1948. It showed how the sled was painted and provided views of its controls behind a curved-plastic cockpit shield. These included five gauges that must have included a speedometer as well as those for fuel and nitrogen pressures, and an electrical ignition switch. Steering was managed by a wheel linked by cable to the rear runner and an aircraft-type rudder. The colors are stunning: overall, a bright glossy red with sleek yellow trim patterns.

On 3 March 1947, Buck Pellington donned an aviator's helmet with goggles, climbed into the cockpit, and took off on icy Lake Hopatcong, shared by both Sussex and Morris counties in northern New Jersey. A day later, the *New York Times* and other papers reported that there had been 250 spectators and that Buck had reached a top speed of 95 mph, which could have been higher if it were not for rough patches of snow.[21] Buck Pellington's brief dash in RMI's rocket ice sled held such appeal that it even entered the popular culture of the day. It was featured in April 1948 in a cartoon-strip- format Camel cigarette advertisement that must have had wide exposure in the United States. A color drawings depicts Pellington seated within the sled. The caption reads: "The hottest thing on ice!"[22]

Far less is known about a possible RMI test of a pulsejet on an ice sled, which would have occurred earlier, during the 1944–1946 period when they were fully occupied with pulsejets. It is certain that a pulsejet was mounted on such a sled. The main proof of this foolhardy project is a photograph in the

Fig. 10.4   RMI rocket-propelled ice boat, 1947. At left, Ernst "Buck" Pellington; at right, Lovell Lawrence Jr. (Courtesy Frederick I. Ordway III collection, U.S. Space and Rocket Center, Huntsville, AL.)

Frederick I. Ordway III RMI collection at the U.S. Space and Rocket Center archives. In addition, NASM has an RMI resojet in its collection (Cat. #1977-1238), and according to a note from the donor (RMD) in the accession file, it was "used on [an] experimental ice sled constructed by H.F. Pierce, 1944."[23] It seems safe to say that this project does date to 1944 and is attributed to Pierce. The solitary photograph depicts a far cruder sled than the one used by Pellington. It is also seen on a frozen lake—no doubt, somewhere also in northern New Jersey—with trees in the background, although no individuals are in sight. The sled shows the horizontally mounted pulsejet, and nearby is a large, upright propellant tank for the pulsejet fuel or compressed gas for feeding in the propellant. But as Pierce and other likely principals have long since passed, the outcome of this odd project may never be known.

JUMP BELTS: SOLID-PROPELLANT TYPES ("HOT GAS")

A series of Buck Rogers-type rocket belts appeared among other, initially exciting and promising, but later luckless miscellaneous RMI projects. They represented another case where the basic concept was not new with RMI. The belt's origin indeed lies with Buck Rogers, the fictional futuristic space hero, and it was first depicted on the cover of the August 1928 issue of the science fiction pulp magazine *Amazing Stories*. Thereafter, the belt frequently appeared in the nationally circulated "Buck Rogers" comic strips and later in films like the old 1949 Republic production, *King of the Rocket Men*.

The whole idea was quite captivating, of course—instantaneous rocket power strapped to one's back that offered absolute freedom and mobility to rocket from here to there at the touch of a button. As RMI and other aerospace companies were to learn from hard experience, however, it was not easy to convert the concept into reality, and there were also several unforeseen weaknesses in the apparatus when put into operation. Regardless of these drawbacks, military and firefighters were naturally intrigued with its great potential for use in their respective occupations.

The idea of the belt was first championed at RMD by engineer Alexander N. Bohr, who happened to be a nephew of the famed Danish nuclear physicist and Nobel Prize winner, Neils Bohr. Bell Aircraft had accomplished much pioneering work on their version of it during the late 1950s, as did RMD, but it is not known if there was any influence from Bell in RMD's effort, and the two systems did show marked differences. On 18 June 1958, Alexander Bohr applied for a U.S. patent, granted on 19 June 1962 as No. 3,039,718 for a "Jet Device" and assigned to Thiokol.

Even before he filed his application, Bohr and two young fellow RMD team members—Harry Burdett Jr. and Raymond E. Weisch Jr.—had already worked out the details. They had gone as far as building and test flying their system in the woods around Lake Denmark. RMD had also assigned a most appropriate name to the project: Project Grasshopper.

Using off-the-shelf hardware, Bohr's version differed considerably from that of Bell Aircraft's, which was based upon hydrogen peroxide. He used solid-propellant canisters that once ignited, produced hot exhaust gases that flowed out of a tiny nozzle facing downward. The designated operator of the belt, Weich, wore a kind of literal belt around his waist. Attached to it were five equidistant pockets, each containing a canister. Each canister generated a short exhaust blast of 2 seconds. But the total duration could be increased up to 10 seconds, depending upon how many canister knobs the operator would pull. Normally, neither RMI nor its successor, RMD, were involved with solid-propellants early in their histories, but this is one instance where they did employ them.

The RMD belt was designed not for rising vertically, like the Bell Aircraft version, but more for hopping or leaping, actions deemed more practical for soldiers wishing to jump over ravines and ditches on the battlefield, or over water. One canister enabled a person to easily run about 25 to 35 yards, or to

Fig. 10.5 Soldier wearing solid-propellant canister model of RMD's JumpBelt. (Courtesy Orbital ATK.)

make a substantial jump—rather than an actual, more sustained ascent. Consequently, the papers had a field day in later publicizing the device. Typical stories ran with colorful titles like: "Superman Rocket Near?," "Up! Up! And Away!," and "Man's First Leap Toward Free Flight."

JUMP BELTS: GASEOUS TYPES

Behind the scenes in secret at RMD, there simultaneously evolved another, more complicated version of a jump belt that was based strictly upon the quick expulsion of a gas. At first, this gas was compressed nitrogen contained in a pair of bottles strapped to the back of a special harness worn by the operator; this type was also referred to as the "cold gas" model, but to further confuse matters, the later hydrogen peroxide versions were also known as "hot

Fig. 10.6   Artist's concept of a paratrooper deploying a JumpBelt, circa 1958. (Photo, courtesy Orbital ATK.)

propellent" types. At any rate, when activated by handheld controls, the gas passed through a fine control valve until it exited through a set of very small downward-pointing nozzles.

Thus, RMD simultaneously developed solid-propellant (canister) and gaseous propellant models of what RMD officially termed the "JumpBelt [sic.]"; this was their registered trade name, to distinguish it from competing similar devices by companies such as Bell Aviation or Aerojet. Another term adopted was the "Small Rocket Lift Device" (SRLD).

Although this technology seems very simple from a 21st-century perspective, these developments (particularly for the cold gas model) were really quite involved and ran into a number of technical hurdles. Among parameters that the developers had to contend with were thrust-to-weight ratio issues; correct center of gravity, or balances, to work out; proper harness designs; nozzle alignments; efficient valve design; anti-sloshing problems; and the development of handheld triggers. Consequently, there was an array of different arrangements, including three-tank versions and three-nozzle versions, as well as adjustments to be made with the appearance of the hydrogen peroxide types that replaced the original nitrogen models. SLRD technology became a field unto itself.

Also, the various complicated data sheets on these products include a potentially wide range of uses for them, far beyond mere "leaping" or "hopping." They encompassed everything from skimming water at high speeds, facilitating brake-parachute drops of men and equipment, and low-level bailouts from disabled aircraft to uses by emergency rescue squads and assisting vehicles over difficult terrain.

In the meantime, the solid-propellant canister types also underwent significant improvements and refinements. For instance, the later canisters had more

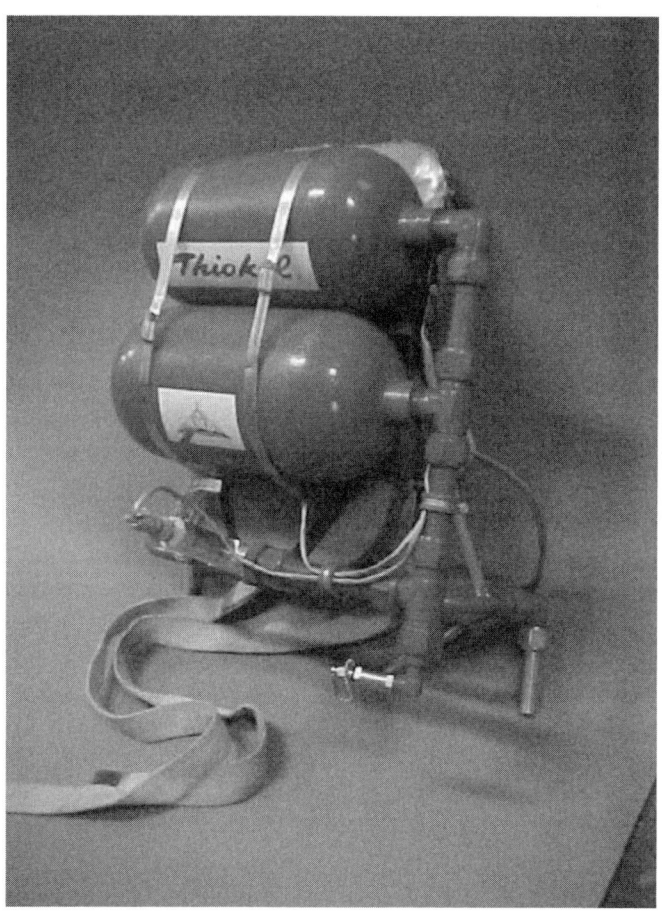

Fig. 10.7   Back of RMD hydrogen peroxide gaseous-type JumpBelt, in the collection of the National Air and Space Museum (Cat. #1964-0698). (Courtesy Smithsonian Institution.)

energy content and could be made impervious to being pierced by enemy rifle or other gunfire. Some of the later model names are indicative of just how specialized they had become: the *Flybelt*, the *Assault jump-Belt*, the *para-Belt*, the *escape-Belt*, *Porta-Belt*, and the *glide-Belt*.

The downside of the basic rocket belt, encountered also by Bell Aircraft, was that this apparent dream invention simply failed to live up to the high expectations for it. Its most significant weakness was an extremely short propellant-supply container, which affected operational duration—average durations of a standard Bell belt lasted only about 20 seconds; for RMD's version, it would have been far less. This also translated into disappointingly limited performance capabilities overall. In RMD's case, the fundamental flaw of limited operational duration, as well as their very short jump heights, applied to both solid- and gaseous-propellant models. These heights ranged from a foot to 6 ft 5 in. in the first tests of the gaseous hydrogen peroxide model. On top of this, the belt operators faced obvious military operational vulnerabilities during combat situations. Bluntly put, operators in such situations could be entirely exposed to enemy fire during their all-too-brief missions. Their quick duration times were also a drawback for peacetime operations. RMD's JumpBelt was demonstrated to the U.S. Army between June 1959 and February 1960, with a much-publicized free-flight demonstration conducted at Fort Eustis, Virginia, in the latter month. But the bottom line is that none of the military services came to adopt the belts as standard equipment, either from Bell or RMI. Today, more than half a century after the first of these devices materialized, even the most sophisticated rocket belts or jump belts or rocket backpacks, as they are variously called, serve as little more than movie-stunt devices or football halftime-show pieces. Notably, the use of such a device thrilled audiences in the 1965 James Bond film *Thunderball.*

AIRCRAFT TAKEOFF CATAPULTS

Both RMI and successor RMD also pioneered in the completely different area of aircraft catapults for very quick plane takeoffs from aircraft carriers, using the rocket for pressure generation. Gas generators for these catapults were to furnish the operating gas from a sort of very controlled, high-pressure rocket exhaust in place of exhaust from standard steam or other gas-activated catapult systems.

This activity started at RMI as early as May 1946, when they received a Navy contract to study the application of gasoline and LOX to produce high pressure for catapults. At that time, the operating gas was obtained from the rapid combustion of smokeless powder, and it was very difficult to achieve proper launching control. It was therefore felt that the controlled combustion of liquid propellants could potentially offer far greater advantages. These were subsequently proven in tests and from then on, RMI evolved several systems during the year for the Navy, although they continued to remain exploratory. These devices are variously referred to in the early Navy study contracts as "catapult power plants" or "gas generators" for Navy catapults.[24] Additionally, the catapult project promised to be a mammoth and most-profitable venture, and by 1951 it was valued at $500,000. According to RMI business forecasts at the time, this activity was hoped to even surpass the expected profits of their regular rocket work. By this juncture, RMI had already constructed a catapult simulator. The simulator was probably one of RMI's earliest applications of a computer for both design and simulation testing—in this instance, toward the realization of a rocket engine-based aircraft catapult.

The rocket approach for catapults had gained a lot of positive interest in naval circles because the Navy was now looking ahead to the onset of far heavier and faster carrier aircraft (jets in particular), and the ordinary steam catapult did not seem capable of accommodating these aircraft, because steam-based catapults could not maintain the expected far-higher pressures. For a variety of reasons then, RMI steadily evolved a succession of more complex catapult power-plant designs and different propellant choices that included compressed air/diesel fuel-and-water mixtures to achieve both improved performance and greater safety.

Several entries from Wyld's neatly handwritten daily RMI *Work Diary* for 1952–1953 show that he was then deeply involved in the catapult project, not only busily calculating this and that for the system, but inspecting and reporting on catapult trial runs—evidently, on the simulator—at conferences with other RMI catapult team members.[25] In addition to the simulator, an RMI catapult power plant was tried on the deck of the aircraft carrier USS *Leyte*, CV-32, in the spring of 1952. This must have been a scale model, although details are wanting and it is not known if it was a working model. Then, as reported in the May 1955 issue of *The RMI Rocket*, the company's house organ, its physics laboratory was "carrying on work on a quarter-scale catapult power plant which has been under development for the last eighteen months."[26]

It is also known that around this time RMI had access to a huge REAC analog computer at Rensselaer Polytechnic Institute in Troy, New York. By a special arrangement, Rensselaer's REAC was used to not only assist in working out the design of the full-scale catapult, but to study a variety of other subjects for RMI—from pressure regulators and complete rocket-motor systems to ballistic problems tied to their regular missile and other rocket work.[27]

By the spring of 1957, a great deal of progress had been made in the catapult program in which the catapult was now officially named the ICCP, for Internal Combustion Catapult Powerplant. According to the April issue of *The RMI Rocket*, a full-sized operable ICCP was "nearing completion at RMI" and was to be installed at the Naval Ship Installation Test Facility at the Naval Air Station at Lakehurst, New Jersey.[28] The ICCP was to undergo "a series of tests which will include the launchings of dead loads and possibly the launchings of 'live' aircraft."[29] The testing phase was to take two years.

Bernard Perlman was RMI's program manager for the ICCP and Edmund Beretvas was chosen as the RMI field engineer at Lakehurst. Eventually, an entire RMI crew would be stationed at Lakehurst. The September 1957 issue of *The RMI Rocket* ran a short picture story of the construction of the massive ICCP installation that was well underway. The pictures included the "270 ft long concrete trough into which the catapult tubes will be installed and will be decked over with a surfaced [aircraft] runway."[30] Also shown was a huge "710 cubic feet air receiver [tank] designed for 1500 pounds per square inch [pressure]"that was "to supply the catapult powerplant during combustion."[31]

In the spring of 1958, Lawrence's brother Bob, RMI's director of finance, reported that the catapult program had accounted for 9% of the company's sales the previous year and would approach 10% in 1958. The sales comprised revenues from the Navy contracts for the ICCP.

Then, *The RMI Rocket* for April 1958 published a long article by Charles E. McKnight, titled "RMI Full-Scale Catapult," that provided a lot more details. For instance, the ICCP at Lakehurst was designated the C-14, and was actually the "second full-scale catapult project. The first full-scale model was built at Lake Denmark and achieved a remarkable record of successful [test] firings."[32] It is further revealed that even that was "not the first catapult at RMI. It grew out of a project successfully pioneered by [RMI's] Engineering and Physics Departments working on a quarter scale model."[33] The remainder of the article goes into an analogous description and capabilities of the ICCP.

As for the C-14 designation, this was strictly the Navy's designation and referred to a type of catapult, rather than a model. In this case, the C-13 referred to a more conventional steam catapult, whereas the C-14 employed a completely different propulsion system using internal combustion.

The aircraft-launching catapult is described by McKnight as "essentially a huge 'gun' with an airplane attached to the 'bullet.' When this gun fires, the 'bullet' drags the plane faster and faster until it reaches launching speed. The 'bullet,' or piston, as it is called in the catapult, detaches from the plane and enters a brake where it is stopped. It is then returned to launching position and a second aircraft is attached."[34] But the giant C-14 model at Lakehurst was built with twin "barrels," or catapult tubes, each 18 in. in diameter and 250 ft (actually, 248 ft) long. "They can launch one plane every 30 seconds," McKnight explains, but he also carefully points out that "The tubes are not part of the RMI project."[35] As a matter of fact, the C-14 power plant was used with conventional steam catapult tubes. RMI's contract, McKnight adds, dealt with only the gas-generating equipment, or power plant.

The ICCP now employed compressed air and JP-5 jet fuel, but fired as a controlled rocket, and produced a minimum thrust of 19,000 lb and a maximum of 360,000 lb, or up to 15,000 HP. This was enough to launch a 100,000-lb aircraft to an end speed of 125 kn, or 143.8 mph. The compressor plant measured 25 ft wide by 23 ft long by 21 ft high. Overall, RMI's C-14 catapult was considered smaller than a conventional steam catapult of the day, although it could boast double the launch capacity. Far more was expected by RMI's engineers, who envisioned the adaptation of an ICCP for zero-length launches of underwater missiles from submarines to "meet an operational need of the nuclear carrier age."[36] All therefore seemed to bode very well for this optimistic expectation.

Not too long after, a tremendous breakthrough occurred early in September 1958 when the *Wall Street Journal* announced that the Reaction Motors Division of Thiokol—five months after the merger of RMI with Thiokol that formed RMD—received a Navy contract to produce four catapult power plants "for the nation's first nuclear aircraft carrier."[37] This would be the USS *Enterprise,* the keel of which was laid the same year.

By the close of 1958, a "work horse" model of the ICCP for the *Enterprise* was planned to determine operating characteristics such as optimum fuel mixtures, pressures, and other parameters beside serving as a test vehicle. It was expected that as many as 30 RMI personnel would be assigned to Lakehurst where they were to operate out of their own on-base separate facility.

The future on the catapult front thus looked exceptionally bright for RMD. In the following year, on 11 May 1959, Bernard Pearlman and two coworkers involved in this enormous undertaking applied for a patent for an automatic pressure control for a gas-generating chamber for catapults—granted on 11 June 1963 as Patent No. 3,092,965, and assigned to Thiokol.

In the same month as the patent application, just a little after 2:00 p.m. on 26 May 1959, a 17,500-lb Grumman F9F Cougar jet fighter aircraft piloted by Cmdr. Ray C. Tylutki became the first successful live launch of a plane from RMD's Combustion Catapult at Lakehurst. It shot into the air at about 159 mph. *The RMI Rocket* proudly publicized this event in its June 1959 issue, and the story also made the papers.

Aviation Week for 8 June 1959 reveals other important facts. For one, following Cmdr. Tylutki's successful demonstration, there was a second successful launch achieved by Lt. Cmdr. John Schaeffer flying a 27,500-lb McDonnell F3H-2M Demon that was accelerated to 145 kn (166.8 mph) during its run. As impressive as these runs were, the magazine comments that "The two demonstrations did not come close to taxing the system's capability."[38] *Aviation Week* additionally states that "Reaction Motors Division...will deliver four units [of the catapult power plant] to the Navy late this year" and that the new catapult "is intended for use on new super carriers."[39]

But as explained years later by Edward "Ed" C. Govignon, who had joined RMI in 1957, "The catapult contract was [indeed] a major contract…The first delivered [RMD] catapult was installed [on the *Enterprise* but was] then actually removed. There were two reasons—(1), we had a major unexplained explosion in [a] test and (2), the nuclear steam generator [of the *Enterprise*] proved capable of supplying required steam and thus no [other] external power source was needed."[40] This appears to have quickly ended years of intense development on what once had been considered a highly promising area for both RMI and RMD, although there is still more to this complex and important story.

Francis Duncan's *Rickover and the Nuclear Navy* provides the insight that during the earlier design phase of the nuclear-powered *Enterprise*, both Adm. Hyman Rickover, the "Father of the Nuclear Navy," and Milton Shaw, one of his leading engineers, had worried about holding the ship "hostage" to the untried internal-combustion catapult development.[41] Therefore, Rickover won an agreement with the Bureau of Naval Weapons—the successor to the BuAer—to accommodate either steam or the internal-combustion arrangement in the ship. As matters transpired, the testing accident happened, although the Bureau of Naval Weapons felt this was only a temporary setback, and still expressed confidence in the success of the new system. The installation plans therefore went forward, but by July 1960, the ICCP was failing to meet expectations and Shaw "found it increasingly difficult to get details of the project."[42] He did not elaborate on this. At this late juncture, the Bureau proposed, but did not agree, to switch to a compressed-air type catapult. *Enterprise* was commissioned right on schedule, on 24 September 1960 at Newport News, Virginia, with the ICCP aboard, yet two days later, the Bureau arrived at a decision to make the switch to steam. As Govignon said, the ICCP was subsequently removed, although the ship retained part of the original ICCP equipment. Rickover, said Duncan, afterwards felt vindicated that if the compressed air or ICCP would have been kept and had been found wanting, "the blame would have been placed upon the nuclear propulsion plant and himself."[43]

As for the intended delivery by RMD of four sets of the ICCP to the Navy, there is no indication this was followed through, and the Navy did not choose to adopt it for any of their carriers. The C-13 steam-type catapult was kept, although continually upgraded, particularly after the opening of the computer revolution.

ROCKET-POWERED AIRCRAFT EJECTION SEATS

About 1954, RMI also tried to enter the rocket seat-ejection business. Aircraft ejection seats that used compressed air first appeared in World War II. Then, in late 1944, Germany's Heinkel He 162A became the first operational military jet featuring a new type of ejection seat fired by an explosive cartridge.

Naturally, the advent of jet aircraft with their much faster operational speeds had necessitated the need for an extremely rapid-acting ejection seat in the event of emergencies. A great deal of development of these seats followed after the war, primarily with explosive-cartridge types; these were found—often from tragic experiences—to have been extremely risky. The rocket ejection seat with its more gradual acceleration thus came into use, and in 1958, the first U.S. aircraft fitted with a rocket seat was the Convair F-102 Delta Dagger jet fighter.

Meanwhile, by 1955, RMI cooperated with the Army's Frankford Arsenal in Philadelphia to arrive at their own solid-propellant T-16 "dual catapult system" for "personnel seat capsule ejection."[44] RMI must therefore have been one of the earliest American developers, or rather, codevelopers, of rocket-ejection seats. According to Jim Tuttle's *Eject! The Complete History of U.S. Aircraft Escape Systems*, it was Frankford Arsenal that had contracted with RMI to develop the catapult. By the spring of that year, RMI announced that six of the systems were completed for Republic Aviation, and were to be shipped to Edwards Air Force Base for testing under flight conditions from Boeing B-47 bombers. This system was a "heavy capsule" using two catapults for the downward ejection of crew members. It was also mentioned that there "will eventually be a separate capsule for each man."[45] Yet, Tuttle also remarks that the RMI-Frankford Arsenal program was terminated prior to a successful test, although he does not say exactly when this happened, nor why.

Judging from an RMI advertisement that appeared a year later in *Aviation Age* and included a dramatic depiction of a seated pilot being ejected upward, it is possible that RMI still entertained hopes that its product was viable. After this, nothing more is heard about their progress in the rocket ejection-seat field. For certain, competition for developing these seats was then very stiff. Most curiously, Tuttle adds that Gerald E. Hirt, who left Frankford Arsenal in 1954 and joined Talco Engineering Co. of Hamden, Connecticut, "was the principal inventor of the rocket catapult, with a patent issued to Talco. The first [U.S.] rocket catapult [and manufactured by Talco]…was first used to upgrade the U.S. Air Force Convair F-102 Delta ejection seat."[46] RMI therefore seems to have been indirectly connected to the history of America's first rocket-seat ejection catapult.

HYPROX STEAM GENERATOR

Perhaps as a byproduct, or influenced by their work with aircraft launching catapults (not to be confused with aircraft ejection-seat catapults), RMD also moved into the steam generator field. But former RMD administrator Edward H. Seymour simply explained that the company merely took up steam generators "as a result of wanting to perform altitude-testing on its smaller engines."[47]

The normal way of carrying out altitude testing," Seymour went on, "was to use stream air ejectors to lower the pressure to the desired simulated altitude."[48] Eventually, RMD engineers at their Lake Denmark rocket engine test facility "came up with the idea of decomposing peroxide, adding fuel gas to burn with oxygen and using this as the steam generator...So we built a set-up like this and took it over to the test stand. A minor business soon got under way and over the next year or so we built steam-generating and ejector pump setups."[49] This system was marketed under the name of Hyprox; the trademark was filed on 15 March 1963. Contrary to Seymour's characterization of the steam generator becoming "a minor business," however, the Hyprox product reportedly created worldwide interest and became quite successful commercially. It is beyond the scope of this book to go into these details from the RMD era.

Hyprox systems were utilized not only by RMD itself, but by clients like the Jet Propulsion Laboratory, Bell Aerosystems, the White Sands Proving Grounds, and the NASA Marshall Space Flight Center. The last even requested a system for the Project Apollo's Saturn V program. In February 1965, a Hyprox was shipped by RMD's Special Projects group to Marshall. There, it was to be used to check out the Saturn S-IVB (the Saturn V third stage) attitude-control engines. Operationally, the control engines on this section of the vehicle fired at about 20-mi altitude. Hyprox was scheduled to simulate this altitude in mid-April, although the results are unknown.

ROCKET TRAIN BRAKES TO ROCKET DRILLS

There were an untold number of other miscellaneous RMI and RMD projects—or near projects—that were more directly linked to rocket propulsion. In the spring of 1947, for instance, the American Locomotive Company contacted RMI to explore the possibility of a unique "rocket brake" that could potentially very quickly halt a speeding train to avert a calamity. This matter was turned over Wyld, who made his inevitable calculations. When he determined that a thrust of some 120,000 lb for 10 s would be required, the request was obviously turned down.

Another potential "exotic" project was a rocket to be used for drilling into the ground, a rocket-age version of the pneumatic drill, that was considered by RMI in late 1951. At about the same time, a rocket for the quick firing of boilers was also discussed, and the next year, RMI was testing a mine-clearing device for the Army. None of these ideas went any further.

ROCKET SLEDS

The application of large-scale RMI liquid-propellant rocket motors for propelling a supersonic rocket sled seemed a more viable undertaking. A rocket sled is a test platform, propelled by rockets that upon ignition immediately accelerate the sled along a very long set of special steel rails called "slippers," so that platform very quickly reaches either subsonic or even supersonic speeds. Typically, the sled then suddenly stops by the use of water brakes. The purpose of sleds varies widely. It ranges from testing the effects of extremely rapid acceleration upon living organisms (usually test animals, although some sleds have tested high accelerations upon humans) to testing aircraft ejection-seat deployment at high speeds to testing instruments or other equipment that might be used in missiles or aircraft that operate at extremely high velocities.

The sled rockets can be solid- or liquid-propellant types, although they are usually the former. In either case, their thrusts have to be very great and their durations very quick. RMI was awarded an Air Force contract for such a rocket-sled project in 1956 or 1957. Britain-born aeronautical engineer Harold Davies, who had worked on turbojet engines at the U.K.'s famed Bristol Aircraft Company, and Dwight S. Smith were assigned by RMI to design the propulsion system according to Air Force specifications toward its development. RMI was to deliver two liquid-propellant "pusher" sleds to Edwards Air Force Base, California, where they were to be applied to testing large-size, full-scale aircraft components over a wide range of velocities from subsonic levels up to Mach 2.

Consequently, Davies and Smith went as far as to design their "PP151 and PP152 Power plants," based upon the cancelled XLR-77 rocket motor that had been intended by RMI to be used as the booster for the supersonic Bomarc surface-to-air missile.[50] The Bomarc came to use an Aerojet booster instead. As for the designation of PP, it likely stood for "Pusher Powerplant."

The XLR-77 was extremely powerful and was rated at 50,000 lb of thrust for a very fast duration of a few seconds, typical of a booster. It burned inhibited red-fuming nitric acid (IRFNA) and ammonia, with an additive of lithium to make the propellants hypergolic. But for the sled application, Davies and Smith preferred the less-problematic choice of 90% hydrogen peroxide. The smaller PP151 arrangement, intended for the subsonic velocities, was to be a three-engine cluster of XLR-77s, for a total thrust of 150,000 lb for 8 s. The larger PP152 arrangement, for supersonic speeds, was to mount eight XLR-77s for a sea-level thrust of 440,000 lb for 10 s. The latter is rounded off in the general literature of the day as a 500,00-lb-thrust arrangement.

Even so, the Air Force rocket sleds would have been RMI's most massive and formidable rocket project, although this was not to be. Soon after the design had been laid out by Davies and Smith, the Air Force cancelled the project for some unknown reason and once more—and for the last time—the XLR-77 motor became defunct. It may be that the cancellation was simply due to the fact that solid-propellant rockets offer far greater advantages over liquid-propellant types for rocket sleds. Mainly, they are far less expensive and far simpler than complex liquid-propellant types, especially for multiple sled-run programs.

ROCKET-PROPELLED DRONES

Drones were another area that once looked promising for RMI. But rather than drones as weapons or the hobbyist drones so popular today, these were target drones—employed by the military as rocket-propelled simulated targets flown in missile-firing practices. Here also, solid-propellant systems for target drones were usually the norm due to their extreme simplicity and very low cost. But about 1957, RMI sought to incorporate low-cost liquid-propellant systems, and they did everything they could in their promotional literature to try to sell to potential military customers the advantages of liquid systems over solids. They pointed out that it was possible to vary the speeds of the drones, which was not at all possible with solids. The inference was that liquids offer better overall control of burning times vs. that for solid-propellant units. Superior performance with liquids was also stressed, as well as other arguments for liquids.

Finally, to negate reaction to the expected higher complexity and cost differential of liquids, RMI stated they had developed a series of very simple-to-use storable liquid-propellant units, either as "boosters" or "sustainers"—for longer duration drone cruise flights. To emphasize their economic persuasive points, RMI additionally tried the "cheaper-by-the-dozen" angle. "In lots of 1,000 per year," wrote RMD's Arthur "Art" Sherman at the time, "prices for the sustainer type powerplant [for drones] will run as low as $500 per unit."[51]

By this period, RMI—or rather RMD, upon the merger of RMI with the Thiokol Chemical Corp. in April 1958, as covered in Chapter 11—had clearly borrowed and adapted their pioneering storable or packaged liquid-propellant technology. This technology had been so successful in the Bullpup missile program, toward working out a family of low-cost and reliable drone motors of different sizes, weights, and performances. These were then marketed as being able to fit all drone-mission requirements. These were clever approaches. Yet, despite all the company's ingenious adaptive engineering and salesmanship, there is no evidence that these efforts succeeded, because neither RMI- nor RMD-powered target drones are known to have become operational.

UNDERWATER DEVICES

Oddly, RMD also had an "Underwater Devices" section, but information about their work is scanty. RMD veteran Arthur Sherman recalled only that the company once pursued a serious study and proposal for a rocket-propelled version of the Navy's Mark 48 series of torpedo. Another project, for which RMD received a contract by July 1960 from an unknown organization, was for "a method for converting ocean wave motion to electrical energy."[52] But no other details are known, much less how rocket power might have been applied in this project. Nothing came of either of these endeavors.

Undoubtedly, both RMI and RMD entertained, or were engaged in, still other miscellaneous or exotic projects. They did not overlook nuclear and other advanced propulsion projects or even deep-space exploration prospects. Some of these are discussed in Chapter 11, which also offers postscripts and retrospectives on more than 30 years of RMI/RMD's remarkable pioneering years.

ENDNOTES

1. James H. Wyld, "Rocket Landing Boat," Jan. 1942, in James H. Wyld collection, Archives, U.S. Space and Rocket Center, Huntsville, AL.
2. James H. Wyld, Notes, "Rocket Boat," 29 Jan. 1944, in James H. Wyld collection, Archives, U.S. Space and Rocket Center, Huntsville, AL.
3. G. Edward Pendray, The Coming Age of Rocket Power, Harper & Brothers, New York, 1945, p. 171.
4. Ibid.
5. Ibid.
6. Ibid.
7. Letter from John Shesta to Frank H. Winter, 12 Feb. 1979, in John Shesta file, NASM.
8. Ibid.
9. Pendray, *The Coming Age*, p. 172.
10. Lovell Lawrence Jr., *Daily Log*, Lovell Lawrence Jr. Papers, NASM, box 4, folder 16.
11. Ibid.
12. Ibid.
13. See "Guy Lombardo in Tempo VI, Wins U.S. Motorboat Crown," *New York Herald Tribune*, 19 April 1946, p. 25.
14. Lawrence Jr., *Daily Log*, entry for 1 May 1944.
15. Frank H. Winter and Frederick I. Ordway III, *Pioneering American Rocketry: The Reaction Motors, Inc. (RMI) Story, 1941–1972*, AAS History Series, Vol. 44, Univelt, Inc., San Diego, 2015 p. 369 that also cites the BuAer contract numbers.
16. James H. Wyld, "The Problem of Rocket Fuel Feed," *Astronautics* (American Rocket Society, New York), No. 34, June 1936, pp. 12–13.
17. "Rocket Fans Hear Sounds of Future," *New York Times*, 10 June 1956, p. 80.
18. Letter from Arthur Sherman to Frank H. Winter, 30 Jan. 2014.
19. Ibid.

20. James H. Wyld, Note with sketches, "Stability of Rocket Ice Boat at High Air Speeds," 19 Feb. 1947, in James H. Wyld collection, U.S. Space and Rocket Center," box and file number unknown.
21. See "Jet-Motored Ice Boat…," *New York Times*, 4 March 1947, p. 32.
22. Advertisement, Camel cigarettes, circa 1948, source unknown, in Frank H. Winter personal collection. Because the identical ad is found in the Sunday edition of the *Des Moines Register* (Des Moines, IA) for 11 April 1948, p. 88, and the *Detroit Free Press* for the same date, it is assumed that it also appeared in other papers at the time.
23. Winter and Ordway III, *Pioneering American Rocketry*, p. 365, and originally from NASM, Accession Work Sheet for Cat. numbers 1977-1224 through 1977-1265, 15 June 1977.
24. Winter and Ordway III, *Pioneering American Rocketry*, pp. 400–401; also gives Navy contract numbers.
25. James H. Wyld, *RMI Work Diary*, unpublished document in James H. Wyld Papers, U.S. Space and Rocket Center, Huntsville, Alabama, entry for 22 Sept. 1952; and 1953 entries for: 2, 6, 16–18, 24, 27 Feb.; 6, 14, 16, 24, and 31 March; and 19 April, etc.
26. "Full Scale Catapult to Launch Planes from Carrier," *The RMI Rocket*, Vol. 6, May 1955, p, 3.
27. By the late 1950s, RMI had acquired a succession of their own analog computers for dynamic design analyses and other tasks. For more on the history of the use of computers by RMI, consult Winter and Ordway III, *Pioneering American Rocketry,* pp. 188, 388–389.
28. "ICCP Catapult Goes to Lakehurst for Demonstration Tests," *The RMI Rocket*, Vol. 8, April 1957, p. 3. See also George Christian, "Carrier Deck Gear Development Pushed," *Aviation Week*, Vol. 67, 2 Sept. 1957, p. 91.
29. Ibid.
30. "Catapult Powerplant Installation at Lakehurst," *The RMI Rocket*, Vol. 8, Sept. 1957, p. 8.
31. Ibid.
32. C.E. McKnight, "RMI Full-Scale Catapult," *The RMI Rocket*, Vol. 9, April 1958, p. 5.
33. Ibid.
34. Ibid.
35. Ibid.
36. Ibid.
37. "Thiokol Gets Contract," *Wall Street Journal*, 4 Sept. 1958, p. 2.
38. "Navy Tests Internal Combustion Catapult," *Aviation Week*, Vol. 70, 8 June 1959, p. 89.
39. Ibid.
40. Letter from Edward C. Govignon to Frank H. Winter, 21 Oct. 2013.
    It is difficult to determine the date of the accident. Francis Duncan, in his *Rickover and the Nuclear Navy: The Disciple of Technology*, indicates it took place in October 1955, but this would have been far too early because RMI's ICCP facility was under construction by 1957. It appears the accident more likely took place in 1959 as the *Enterprise* was completing its construction.
41. Francis Duncan, *Rickover and the Nuclear Navy: The Disciples of Technology,* Naval Institute Press, Annapolis, MD, 1980, p. 110.
42. Ibid.
    The Bureau of Naval Weapons had been established in August 1959 upon the merger of the Bureau of Aeronautics (BuAer) and the Bureau of Ordnance.
43. Duncan, *Rickover*, p. 111.
44. "RMI's T-16 Seat Ejection Catapult System Completed," *The RMI Rocket*, Vol. 6, April 1955, p. 1.
45. Ibid.

46. Jim Tuttle, *Eject! The Complete History of U.S. Aircraft Escape Systems,* MBI Publishing Co., St. Paul, MN, 2002, pp. 85–86.

The history of the rocket-propelled aircraft ejection seat is more convoluted than it first appears. One of Hirt's patents—and evidently his first to mention rocket propulsion—was applied for on 22 January 1957, and was granted 18 August 1959 to Hirt and Eugene A. Martin as No. 2,900,150 for an "Ejection Seat Catapult." However, Hirt was preceded by Earl Schuyler "Sky" Kleinhans, the prominent airplane designer, who applied for a patent on 24 May 1946, granted 8 May 1951 as No. 2,552,181 for an "Ejecting Device" that was assigned to Douglas Aircraft. This patent incorporated a "non-explosive propelling type of rocket." U.S. Patent No. 2,552,181, available online.

47. Interview with Dr. Edward H. Seymour by Frederick I. Ordway III, 25 Jan.1986, cited in Winter and Ordway III, *Pioneering American Rocketry*, p. 158.

48. Ibid.

49. Winter and Ordway III, *Pioneering American Rocketry*, pp. 158–159, 401. See also "Small 'Hyprox' Generators Boost High Rate Steam Rate Flow," *Missiles and Rockets*, Vol. 2, April 1962, p. 35.

50. Harold Davies and Dwight S. Smith, "Design Considerations of Two Large Liquid Rocket Sled Pusher Vehicles," *Jet Propulsion*, Vol. 27, Sept. 1957, p. 999.

51. Arthur Sherman, "Rocket Propulsion for Drones," *Astronautics*, Vol. 3, June 1958, pp. 24–26. Consult Winter and Ordway III, *Pioneering American Rocketry,* pp. 381–383, for more details on RMI and RMD's drone propulsion developments, and p. 196, Fig. 6.9, for a picture of their standard drone propulsion unit.

52. "Research Reports Receipt of Seven Contracts," [RMD] *Rocket Weekly News Bulletin*, 8 July 1960, p. 11, in Archives, U.S. Space and Rocket Center, Huntsville, AL.

Chapter 11

RMI AND RMD: POSTSCRIPTS AND EPILOGUE

"Their dedication was fantastic. They put real blood into those projects."

–Ted D. Sjoberg, at 1970 auction of RMD's moveable assets prior to closing the company.

BACKGROUND

Sadly, by the end of the 1960s, RMD and its predecessor Reaction Motors, Inc. had seen their greatest glory days behind them. Those early years had been an incredible ride, replete with outstanding success stories for both of these truly pioneering companies. Their extraordinary achievements also represent major milestones in the histories of aerospace technology and American business enterprise.

Imagine, RMI was very modestly begun by only four men previously engaged in the fantastic—and dangerous—hobby of amateur rocketry during the Great Depression, before they formed their tiny company in northern New Jersey just a few weeks after the attack on Pearl Harbor in December 1941. Then, five years later, they initiated the development of a rocket engine that went on to power the Bell XS-1—the first aircraft to break the sound barrier, in October 1947. In November 1953, the same motor powered the first plane to fly *twice* the speed of sound: the Douglas Skyrocket in November 1953.

In the same period, another RMI engine propelled the Viking rocket as America's first single-stage vehicle to penetrate space, when Viking 4 launched on 11 May 1950 from the deck of the USS *Norton Sound* and reached a peak altitude of 105 mi. Other Vikings set other outstanding records. Viking 7, flown on 7 August 1951, reached 136 mi and beat the old V-2 record for a single-stage rocket. Viking 11 rose to 158 mi on 24 May 1954 and took some of the first photographs from space. The last two Vikings flew as test vehicles for Project Vanguard, America's first rocket designed specifically as a satellite launcher.

From 1958, RMI successor RMD literally soared to far higher glories. RMD completed the development of the most sophisticated rocket power plant of its day, the XLR-99 Pioneer that propelled the X-15, which would fly almost seven times the speed of sound (Mach 7). The same aircraft became America's first "space plane." X-15 flight 90 on 19 July 1963 was the first

X-15 flight of more than 100 km (62 mi)—altitude that made pilot Joe Walker the first U.S. civilian in space; it was also the first flight of a space plane in aviation history. A month later, on 22 August, Walker flew X-15 Flight 91 in the highest flight of the X-15 program at 107.96 km (67 mi). This altitude remained the world record until the 1981 flight of the Space Shuttle *Columbia*.

Beyond this, RMD vernier rocket motors played major roles in the unmanned Surveyor Project spacecraft that helped pave the way for the manned Project Apollo missions to the moon. The verniers enabled Surveyor 1 to make critical midcourse correction maneuvers and helped enable the spacecraft to become America's first to make a soft landing on the moon—and the first craft to land on any astronomical body beyond Earth—on 2 June 1966. Altogether, RMD verniers completed five successful trips to the moon up to the conclusion of the Surveyor program in February 1968. To this day, the 15 northern New Jersey-made verniers on those spacecraft remain on the moon.

RMD also pioneered landmark technical advances, notably their developments of the regeneratively cooled rocket chamber, the spaghetti type of engine configuration, and the gimbaling of liquid-propellant rocket motors as proven in the successful flights of the MX-774 and Viking vehicles. All three of these highly significant developments rapidly spread throughout the rising American rocket industry during the late 1940s, and eventually became incorporated into almost all major rocket engines up through all the stages of the mighty Saturn manned lunar-launch vehicles. For example, the spaghetti type of combustion chamber continued to be used up to the Space Shuttle main power plant and in later-generation large engines. At the same time, it is remarkable that RMI, started back in late 1941 as America's first liquid-rocket company, was responsible for opening the way to these exceptional and history-making technological accomplishments. This book now helps commemorate the RMI/RMD enterprise and their recent 75th anniversary.

Yet, both RMI and RMD had faced their share of struggles throughout their respective histories, notably RMI's attempts to arrive at sound financial management in the early postwar years, encountering incessant noise and damage complaints from testing the X-1 and X-15 power plants, and the lesser-known, abortive attempts by RMI and successor RMD to gain technological footholds in the large-scale rocket engine field, as well as in the areas of nuclear and other exotic propulsion sources.

This section explores some of these and other aspects of the histories of RMI and RMD not covered in the earlier chapters. It includes coverage on the only known major accident in all their years of rocket testing, the merger of Thiokol Chemical Corporation with RMI to form RMD, the advanced chemical-propellant work of RMI and RMD, their forays into nuclear propulsion, and the final fate of RMI's original blockhouse. It concludes with RMD's last known projects, and how the enterprise finally came to be dissolved in 1972.

Fig. 11.1 Early RMI "spaghetti" motor, circa 1940s, on exhibit in NASM's Udvar-Hazy Center, Chantilly, VA, near Dulles International Airport. (Courtesy Frank H. Winter.)

This is followed by an epilogue that more fully rounds out the rich histories of both RMI and RMD, offering a more complete picture of America's first commercial rocket company.

POSTSCRIPT TO RMI'S EARLY FINANCIAL WOES AND THE ENTRY OF LAURANCE S. ROCKEFELLER

As Chapter 5 demonstrates, from the start RMI strove to create—rapidly succeeding—great technological advances in the still-young field of liquid-fuel rocketry. Paradoxically, at the same time during the early postwar years, when RMI's intense development of their Black Betsy engine was about to bring aviation into the supersonic age, the company was hardly on firm financial footing. This precarious situation then got worse.

Initially, in order to retool toward advancing the state-of-the-art rocket technology that RMI pursued, the company expended a large chunk of their revenues in acquiring capital equipment. By doing so, they strangled their net worth. The situation then became far more complicated upon their sudden, unexpected expansion and the move to Lake Denmark in 1946. That is, by the close of 1945, RMI had only 55 employees; post-move by mid-1946, the number more than doubled to 120. The number then rose sharply to 434 by June 1947.

Because of this dramatic increase, however, RMI was barely able to meet its monthly payroll. There were other consequences as well. "During this period of rapid employment," explains the anonymous author of a long article on RMI's financial history that appeared in *The RMI Rocket* for December 1951, "every payroll was a major financial crisis, and after they had been met, there was little money left for vendor payments."[1] The author of this lengthy and comprehensive financial history was very likely Lawrence's brother Robert, who had served as the company's treasurer throughout much of its early history and therefore possessed intimate knowledge of the topic.

It is true, the writer pointed out, that sales for RMI had mushroomed from $49,000 in 1942 to $900,000 by 1946, but purchases of equipment and materials simultaneously grew to "dizzy proportions," and operating losses further aggravated matters.[2] He further reported that at the end of 1946, RMI "had a net worth of minus $92,787 and a discouraging working capital deficit of $180,367."[3] According to a most revealing November 1950 article in *Fortune* magazine, "RMI's books were so frightening that a long series of potential backers took one look and ran."[4] On the plus side, RMI's outstanding technical accomplishments at the time, notably its Black Betsy for the Bell X-1 and its upcoming 20,000-lb-thrust Viking rocket motor, played key and most timely roles in helping spare RMI from total bankruptcy.

RMI's huge operating costs cannot be overlooked as an additional complication to their financial woes during that period. A memo of 28 February 1948 from RMI's executive vice president and general manager Charles W. Newhall Jr. to chief engineer Harry B. Horne Jr. laid out propellant and other costs. Typical tests, with even a 1,000-lb-thrust engine, amounted to $7,900 for "labor and overhead," and $6,500 for propellants, or $14,000 total.[5] For the development of the engine for the Neptune—later renamed the Viking project—the figures were $90,000 for "labor and overhead," $135,000 for propellants, and $100,000 for a test stand, or $325,000 total.[6]

By mid-1947, with the prospect of RMI owing creditors some $600,000, Laurance S. Rockefeller stepped in. Rockefeller and his people had already been spending months on investigations and negotiations going back to at least the previous year with company officers and high-ranking Navy officials, in addition to consultations with banking executives, before actively seeking to help try to solve the problems. Lawrence's *Daily Log* reveals that as early as 9 January 1947, he and John A. "Jack" Pethick, RMI's treasurer, went to New York to the Rockefeller Foundation for a crucial conference with the "Rockefeller group—re: financing."[7]

The reasons for Laurance Rockefeller's involvement with RMI were twofold: First, there was his own personal interest in and support of aviation, of which rocketry was then regarded as a highly promising potential extension—nowadays known as aerospace. Second, he was a leading figure in the pioneering field of venture capitalism. Venture capitalism meant investment in businesses deemed to have high growth potential due to demonstrated growth and innovative technology. In RMI's case, Rockefeller quickly understood that fulfilling this potential required a combination of both his financial management skills and an infusion of very seriously needed capital.

Rockefeller himself had learned to fly as a young man. He had made friends with the likes of World War I American flying ace Eddie Rickenbacker, and he subsequently foresaw the promise in postwar commercial aviation. This led the multimillionaire to become the largest shareholder of Eastern Aviation, which had been headed by Rickenbacker since 1935, as well as to fund the very important post-WWII military contractor McDonnell Aircraft Corp. Along similar lines, Rockefeller must have been chiefly persuaded to invest in RMI's future in light of their work toward developing the power plant for the Bell X-S1 aircraft that would break the sound barrier later in the same year. RMI therefore ideally fitted the innovative-technology criterion of venture capitalism.

It is not known how Rockefeller himself learned of RMI, although he enjoyed the highest connections in the aviation world; perhaps the connection came directly or indirectly through Bell Aircraft. On the other hand, it was general knowledge by this time—inside and outside aviation circles on the East Coast—that RMI was working toward rocket engines for aircraft.[8] But available evidence suggests that Rockefeller learned of of RMI's financial plight through the Navy. In fact, a November 1950 *Fortune* article stated that by 1947 the Navy had become so alarmed over the potential collapse of RMI that they "asked Laurance Rockefeller to come to the rescue," although this cannot be verified.[9]

However matters transpired, Laurance Rockefeller came to help support the financially ailing RMI through a substantial debenture loan of $200,000, on top of his own heavy investment in the company through stock purchases, and also by affording RMI the services of a team of his own financial managers.

On 29 August 1947, Rockefeller addressed a letter to Adm. Alfred M. Pride, the newly installed chief of BuAer, expressing "assurances" regarding the proposed loan and other financial management help being offered to RMI toward fulfilling its contractual obligations with the Navy. This letter, which appears to corroborate the statement in *Fortune*, was presented and discussed at a subsequent RMI board of directors meeting. The terms of the loan and other arrangements were approved soon thereafter, and on 10 September, a three-year, $200,000 loan at 4%—to be paid in full by 10 September 1950—was signed.

Less than a week later, in the early afternoon of 15 September 1947, Lawrence's *Daily Log* reveals an all-hands meeting was held in the RMI cafeteria, in which Charles W. Newhall Jr., an ex-Air Force colonel and now the company's general manager, "talked to RMI employees, explaining [the] Rockefeller Bros. Partnership in this company. His talk [was] preceded by short remarks from Mr. [Laurence P.] Heath [manager of RMI Contracts, Administration, and Service Division] and Mr. Lawrence." The next day, Laurance Rockefeller and his party arrived.[10]

Thus, the arrangements amounted to far more than a loan—they created a partnership with the Rockefeller brothers (that is, Laurance). Beyond this, Laurance Rockefeller also acquired a majority stock position in RMI that by 1951 had reached 44,100 shares, or 43.6%, compared with the grand total of 31,050 shares, or 30.7% for the three original partners.[11] Bowing to the stark reality of a complete company takeover, coupled with the inevitability of an approaching new corporate environment in which he, as a non-professional and former subway ticket taker, simply could no longer work—Pierce submitted his letter of resignation as a "Director" on 8 September.[12] This was announced in the RMI board of directors meeting the following day.

In fact, it was a forced situation, because on 10 September, Lawrence, Wyld, and Shesta each signed an agreement in which they would "remain in the employ of Reaction [Motors, Inc.] and Reaction agrees to employ him until September 1950 (or for the period that any three-year 4% Notes are outstanding whichever period is shorter)."[13]At that point, their positions would be "terminable."[14]

With the Rockefeller partnership with RMI came new operating and financial plans and policies, in addition to the hiring of new upper-level managers to bolster the company's executive and administrative strengths. In addition, as Shesta put it succinctly, Lawrence was "kicked upstairs" and made the chairman of the RMI board of directors, although he retained his title of president, while Newhall served as the actual administrative head who answered to Rockefeller directly.[15]

Already, by the end of 1947, a profit of $33,202 was gained, and the deficit reduced after net sales of $1,447,880. In 1948, a profit of $99,205 was reached with net sales of $3,592,288. Up to the mid-1950s, RMI's net sales averaged about $4.5 million, with average profits of about $110,000 per annum. Similarly, by the end of 1950, RMI had established a net worth of almost $700,000 and a net working capital of more than $500,000. Just five years later, RMI gained a record net sales of $10,298,628, with a net income of almost $200,000, while the number of employees had grown to approximately 1,250. Their net worth had also doubled to more than $1,500,000. Sales and earnings climbed thereafter, amounting to $16,193,944 and $348,348, respectively, in 1956 for instance.[16]

LATER FATES OF THE FOUR FOUNDERS

As for the fates of the four original founders, the first to leave, Pierce, had settled down in California, and even though he left with his 10,000 shares in RMI, he ran into bad luck. By 1948, he had purchased a citrus grove, evidently in Orosi, Tulare County, in the southern part of the state, but a freak frost ruined his orchard. That was the last his former RMI colleagues heard of him for many years. It was later learned that after the loss of his citrus grove, Pierce held positions in various southern California aircraft and/or rocket companies—including Douglas Aircraft by 1960. By the late 1960s or early 1970s, Pierce was living in modest circumstances, having purchased a mobile home based in Apple Valley, near Barstow in the Mojave Desert. About 1972 or 1973, he met a retired, widowed elementary school teacher in Barstow and married her in 1974 (He had earlier divorced his first wife.). By this time he had retired as "an aerospace engineer," according a recently discovered brief obituary.[17] He settled in Barstow, and on 22 February 1979, he died of pneumonia at age 74 in nearby Loma Linda.

Arguably, it was due to Lovell Lawrence's agreement of 10 September 1947 with RMI—or rather, with Laurance Rockefeller—that by late 1950, when the terms of the agreement were met, his position at RMI no longer seemed tenable nor challenging. But it was not until 18 July 1951 that Lawrence resigned as president and chairman of the board, and his letter of resignation was also sent to Rockefeller. Raymond W. Young Jr. then became the second president of RMI. From 1952 to 1953, Lawrence briefly ran his own firm, Lawrence Engineering Company, Inc., set up in Eau Gallie, Florida. His new enterprise made home-living items such as the "Petit-Edo pocket lighter" and the "Clo-lite closet lamp"—products nowhere near as exciting nor challenging as those of his RMI days. Consequently, in 1953, Lawrence left this business and returned to rocketry by joining the Chrysler Corporation of Detroit, where he was named the technical director of their Missile Division. In this capacity he directed the very important Redstone missile development, which became the nation's first operational ballistic missile.

In 1958, with modifications and an upper stage, the Redstone (redesignated as the Jupiter-C) launched America's first successful satellite into space, Explorer 1. Then in 1961, with a spacecraft atop the rocket and now named the Mercury-Redstone 3 (MR-3), it achieved the first flight of an American into space, Alan B. Shepard Jr.

Lawrence was to hold other important positions with Chrysler, eventually becoming their chief research engineer. He died of cancer in Grosse Point, Michigan, at age 56 on 24 January 1971—a highly regarded major pioneer in American rocketry. Notably, in 1950 he had received the American Rocket Society's most prestigious honor, the Goddard Memorial Award for directing the development of the X-1 engine.

But there is more to Lawrence's story than his contributions to rocketry. He had always been a visionary. In 1956, a year before the launch of Sputnik 1, he advocated nuclear-propelled spaceships, and in the same year he presented the paper, "Astro: An Artificial Celestial Navigation System," at a Franklin Institute Symposium in Philadelphia. He also had an article published on this same concept later that year. In essence, this was among the first—if not the first—unclassified treatment of satellites for navigation and a precursor to the Global Positioning Satellite (GPS). The first satellite in the U.S. GPS system, Navstar 1, was not launched until more than 20 years later, on 29 February 1978.[18]

Shesta stayed at RMI slightly longer that Lawrence, resigning effective 27 February 1953 as their vice president of research. An avid skier, he then moved to Stowe, Vermont, and opened up a ski lodge, the Deer Cross Inn, in partnership with another RMI administrator and fellow skiing enthusiast, Laurence P. "Larry" Heath. Shesta died in Morrisville, Vermont, on 13 July 1987 at age 86—the oldest survivor of the four RMI founders.

As for Jimmy Wyld, whose diminutive 90+-lb-thrust regen motor had started it all, he had technically resigned as RMI's chief research engineer on 21 November 1947, soon after the Rockefeller management takeover, although he remained with the company, apparently as a consultant. In 1950, he took temporary leave to work for the Atomic Energy Commission's National Laboratory at Oak Ridge, Tennessee, to assist in the design of a potential reactor small enough for use as aircraft propulsion: their Nuclear Energy for Propulsion of Aircraft, or NEPA project. This contract was extended for another year, and he returned to RMI in 1952, although the NEPA project never materialized. Tragically, however, Jimmy died prematurely of a heart ailment on 1 December 1953, at age 41, at his home at at 16 Pompton Avenue, Pompton Lakes—where RMI had started a dozen years before.

A man of great modesty despite his many achievements, Wyld received several posthumous honors, including the eponymous James H. Wyld Award, established by the AIAA in 1964 and now called the Wyld Propulsion Award. In 1976, he was inducted into the International Space Hall of Fame at Alamogordo, New Mexico, and in 1970 he was accorded his highest honor—a crater on the far side of the moon was named after him.

A CLOSER LOOK AT RMD'S NOISE AND DAMAGE SUITS

As described in Chapter 4, the noise and damage complaints RMI had faced earlier in its history, especially during the full days and nights of test firing their 6,000-lb-thrust Black Betsy, had been bad enough. But the situation became far more pronounced upon the development of the 50,000-lb-thrust Pioneer for the X-15 into the RMD years of the late 1950s. This time, the damage suits reached historic proportions—at least in legal annals. The 25 plaintiffs in a case known as Magnus Berg, et. al. v. Reaction Motors Division, Thiokol Chemical Corp. were residents of fifteen homes in Lake Telemark, a community close to Lake Denmark. By late 1960, the Morris County (NJ) Superior Court awarded a total of $100,000 in compensatory and punitive damages. "As far as can be determined," according to *Aviation Week* for 28 November 1960, "this is the first case in which rocket power (as distinguished from turbojet engine power) has been cited as a damage-causing agent and so the ultimate decision could have important consequences for this country's growing missile and space effort."[19]

The Berg v. Reaction Motors Division case did not end there. The defendants appealed, and the case went all the way up to the Supreme Court of New Jersey. It was argued during 19–20 March 1962 and decided the following day. The court concluded, as summarized by the legal writer R. Bender, that RMD's activities were not found to be negligent, but were "ultra-hazardous" and therefore they were "absolutely liable for property damages caused to homeowners near the tests."[20] RMD was compelled to pay the damages. The "consequences" upon the country's "missile and space effort" were therefore not as severe as the writer for *Aviation Week* had feared, but the Berg v. Reaction Motors Division case was certainly enormously distressful at the time to both the townspeople near Lake Denmark and RMD. RMD also earned the dubious historic distinction of being involved in one of the first legal "space cases." Moreover, Berg v. Reaction Motors Division is often cited in property-damage cases. Some interesting details are likewise learned of the nature of the then-controversial tests, and well as some of the measures taken by RMD to help rectify matters.

For example, the XLR-99 tests were carried out on test stands E-1 and R-2 at Picatinny Arsenal —in the Lake Denmark area—or, as close as approximately 3,500 ft from the center of the nearby village of Lake Telemark in Rockaway Township. When the tests were conducted, the thrusts created air turbulences as well as powerful exhaust flames extending as much as 20 to 25 ft in length and 18 in. to 2 ft around. One of the homeowners justifiably complained that he was disturbed in his sleep, and that the roar and accompanying awful vibrations caused his home "to come apart."[21] When the complaints began, the residents were shown films by RMD's Public Relations Office "about the high importance of the X-15 and the urgent need for the testing."[22] But this did not really help matters, and a Rockaway Township committee then insisted that RMD restrict its major test operations to between 7 a.m. and 8 p.m., and to take other restrictive measures. After all, the 3-minute tests had been run as many as 10 times per 12-hour work day, which meant that much of the testing had been carried out much later in the evenings. At first, RMD complied with this prescribed schedule requested by the township committee.

But RMD was pressured by the Air Force when the XLR-99 project fell far behind schedule and they insisted that RMI actually "step up" the program. To help improve the situation, RMD went as far as to devise and build a "sound suppressor" and undertake noise tests. Even those measures were not enough at the time, however, and the residents still lodged complaints until law suits were filed. In the end, the XLR-99 developmental program most fortuitously matured not long after, with qualifications-testing shifted way across the country, to Edwards Air Force Base in California. All in all, the XLR-99 testing phase had been a thrilling and at the same traumatic experience for all involved with it and affected by it. As later recalled by RMI veteran Al Miller, "The rocket engines made a lot of noise at Rockaway…That's probably what did us in."[23] Matters did not work out quite that badly, but came close.

RMI'S MOST SERIOUS ACCIDENT: EXPLOSION OF THE "SUPER P" ENGINE

A far more serious episode in RMI's history was the accident that occurred late in 1957 during the testing phase of their XLR-40 rocket engine—also known as the "Superformance," or "Super P" engine—in a Navy Vought F8U-1 Crusader carrier-based jet interceptor. This throttleable, 3,500–8,000-lb-thrust range spaghetti-type motor, designed by Robertson Youngquist, used 90% hydrogen peroxide and JP-4 as the propellants, with the JP-4 jet fuel supplied from the main aircraft tanks, and the peroxide tanks designed to fit into cavities in the plane. The motor was to be imbedded in the vertical tail root of the aircraft. The augmentation rocket motor was called Superformance for the following reason.

By the late 1950s, there was a very real concern that the Soviets would soon have bombers capable of cruising to more than 60,000 ft. Chance-Vought and other aircraft firms therefore sought a means to reach and maneuver at such altitudes, and RMI devoted considerable study to the problem that led to their development of the XLR-40 engine to provide short-duration thrust augmentation at high altitudes. The sudden thrust augmentation from this built-in adjustable booster was thereby designed to give the plane—or rather, a later model of the aircraft—an extra superperfomance boost when needed. RMI was subsequently awarded a $15 million contract for the XLR-40.

But at 11:20 a.m. on Monday morning, 16 September, during routine work on the engine at the Naval Air Rocket Test Station (NARTS) about a mile from Lake Denmark, there was a sudden terrific explosion, killing Chance-Vought test engineer Herbert. L. Bell and leaving six others badly burned. The burn victims also suffered from puncture wounds. As observed at the time, this was the first fatality in RMI's 16-year history. The local newspapers, as well as the *New York Times,* the *Boston Daily Globe* and undoubtedly other papers, covered the disaster as best they could given that the project was a highly classified

Fig. 11.2   XLR-40 rocket engine. (Courtesy Harry W. Burdett Jr.)

one. Both RMI and the Navy imposed restrictions on what could be reported.[24] On the Navy side, a detachment of 150 marines had been quickly dispatched to put the 300-acre testing site under a security lid. Consequently, the press stories were largely confined to identifying the victims and could not go into the nature of the project, nor exactly what happened. An RMI spokesman could only tell the reporters that the engine was "an experimental type" and that the equipment—which meant the static test stand at the site—"was only slightly damaged."[25]

It is now known that up to the time of the accident the engine was fully developed and was approaching its final delivery to Chance-Vought. Modification had already started on two earlier F8U-1 airframes. The only known surviving Chance-Vought technician involved in the accident, Glenn Repp, provides with the gist of what happened.

Repp was Vought's engine test engineer who had had experience not only with jet engines, but briefly as well with liquid-propellant rocket motors at Rocketdyne "Everyone," he recalls, was "afraid of hydrogen:

> I had made several training trips to RMI for peroxide training, and brought several of our fuels technicians with me. Our run [static test] cell was nearly complete…On Friday [13 September]…we had done a motor run and experienced some kind of leak in the pump plumbing. It was at the

end of the day and so it was decided we would let things rest until Monday...On Monday we came back to the test cell to find and correct the leak. Our safety officer and I were in the cell, perhaps 20 feet from the motor, observing. Work was being done on the motor by an RMI tech[nician] and two of my techs[,] Herb Bell and [Robert D.] Don Tweedy. Herb was hunched over the motor loosening a B-nut when the explosion occurred. He took the full blast...The safety officer and I were fortunate. We were far enough from the blast and close enough to the safety shower.[27]

As explained by Repp, a "B-nut" is simply a threaded part of (aircraft) plumbing. It takes three pieces of plumbing to join a hydraulic line.

Repp also briefly summarized RMI's subsequent findings: "RMI ran a lot of tests and determined...that a pump seal had leaked and over the weekend."[28] The JP-4 and peroxide, he also continued, had "mixed together forming a shock sensitive mixture (or heat sensitive) and action of twisting the B-nut was enough to set it off."[29] During their investigations, RMD thus discovered the shock/temp sensitivity of this propellant combination. In addition, they made major redesigns of the turbopump—probably including more secure pump seals. This was to no avail since, as Repp concluded, "The Navy cancelled the [RMI] thrust aug[mentation] program."[30] As far as is known, the terrible Super P experience was the only known fatality throughout the entire course of the histories of both RMI and RMD.

EFFORTS TO BUILD A LARGE-SCALE ENGINE

From time to time there had been a number of nonfatal accidents at both RMI and RMD. One case is especially significant historically because it had to do with RMI's secret efforts to build and test their largest rocket engine ever. According to the late RMI veteran Harold S. "Sam" Bell, "The 100,000 lb. thrust engine was an in-house, 'can-do' operation ostensibly to let us compete with the California big boys [that is, Rocketdyne and Aerojet]. It was built in secrecy and only a few employees were even aware of its existence. The first firing was a disaster and no further units were built. Harold Davies... was involved in this program."[31] Davies corroborated this account, although he said it was a 150,000-lb engine. The year or years of this program are unknown, although there are some strong clues as to the period.

In another letter, Bell also remembered that "the chamber stood six or seven feet and had a tube bundle [that is, spaghetti configuration] similar to the LR-99 [another designation of the XLR-99] that was welded and micro brazed. This effort followed the LR-99 program...RMD could never have gotten local [New Jersey] permission to develop such a large engine. We had trouble enough with the LR-99."[32] Therefore, it is fairly certain this must have happened by the early 1960s, when the XLR-99 was completed and became operational.

Davies provided finer details. He said that that this was a company-funded project because they "became concerned that it was being left behind by Rocketdyne and Aerojet in large engines."[33] Henry A. "Buzz" Barton was in charge, while Davies headed up the design, with Haakon O. Pederson working on the large turbopump. The propellants were LOX and JP-4. The pump ran around 8,000 rpm and tested well with water in place of the propellants. "The pump tested well with water," Davies continued, "and in later tests with propellants. The first [and only] hot test of the entire engine was run on the large Navy [static] test stand across the road from [the] RMI 'C' [testing] area [at Lake Denmark]. The chamber broke up on starting. The pump was not damaged and we used it for many years as a convenient power source."[34]

Just recently, another significant piece of evidence on this project has surfaced. It is a photograph of the engine and the pump together and an accompanying typed data sheet, both found in an untitled light-brown leather three-ring binder full of other RMI photos and documentation in the library of the New Jersey Aviation Hall of Fame at Teterboro. Unfortunately, the data sheet gives neither a date nor a designation, but does identify the engine as a "120,000 to 240,000 Pound Thrust Liquid Propellant Rocket Engine [sic.]."[35] Davies's memory may have erred on the thrust of the motor, or he may have recalled the initial thrust rating that was perhaps later upgraded. At any rate, among other data on this sheet is that the engine weighed 515 lb, the turbopump weighed (an additional) 495 lb, and the turbopump speed was rated from 6,500 to 8,100 rpm; those numbers do agree with Davies's recollection on that point.

Even earlier, by 1955, according to historians Constance McLaughlin Green and Milton Lomask, RMI "was working on a motor with a 75,000-pound thrust," perhaps to try to adapt it to the proposed satellite launch vehicle for the Vanguard project. But General Electric offered what "looked liker a better choice," and GE got the contract.[36] Nothing more is known of RMI's 75,000-lb-thrust effort, although it is possible it was a further upgrade of their 50,000-lb-thrust Super Viking engine.

There is also some evidence that by the same year, RMI had already built experimental parts for a 120,000–250,000-lb-thrust engine, depending upon the chamber operating pressure.

Years before Project Vanguard was conceived, RMI already had its sights set on a really large engine. An RMI interoffice memo dated 5 February 1951 and titled "Possible Air Force Projects" from Charles W. Newell to Raymond W. Young—then vice president of engineering—offered a number of suggestions. Among them was the proposed "Development of a high performance, high specific impulse engine between 20,000 lbs and 500,000 lbs for use in the long range surface to surface missile."[37] Hence, RMI—and its successor, RMD—had probably *always* harbored a desire to truly get into the very big rocket engine market, although they never attained that goal.

At this point it is useful to offer further details on another aspect of RMI history already encountered throughout this book—RMI's merger with Thiokol.

THE MERGER WITH THIOKOL

The Thiokol Chemical Company was founded in 1929. Its initial business was the manufacture of a range of synthetic rubber and polymer sealants. Thiokol's name was derived from the Greek words for words for sulfur (*theion*) and glue (*kolla*)—the chemical basis for their synthetic rubber. Thiokol thus became a major supplier of liquid polymer sealants during World War II. After the war, when scientists at the Jet Propulsion Laboratory discovered that Thiokol's polymers made ideal binders for solid rocket fuels, Thiokol moved into the new field of composite rocket propellants and became a leading developer and manufacturer of solid-propellant rocket motors, initially for the highly successful Falcon air-to-air missile. Thiokol went on to make its own major breakthroughs in the technology, the most outstanding of which was the large-scale, long-duration-burning, solid-propellant rocket motor that resulted in such key aerospace developments as the Polaris missile and the SRBs (solid-rocket boosters) for the later Space Shuttle.

By 1957, Thiokol—now a large corporation—had annual sales of approximately $30 million. A merger between it and RMI seemed appropriate to the directors of both concerns, as well as to the Olin Mathieson Chemical Corp. and Laurance Rockefeller. The latter two controlled more than two-thirds of RMI stock. A merger was thus proposed to Thiokol stockholders at a meeting at Trenton, New Jersey, on 17 April 1958. By an overwhelming margin, the stockholders voted for the proposal. The merger plan was subsequently brought up at the regular RMI board of directors meeting held on 24 January 1958 in room 5600 at 30 Rockefeller Plaza in New York City.

The merger plan became similarly approved by an enthusiastic majority of RMI stockholders and the two companies were finally officially merged at midnight on 30 April 1958. The firm was now known as the Reaction Motors Division (RMD) of Thiokol. RMI's financial and technical alliance with Olin Mathieson was also dissolved in order to more easily and speedily integrate the new division into the Thiokol organization. For the first two years following the merger, Ray Young would serve as the general manager of the new division and its de facto head. After this, and for most of the 1960s, this post was assumed by Dr. Edward H. Seymour, who had joined RMI in 1956 as their manager of preliminary design and later moved up to become director of research.[38]

Regarding the Olin Mathieson connection, this other giant chemical corporation was impressed by RMI's technical skills and accomplishments and in October 1953 they had bought up 50% of RMI stock. Furthermore, Olin Mathieson wished to pursue its own technical and business association with RMI in the field of rocket propulsion. On 8 December of that year, after the untimely death of Jimmy Wyld on 1 December and the resignation of two other members of RMI's board of directors, including RMI executive vice president Charles W. Newell, three representatives of Olin Mathieson were elected to the board of directors. Olin Mathieson thereby acquired control of RMI. Newell had left RMI because he had been named president of Flight Refueling, Inc. Among the technical connections between Olin Mathieson and RMI, in 1955 they both established a joint program of shared applied research and development that additionally involved the Marquardt Aircraft Company. This program became known as the OMAR Technical Liaison Committee, after the first letters of the names of each of the member companies. Among other arrangements on the business side, Olin Mathieson and RMI successfully teamed up in acquiring Air Force contracts for joint propellant studies.

EXPANDED FACILITIES

Upon the merger of Thiokol with RMI came rapidly expanded new facilities. When RMI first occupied part of Lake Denmark back in 1946, the company had also rented buildings in adjacent Rockaway. In 1949, RMI purchased industrial property there and set up their headquarters at Elm Street and Stickle Avenue, while their research laboratories and engineering department were situated about five miles away; full-scale testing continued at Lake Denmark. A few years later, they built a plant with Navy support in nearby Denville. That plant opened in early October 1955. Upon the merger, Thiokol purchased additional Denville facilities, although headquarters (now RMD) remained at 100 Ford Road in Denville until the end. With the merger—or rather in anticipation of it—Thiokol had also acquired a production plant at Bristol, Pennsylvania, where RMD carried out most of its production, principally of the Falcon III and later Bullpup motors.

GROWTH IN PROPELLANT RESEARCH

Inevitably, another result of the merger with the huge Thiokol chemical concern was that RMD became far more propellant oriented. Shortly after the merger during that spring of 1958, RMD managers requested about $4 million from Thiokol headquarters to invest in test facilities to prepare them to enter the field of liquid-hydrogen technology. But the large amount, coupled with reality that competitors Pratt & Whitney, Rocketdyne, and Aerojet were already firmly established in this field, caused the request to be turned down.

This disappointment neither halted nor hindered the Division's considerable efforts in other high-energy propellant research and development—nor did it hinder their investigations into potential nuclear and plasma propulsion systems.

In their earliest pioneering years, RMI was of course largely confined to basic propellants of LOX/gasoline, LOX/alcohol, LOX/nitric acid, and similar combinations. But the addition of the distinguished Austrian chemist Dr. Paul F. Winternitz to their staff just after the war considerably widened their propellant horizons. By the same year of 1946, he introduced RMI to liquid ammonia and undertook among the country's first studies on the exotic and high-energy borane and diboranes. By the 1950s, in coordination with Olin Mathieson, RMI became engaged in high-performance solid-propellant research. This led to RMI forming a separate Applications Engineering Department for Solid Propellant Rocket Power Plants in 1957; William M. Davidson was named to head the department. This was followed in August 1959 by RMD's initiation of a substantial program on hybrid propellants: combination solid and liquid fuel systems. That same year, increased research was carried out with solid boron fuels, in addition to energetic hydrazine liquid propellants.[39] By the early 1960s, complex work was also underway to arrive at far more powerful fluorine-based oxidizers and included extensive test firings using small rocket motors—much of this advanced propellant research carried out with the very farsighted aims of potentially developing thrusters for long-term deep-space missions. More than merely exploratory, these propellant studies were also often undertaken as very lucrative contracts granted by NASA and other entities. In 1963, NASA was even considering hybrids for use in lunar landers.

Such exotic propellants were often unpredictable to deal with and even downright dangerous. As Arthur "Art" Sherman recalled: "While witnessing one of the unsuccessful tests with chlorine trifluoride...I became known as the fastest man in the test area while wearing one of these [a Scott air pack]...We did live dangerously, but by and large safely and the idea of working in totally virgin territory such as this was fascinating."[40]

Atomic rocket propulsion—better known as nuclear rocket propulsion—always seemed to be the ultimate answer to deep spaceflight.

Atomic Propulsion

As far back as mid-January 1947, Jimmy Wyld enjoyed a stimulating "preliminary discussion" with Lawrence, Shesta, Dr. Winternitz, and a few others on the possibilities of an atomic-powered aircraft in which RMI was to "be in charge of constructing the motor."[41] Nothing came from this, although Wyld continued to be fascinated by the prospects.

It was not until 1956 that RMI started a modest program that tasked just four engineers to work on the theoretical possibilities of nuclear propulsion.

It was soon to come under the leadership of physicist Dr. John J. Newgard. By 1959, RMI/ RMD's nuclear "staff" had produced at least 17 technical papers on the topic. Other leading members of RMD's small nuclear-technology group around this time were Myron M. Levoy and Robert "Bob" P. Helgeson, although the latter left in 1963 to work for NASA as head of their Nuclear Rocket Development Station at Jackass Flats, Nevada, for Project Rover.

The bulk of RMI's and RMD's nuclear work was on the theoretical side. Yet RMD did develop some modest nuclear hardware. One example was a Control Rod Servo, developed by 1958, for the Atomic Energy Commission's (AEC) "Little Eva," a small enriched-uranium "critical assembly" for testing nuclear components. They had also participated in the design and development of controls for Kiwi-A—the NASA/AEC nonflight, static-test, nuclear rocket-engine test rig in Nevada that came to demonstrate the feasibility of gaseous-hydrogen nuclear rocket propulsion from 1959 to 1960. In addition, RMD had contributed toward the planning of the NASA/AEC facility at Jackass Flats.

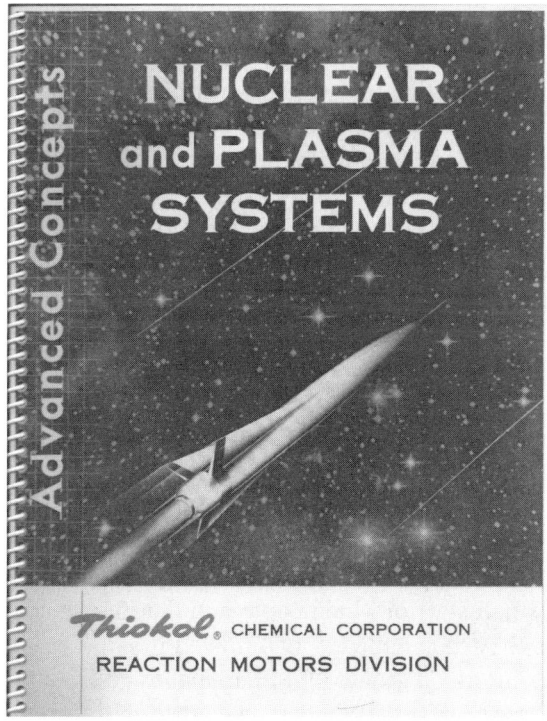

Fig. 11.3   Cover of RMD report on proposed nuclear and plasma propulsion systems, circa 1960. (Courtesy Orbital ATK.)

But the group's most notable theoretical treatment was Newgard and Levoy's most ambitious 1958 concept for a 150 ft long, 15 ft in diameter, nuclear-propelled, deep-space spacecraft powered by a solid-core reactor. Various design configurations and different performances were also worked out. An abridged version of the study appeared in *Scientific American* for May 1959. One evening per week, Newgard also delivered well-received lectures on nuclear propulsion in RMD's cafeteria. In addition to these activities, by the end of the 1958 Newgard and Levoy presented a paper on their nuclear space-propulsion concepts at the 10th International Astronautical Federation (IAF) Congress, held in London. Thiokol's upper-echelon management must have been highly encouraged by the combination of Newgard and Levoy's farsighted conceptual study and the international recognition it received, in addition to the obvious rising enthusiasm about the topic among the RMD staff and the beginnings of RMD experience in nuclear-propulsion engineering, however modest at this point. All these factors must have convinced RMD that far larger prospects in the nuclear field were close at hand.

Consequently, the *New York Times* for 8 June 1960 ran a small and curious item, datelined Thiokol's corporate headquarters in Ogden, Utah, reporting that Dr. Harold J. Ritchey, Thiokol's vice president for operations, had announced the formation of a new "nuclear unit...to service prime contractors...for propulsion development."[42] This unit was to be "headed by W.M. Davidson, formerly of the Reaction Motors Division."[43] Other sources call the nuclear unit a division. Closer examination reveals, however, that it hardly can be described that way; rather, the "unit" consisted of just a handful of people stationed in a small office in an isolated two-story building off U.S. Highway 46 in the township of Parsipanny-Troy Hills, New Jersey, about eight miles from RMD's headquarters at Denville. Despite this, the local 1960–1961 telephone directories do, indeed, list this entity as Thiokol's "Nuclear Div.[ision]."[44]

It was revealed that Davidson is more properly identified as William M. Davidson, who was more of a "marketing person," rather a technical one, according to Sherman.[45] Hence, this small office was to serve strictly as a marketing operation to first seek out contracts, which, if successful, could perhaps lead to a true division. Davidson must have had similar marketing responsibilities back in 1957 when he headed RMI's newly opened Applications Engineering Department for Solid Propellant Rocket Power Plants. But the nuclear field was vastly more complex and demanding.

It is true that Bob Hegelson, a leading authority on nuclear propulsion, was on Davidson's small staff of nine people and that the "Nuclear Technology Department" as they then called it, had moved in late December 1961 from their cramped quarters off Route 46 in Parsipanny to Denville. It is likewise true that by August 1960 they had managed to gain a half-million-dollar shared contract with three other firms, awarded by the Air Force Flight Test

Center at Edwards Air Force Base to provide sound system data to contractors considering nuclear rocket propulsion. Outside of this additional propulsion study, however, nothing further is known of RMD's "Nuclear Development Department," or "Nuclear Development Center," as it was also called. Nor is anything known of the fate of the Air Force contract. It appears that the RMD's "Nuclear Division" was thus short lived. Without a doubt, anticipated nuclear propulsion contracts for actual hardware development never materialized.

A more bona fide attempt by RMD to directly enter the nuclear rocket propulsion field was their planned Nuclear Technology Department, promoted in 1961. However, this may have been an effort to enlarge the previous Nuclear Energy Department formed by Thiokol headquarters to more fully integrate it into RMD. RMD produced a very comprehensive booklet that outlined their objectives and proposed department sections, and included biographical sketches of the ten projected section leaders, headed by Harry M. Bowman. Their main objectives were to perform engineering analyses of proposed nuclear rockets; undertake feasibility studies of the same; develop prototype nuclear rocket engines; and establish manufacturing specifications for the same.

However, an examination of the individual biographies of the projected staff members of the Nuclear Technology Department shows that while these men had impressive experience on the chemical rocket side, with just one exception none were really experienced in the nuclear field. And although there had been RMD staff enthusiasm for atomic-propulsion deep-space mission studies, RMD's bid for the joint AEC-NASA Nuclear Engine for Rocket Vehicle Application (NERVA) did not succeed for a number of reasons— technical, economic, and political—although the company did gain some component work on that project. It seems that many in the aerospace community, including Thiokol, had misgivings about the future of the existing NERVA program in the first place. As it turned out, the multimillion-dollar NERVA program did experience several outstanding technological successes, even though it was eventually cancelled in 1972, principally due to a growing lack of political support that led to severe funding cuts.

As for RMD's 1961 proposal to set up their own Nuclear Technology Department, there is no proof that this plan ever came to fruition. It is said that by 1961 RMD already had some 50 engineers engaged in some capacity on nuclear projects. But as noted, Davidson's "unit" was quite small and there is no evidence there ever really was a large-scale nuclear-propulsion department or section in the histories of either RMI or RMD. Rather, RMD staff members involved in nuclear research simply seem to have carried out this work as part of their assignments in other departments, like their Physics Department that regularly concentrated on advanced projects. In the cases of Hegelson and Levoy, and evidently Newgard as well, for instance, they had been in the Advanced Planning Department, and the latter two had both undertaken theoretical studies and not hardware development. Also, Newgard, the leading nuclear-propulsion authority at RMD, had left around this time to help start his own firm, Electro-Nucleonics, Inc., at nearby Caldwell, New Jersey. The firm went on to pioneer nuclear technology as applied to medicine.

PLASMA AND ION RESEARCH

Far less is known of RMD's involvement in plasma and ion research, which also took place during this time. Robert W. Ellison of the Applied Physics Department and Raymond E. Weisch, Jr. of their Special Projects Unit had co-authored a massive, 247-page 1958 report entitled, *Research Leading to a Heavy Ion Propulsion Unit* that advocated RMD's development of "ion thrust devices." Ion devices are thrusters that use electrically charged molecules, or ions, as the propellant. These forms of propellants have extremely high specific impulses compared with chemical propellants; the higher the specific impulse the higher the velocity of the rocket or craft, especially in airless—and there-fore frictionless—space. Ion engines can only be used in space because a large amount of electric power is needed to generate the ions. The ion thruster, or ion engine, produces extremely minute levels of thrust that come out as exhaust streams—but with extremely long durations of thrust for the amount of propel-lant—and eventually exceptionally high velocities for the spacecraft. Ion engines are therefore theoretically ideal for very long, deep-space voyages.

By late 1959, a special shielded laboratory funded by the Air Force Office of Scientific Research also existed at RMD's Denville plant and was devoted to ion research. Myron Levoy recalls seeing a kind of "rig" in "some labora-tory" at the Denville plant during that period in which "you could even see a glow in the back of the thing" that suggested ion propulsion.[46] It turns out that this laboratory was in RMD's Physics Department and was then responsible for not only key areas like basic combustion research, but also plasma physics and propulsion investigations. The latter entailed the use of a working ion engine, in addition to advanced plasma propulsion investigations.

Yet, for all their intense interest and work toward these very advanced and exotic forms of propulsion, they did not lead to any viable programs. Rather, both RMI and RMD were far better known and remain revered as outstanding pioneers in chemical rocket propulsion.

RMD'S FINAL YEARS

As for RMD's final years, after the booms of the multiple Bullpup contracts and closeout of the Surveyor vernier program, these were sadly marked by diminishing backlogs and no new major programs. RMD sales had reached a record $35.7 million in 1963 then declined steadily, to $30.4 million in 1965 and $27.7 million in 1967, while in 1968 they plummeted to $7 million. The last was attributed to the phasing out of production of Bullpup B motors and heavy expenditure for the development of the Condor motor, although it was terminated in the autumn of 1967; a funding loss of $5.2 million for research and development from 1958 to 1969 aggravated the situation. With the simultaneous phaseout of several programs, including the C-1 and Surveyor verniers, Division employment also dropped dramatically from 1,400 at the start of 1967 to about 530 by year's end. At the beginning of 1968 it was worse, at 298. This was also a time of severe Department of Defense budget cuts that particularly affected missile liquid-rocket technology, as well as the winding down of NASA's Project Apollo.

During RMD's later years, they were involved in minor successes, such as additional contracts for Lance missile motor improvements. There were also rocket components and especially valves, another one of their mainstays. In fact, by 1966 RMD had secured a lucrative valve contract from NASA for Project Apollo's Saturn 1B vehicle—a predecessor to the manned Saturn V launch vehicle. Another Saturn component was a "universal ball joint," while a more unusual one was their Hyprox steam generator that was shipped to NASA's Marshall Space Flight Center at Huntsville in 1965 to be used to simulate a 20-mile altitude to check out the workability of the attitude-control engines of the S-4B—the second stage of the Saturn 1B. RMD additionally produced vibration monitors for the Saturn V and Apollo Service Modules to monitor combustion stability and vibration levels in this hardware.

Incidentally, RMD's (and RMI's) staging and other valve work went back for many years. For instance, RMI's valve developments were started in the mid-50s for the Atlas, eventually covering Atlas A to D—as well as Titan missiles from Titan I up to the Titan III. The valve work was thus very extensive and also came to encompass space-launch applications for these vehicles. By late 1959—as an extension of their work on the Atlas ICBM missile valves—RMD developed and supplied disconnect valves for the Atlas booster for America's first manned space program, Project Mercury. Disconnect valves are hermetically sealed and squib actuated. They permit entry of fuel and oxidizer into the first and second stage engines. Similar valves were used in Atlas boosters for Project Gemini, as well as the unmanned Agena Target Vehicle (ATV) that enabled the Gemini astronauts to practice orbital docking and rendezvous techniques in preparation for the coming Apollo program. On 17 May 1965, astronauts Gus Grissom and John Young sent a letter of thanks to RMD for the flawless operation of the six prevalves, all of different configurations, that were used in both stages of the modified Titan booster for the Gemini-Titan GT-3 mission.

RMD did attempt to enter Project Apollo itself in a more major way. They went as far as to have a full-scale wooden mockup made of the Apollo LEM (Lunar Excursion Module) descent engine by the firm of Lester Associates of Thornwood, New York, premier scale-model specialists. They then shipped the model early in 1963 to Grumman Aircraft, the chosen main contractor for the LEM, to enter the bidding as a subcontractor for this key component. But they were beaten out by Space Technology Laboratories, Inc. (STL). Nonetheless, RMD was still proud of this effort and afterwards kept the mockup that may have been displayed in their company "museum"—evidently a collection of artifacts not only to commemorate their achievements, but useful for drawing upon for trade shows.

RMD's Last Project: ARE

But as far as we know, the very last project of RMD was the ARE project, or Advanced Research Engine. As Harvey Fox, who worked on the project, remembered, "Since I had to finish the final report for the ARE project, I had to stay at the [RMD] site for a few weeks after essentially everybody else had left. The few of us that were left behind had to move into a quonset [sic.] hut in the parking lot."[47] The ARE, he added, involved a very short—two seconds or less—duration motor that burned chlorine trifluoride as the oxidizer and MAF (mixed amine) fuel and "the object was to optimize injector configurations, etc. The tests were short to keep the copper thrust chamber from melting…I have many fond memories (and not so fond) memories of those chlorine trifluoride tests."[48] As seen earlier, Art Sherman has his own "not so fond" memories of chlorine trifluoride.

However, William "Bill"/"Billy" Arnold, who joined RMI in 1946 and for years had been the company's field service representative, made the cogent point that the usage of XLR-11 engines in NASA's lifting bodies up to 1975 (covered in Chapter 5) was technically RMD's final project. Beyond this, Arnold worked on the proposed, though never realized, lifting body followup, the X-24C program. He added that the XLR engines still bore the Reaction Motors name and he was still considered the Reaction Motors field representative, receiving paychecks from Thiokol up to 1976. He therefore contended that he was the last Reaction Motors employee, even though he was paid by Thiokol past the end of their Reaction Motors Division in 1972.

Arnold also worked at Edwards Air Force on a little-known related project until 1979 that incorporated both the XLR-11—using LOX and liquid ammonia as the propellants—and the XLR-99 toward the next step in piloted aircraft later called the National Aerospace Plane (NASP). The next step involved potentially exploring flight from Mach 5 up to Mach 10. For this exploratory work, however, Thiokol's Elkton, Maryland, plant sold all the rights to these engines for $10,000 to Aerojet—and Arnold then worked for Aerojet.

End of the Company

By the end of 1969, with RMD's backlog at a minimum and no new major projects in sight—and projected operational losses linked to the failing aerospace marketplace that was due in part to the wind-down of Project Apollo—Thiokol reached the decision to terminate their participation in the liquid-propellant field. RMD would be phased out. In its 12 August 1970 issue, the *New York Times* ran a lengthy and emotional piece on the public auction that had closed the previous day, liquidating the Division's last "moveable assets."[49] The auction was held in a building at the Denville plant. Assets consisted of electronic and chemical laboratory equipment, shipping pallets, and the like. The auction attracted some 300 bidders from perhaps 30 states, from Florida to California. The auction also drew curious company old-timers, like Ted D. Sjoberg who had worked on the X-15 engine. "Their dedication," he said nostalgically of his former coworkers, "was fantastic. They put real blood into those projects."[50] By the time of the auction, the article adds, RMD employed only about a dozen executives, guards, and maintenance people "to wind things up."[51]

The end came in June 1972, or more than 30 years after RMI was founded. Thiokol continued on, and in 2001 ATK (Alliant Techsystems) acquired Thiokol in a merger that became ATK Thiokol Propulsion. In February 2015, ATK merged with Orbital Sciences Corporation to became Orbital ATK.

FINAL POSTSCRIPT: THE OLD *RMI* "BLOCKHOUSE"

As a final postscript, there is the story of RMI's old Franklin Lakes "Control Room," or "blockhouse" as it came to be called in more modern terminology.The blockhouse existed for many years after its abandonment in 1944. In its afterlife, it became a garden shed for a house with the address of 936 Dogwood Trail. The other structures of the former small test complex, including the roll-over shed that protected the test stand, storage sheds, concrete walkways throughout, and the test stand itself, had long since disappeared. Then, in 1977, the site and its historical significance came to the attention of archeologist Edward J. Lenik of the Archeological Society of New Jersey, who conducted further research, including consulting the author, then with NASM, to verify its historical importance.

Consequently, on 6 June 1979, the modest building—now minus a few of its viewing ports—was listed in the National Register of Historic Places.[52] Alas, some years later some of the neighbors considered the building an eyesore and they pressured the owner, who eventually had it unceremoniously demolished in 2005. This was a tragic loss for the history of New Jersey and for U. S. aerospace history, although thanks to another neighbor, a solitary cinder block of the quaint old building was saved and still exists. It is presently in the hands of a private collector.[53]

Fig. 11.4   Former RMI blockhouse at Franklin Lakes, NJ, used 1942–1944, photo by Edward J. Lenik, 1979. (Courtesy Ronald J. Dupont Jr.)

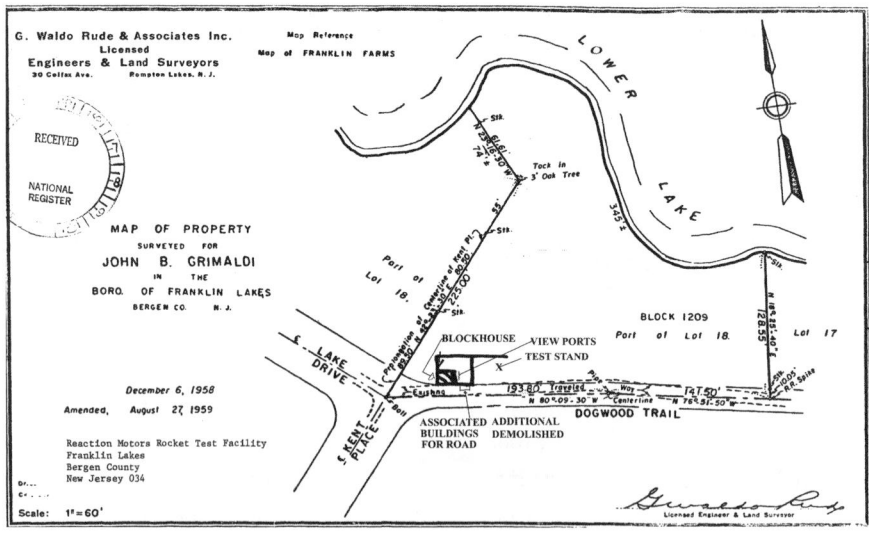

Fig. 11.5   Property map, 1959, with added annotations showing the location of RMI's blockhouse at Franklin Lakes. (Courtesy Jack Goudsward.)

EPILOGUE

The single cinder block of RMI's first control room is by no means the only remains of America's first rocket company. A most impressive number of both RMI and RMD motors, as well as an assortment of other parts, are in the collections of NASM and other museums.[54] Several of the more prized objects are on exhibit. Among these are the large and well-used X-15 engine that may be very easily viewed as installed in the aircraft itself that hangs from the ceiling of the museum's Pioneers of Flight gallery. A set of four of the venerable XLR-11 engines of the type that powered Chuck Yeager's Bell X-1 protrude from the back of the M2-F2 Lifting Body that is suspended from the ceiling of the adjacent Space Race gallery. The original X-1 is also in Pioneers of Flight, although it is not very easy to spot the four imbedded nozzles of this engine—originally designated the 6000C-4—in the aft section of this world-famous plane. But the four XLR-11 nozzles at the rear of the Douglas D-558-2 Skyrocket, on the second floor of NASM, a short distance from the X-15, are distinctly visible.

In the Space gallery is found an upright-mounted Viking rocket containing original parts from a flown Viking; it includes the vehicle's engine, which is visible through a Plexiglas covering. At the museum's Steven F. Udvar-Hazy Center in Chantilly, Virginia, near Dulles International Airport, is an engineering model (designated the S-10) of the Surveyor spacecraft, which is fitted with a set of three RMD vernier motors. This spacecraft model, used in thermal control tests and representing Surveyor 3, is presently in the Mary Baker Engen Restoration Hangar at Udvar-Hazy. Likewise at Udvar-Hazy, a rare late 1940s specimen of RMI's spaghetti-type rocket chamber rests in the smaller of the exhibit cases in the rocketry display.

There are two cutaway examples of the C-1 engine—also called the "Radiamic engine—one in the How Things Fly gallery in the main NASM building in downtown Washington, D.C., and the other in the long "storefront case" in the Rocketry exhibit in the McDonnell Space Hangar at Udvar-Hazy. In this same case, on the bottom shelf, sits an XLR-48-RM-2, or Patriot rocket motor for the Corvus missile. Also at Udvar-Hazy, suspended from the ceiling along the right side aisle of the rocketry exhibit, are wonderfully restored examples of early postwar Gorgon 2A and Lark missiles, although the early RMI motors within them are not visible. However, there is a cutaway Lark engine in the small display case close to the spaghetti motor mentioned above.

Finally, on the second floor of the main NASM building, in the Barron Hilton Pioneers of Flight gallery, is the bright-red ARS Test Stand No. 2 that had been loaned by the American Rocket Society for RMI's first static tests. Part of this exhibit also features the all-important Wyld Serial No. 1 rocket motor that started RMI, and the Wyld Serial No. 2 variant that was among RMI's very first rocket motor products. RMI and RMD objects are also to be

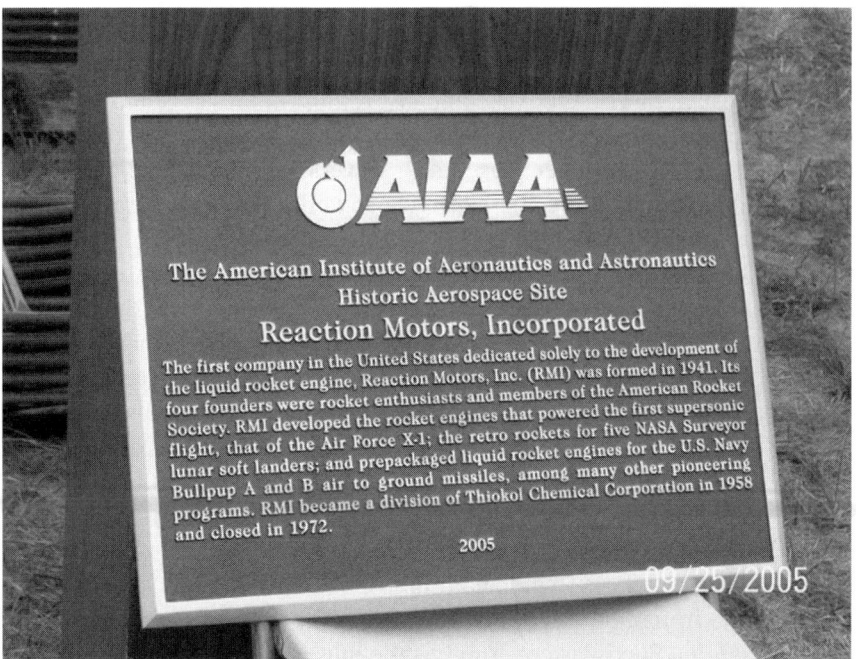

Fig. 11.6   AIAA historic marker at 100 Ford Road, Denville, NJ, next to former RMI/
RMD plant. (Courtesy Ken Montanye.)

found in other museums across the United States as well as in some international museums.[54]

Clearly, this amazing amount of representation of both RMI and RMD
artifacts in museums—particularly in the National Air and Space Museum,
one of the world's most visited museums—is a solid testament to the enormous legacy of Reaction Motors. Moreover, the objects represent a wide
spectrum of phases in the histories of both RMI and RMD. This is not to leave
out several historic markers at both RMI and RMD historic sites: one near the
old RMI Dunn barn plant, two by the RMI/RMD former plant at Denville
(both dedicated by AIAA); one by buildings used by RMI/RMD at Lake
Denmark, Picatinny Arsenal (also dedicated by AIAA), and another by the
original testing site at Lake Denmark. All these underscore that America's first
liquid-fuel rocket company and its successor were undeniably substantial
contributors to some of the greatest milestones in the history of both aeronautics and astronautics.

ENDNOTES

1. Anon., "Hobby Discussion Led to RMI Birth," *The RMI Rocket*, Vol. 2, December 1951,
   p. 3.
2. Ibid.

3. Ibid.
4. Anonymous, "Rockets," *Fortune* magazine, Vol. 42, Nov. 1950, p. 129.
5. Memo from C.W. Newell Jr. to H.B. Horne Jr., RMI, Feb. 1948, copy in "Reaction Motors, General Correspondence, Corporate History" file, NASM.
6. Ibid.
7. Lovell Lawrence Jr., *Daily Log*, Lovell Lawrence Jr. Papers, NASM, box 4, folder 18.
8. See for example, "Rocket Business Soars, in the *Newark Evening News*, 1 Sept1946, p. unknown.
9. Anon., "Rockets," p. 29.
10. Lawrence Jr., *Daily Log*, box 4, folder 19.
11. "Distribution of Stock By Major Groups," as of 31 Jan. 1951, Lawrence Jr. Papers, box 5, folder 5.
12. H. Franklin Pierce, letter of resignation, Lawrence Jr. Papers, box 5, folder 5.
13. "Agreement of with Reaction Motors, Inc.," [made with Lawrence, Shesta, and Wyld], 10 Sept. 1947, in Lawrence Jr. Papers, box 5, folder 5.
14. Ibid.
15. Shesta, "Reaction Motors, Inc.," in Frank H. Winter and Frederick I. Ordway III, *Pioneering American Rocketry: The Reaction Motors, Inc. (RMI) Story, 1941–1972*, AAS History Series, Vol. 44, Univelt, Inc., San Diego, 2015, p. 76.
16. These figures are from "Comparative Analysis [of RM] Net Sales – Net Earnings – Years 1942 Through 1955," in "Reaction Motors, Inc., Administrative, General" file, NASM; Anon., "Hobby Discussion," p. 3; and Abbot, Baker & Co., N.Y., Memo from the [Investors'] Research Department, re RMI, 14 Nov. 1956, in Willy Ley Papers, NASM, box 48, folder 1; RMI, *Annual Report*, 1956, n.p.
17. Obituary, "Hugh F. Pierce, Barstow," *San Bernardino County Sun* (San Bernardino, CA), 24 Feb. 1979, p. unknown.
18. Consult Eric D. Eason and Eric D. Frazier, *GPS Declassified: From Smart Bombs to Smart Phones*, Potomac Books, University of Nebraska Press, 2013, especially p. 43 on Lawrence's concept, although there is no indication it had an input into the development of the later GPS system. Lawrence's article is "Navigation by Satellites," in *Missiles & Rockets*, Vol. 1, Oct. 1956, pp. 48–52.
19. Michael Gaffe, "Rocket Noise Suit May Set Precedent," *Aviation Week*, Vol. 73, 28 Nov. 1960, p. 30.
20. R. Bender, *Space Transport Liability: National and International Aspect,* Kluwer Law International, The Hague, Netherlands, 1995, p. 135.
21. Winter and Ordway III, *Pioneering American Rocketry*, p. 432.
22. Ibid.
23. Al Miller quoted in Sharon Sheridan, "Having fun was part of the [RMI/RMD] job description," *The Star-Ledger* (Newark, NJ), 19 June 1997, p. 2.
24. See "One Dead in Jersey as Rocket Blows Up," *New York Times*, 17 Sept. 1957, p. 26, and "One Killed, 6 Badly Burned as Rocket Engine Explodes," *Boston Daily Globe*, 17 Sept. 1957, p. 7.
25. "Engineer dies, 6 hurt in N.J. engine blast," *Newark Star-Ledger*, 17 Sept. 1957, p. 1.
26. E-mail from Glenn Repp to Ken Montanye, 28 Aug. 2016.
27. Ibid.
   "Don" was obviously a nickname for Robert D. Tweedy.
28. Ibid.
29. Ibid.
30. Ibid.
31. Letter from Harry S. Bell to Frank H. Winter, 26 May 2001.

32. Letter from Harry S. Bell to Frank H. Winter, 20 Aug. 2000.
33. Letter from Harold Davies to Frank H. Winter, undated, copy in "Reaction Motors, Inc." file, NASM.
34. Ibid.
35. Data sheet, "120,000 to 240,000 Pound Thrust Liquid Propellant Rocket Engine," in Anon. and unpublished, collection of Reaction Motors, Inc. photographs and data sheets, in three-ring binder in Aviation Hall of Fame Library, Teterboro, NJ, n.p.
36. Constance McLaughlin Green and Milton Lomask, *Vanguard: A History*, Scientific and Technological Information Division, NASA Historical Series, NASA, Washington, DC, 1970, p. 69.
37. Memo from R.W. Young to G.W. Newell Jr., "Possible Air Force Projects," Lawrence Jr. Papers, box 7, folder 7.
38. For typical coverage on the merger see "Thiokol, Reaction Motors Boards Agree to Merger," *Astronautics*, Vol. 3, March 1958, p. 42; "Holders of Thiokol, Reaction Motors Vote to Merge Companies," *Wall Street Journal*, 18 April 1958, p. 18; and "Thiokol Merger Wins Approval," *Washington Post*, 19 April 1958, p. 18.
39. See, "Thiokol Finds Promising Approach To Solid Boron Rocket Propellant," in *Aviation Week*, Vol. 70, 15 June 1959, p. 37, and on the same page, "Hydrazine Fuel."
40. E-mail from Art Sherman to Frank H. Winter, 23 April 2014.
41. James H. Wyld, RMI Memo, 17 Jan. 1947, on "Use of atomic power for jet propulsion," copy in "James H. Wyld" file, NASM.
42. "New Unit at Thiokol: Nuclear Group Formed for Research on Propulsion," *New York Times*, 8 June 1960, p. 55.
43. Ibid.
44. Morristown, NJ, and vicinity, telephone directories, 1960–1962, various pages.
45. E-mail from Art Sherman to Frank H. Winter, 6 Feb. 2014.
46. Telephone interview with Myron Levoy by Frank H. Winter, 17 April 2013.
47. E-mails from Harvey Fox to Frank H. Winter, 23 Jan. 2015 and 24 June 2016.
48. Ibid.
49. Robert Walker, "Sun Sets on a Pioneer Space Age Company," *New York Times*, 12 Aug. 1970, p. 55.
50. Ibid.
51. Ibid.
52. See "Landmark Status Asked for Rocket Site of the 1940s," *New York Times*, 19 Aug. 1977, p. 45.
53. As stated in Chapter 3, the remaining cinder block is in the possession of RMI/RMD collector and authority Ken Montanye of Butler, NJ.
54. Some of these artifacts are alluded to in this book. Consult also Appendix 2 in Winter and Ordway III, *Pioneering American Rocketry*, pp. 317–324, for an annotated list of RMI/RMD objects within NASM's collections that includes the present whereabouts of some of these objects on loan to other museums. Appendix 3 specifically covers the known locations of other XLR-11 and XLR-8 engines in the collections of other museums while Appendix 4 covers the locations of the XLR-99 (X-15) engine in the collections of other museums.

INDEX

SUPPORTING MATERIALS

Many of the topics discussed in this book are examined in other AIAA publications. For a complete listing of titles in the Library of Flight, as well as other AIAA publications, please go to AIAA's electronic library, Aerospace Research Central (ARC), at arc.aiaa.org.

Visit ARC frequently to stay abreast of product changes, corrections, special offers, and new publications.

AIAA is committed to devoting resources to the education of both practicing and future aerospace professionals. In 1996, the AIAA Foundation was founded. Its programs enhance scientific literacy and advance the arts and sciences of aerospace. For more information, please visit www.aiaafoundation.org.